Adaptive Control of
Robot
Manipulators
A Unified Regressor-Free Approach

Adaptive Control of
Robot
Manipulators

A Unified
Regressor-Free
Approach

An-Chyau Huang
Ming-Chih Chien
National Taiwan University of Science and Technology, Taiwan

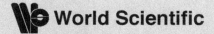

World Scientific

NEW JERSEY · LONDON · SINGAPORE · BEIJING · SHANGHAI · HONG KONG · TAIPEI · CHENNAI

Published by

World Scientific Publishing Co. Pte. Ltd.

5 Toh Tuck Link, Singapore 596224

USA office: 27 Warren Street, Suite 401-402, Hackensack, NJ 07601

UK office: 57 Shelton Street, Covent Garden, London WC2H 9HE

British Library Cataloguing-in-Publication Data
A catalogue record for this book is available from the British Library.

ADAPTIVE CONTROL OF ROBOT MANIPULATORS
A Unified Regressor-Free Approach

ISBN-13 978-981-4307-41-3
ISBN-10 981-4307-41-6

Printed in Singapore.

To Our Families

Preface

The traditional computed torque design is only adequate for the control of robot manipulators with precisely known dynamics. In the industrial environment, however, an accurate robot model is not generally available, and most robots are limited to be operated under slow motion conditions so that some system dynamics can be ignored. When performing the tasks with precise tracking of fast trajectories under time-varying payloads, several considerations such as the joint flexibility and actuator dynamics are unavoidable. This will generally lead to some extremely complex robot model which greatly increases the difficulty in the controller design. What is worse is that estimation of the system parameters in this complex model becomes more challenging. It is reasonable to regard some system dynamics as uncertainties to simplify the modeling tasks. The robust controls and adaptive designs are then utilized to deal with these uncertainties. However, the former needs the knowledge of the variation bounds for the uncertainties, while the later requires the linear parameterization of the uncertainties into a known regressor multiplied by an unknown constant parameter vector. When the system contains time-varying uncertainties whose variation bounds are not given (defined later as general uncertainties), both the robust control and adaptive design are not feasible in general.

In the conventional adaptive control of robot manipulators, the robot model is assumed to be linearly parameterizable into the regressor form. But the derivation of the regressor matrix is tedious in most cases, and computation of the regressor matrix during each sampling period in the real-time realization is too time-consuming. This suggests the need for some regressor-free adaptive designs.

The aim of this book is to address recent developments of the unified regressor-free adaptive controller designs for robot manipulators with consideration of joint flexibility and actuator dynamics. The unified approach is still valid for the robot control in the compliant motion environment. The main tool used in this new design is the function approximation technique which represents the general uncertainties in the robot model as finite combinations of basis functions weighted with unknown constant coefficients.

The book has been written as a text suitable for postgraduate students in the advanced course for the control of robot manipulators. In addition, it is also intended to provide valuable tools for researches and practicing engineers who currently employ the regressor-based algorithms but would benefit significantly from the use of the regressor-free strategies.

We would like to thank all of our colleagues and students with whom we have discussed the basic problems in the control of robot manipulators over the last few years. Without them, many issues would never have been clarified. This work was partially supported by the National Science Council of the Republic of China government and by the National Taiwan University of Science and Technology. The authors are grateful for their support.

An-Chyau Huang,
Ming-Chih Chien

Mechanical Engineering Department
National Taiwan University of Science and Technology

Contents

Chapter 1

Introduction

Robot manipulators have been widely used in the industrial applications in the past decades. Most of these applications are restricted to slow-motion operations without interactions with the environment. This is mainly due to limited performance of the available controllers in the market that are based on simplified system models. To increase the operation speed with more servo accuracy, advanced control strategies are needed. Consideration of the actuator dynamics in the controller design is one of the possible ways to improve system performance. Although some of the industrial robots are driven by hydraulic or pneumatic actuators, most of them are still activated by motors. Therefore, similar to majority of the related literature, we are only going to consider electrically driven (ED) robot manipulators in this book. On the other hand, explicit inclusion of the joint flexibility into the system dynamics can also improve the control performance. Since the robot dynamics is highly nonlinear, consideration of these effects will largely increase the difficulty in the controller design. Besides, the robot model inevitably contains uncertainties and disturbances; this makes the control problem extremely difficult.

In this book, we would like to consider the control problem of robot manipulators with consideration of actuator dynamics, joint flexibility and various system uncertainties. These uncertainties are assumed to be time-varying but their variation bounds are not available. Due to their time-varying nature, most traditional adaptive designs are not feasible. Because their variation bounds are not known, most conventional robust schemes are not applicable. The main strategy we employed in this book to deal with the uncertainties is based on the function approximation techniques (FAT). The basic idea of FAT is to represent the uncertain term as a finite combination of known basis functions so that proper update laws can be derived based on the Lyapunov-like design to give good performance. Since the FAT-based adaptive design does not need to represent the system dynamics into a regressor form, it is free from computation of the regressor matrix. Derivation of the regressor matrix is well-known to be very tedious for a robot manipulator with more than four joints. The regressor-free strategy greatly simplified the controller design process. Because, for the traditional robot adaptive control, the complex regressor matrix has to be updated in every control cycle in the real-time implementation, the regressor-free algorithm also effectively simplified the programming complexity.

When the robot end-effector contacts with the environment, the controllers designed for performing the free space tracking tasks cannot provide appropriate control performance. In this book, the renowned impedance control strategy will be incorporated into the FAT-based design so that some regressor-free adaptive controllers with consideration of the actuator dynamics for the robot manipulators can be obtained.

To have a better understanding of the problems we are going to deal with, Table 1.1 presents the systems considered in this book. The abbreviations listed will be used throughout this book to simplify the presentation.

Table 1.1 Systems considered in this book

	Systems	Abbreviation
1	Rigid robot in the free space	RR
2	Rigid robot interacting with the environment	RRE
3	Electrically-driven rigid robot in the free space	EDRR
4	Electrically-driven rigid robot interacting with the environment	EDRRE
5	Flexible-joint robot in the free space	FJR
6	Flexible-joint robot interacting with the environment	FJRE
7	Electrically-driven flexible-joint robot in the free space	EDFJR
8	Electrically-driven flexible-joint robot interacting with environment	EDFJRE

Free space tracking of a rigid robot

It can be seen that systems from 1 to 7 are all special cases of 8. However, it is not appropriate to derive a controller for the system in 8 directly, because starting from the simple ones can give us more insight into the unified approach to be introduced. Let us consider the tracking problem of a rigid robot in the free space first. It is the simplest case in this book and several control strategies can also be found in robotics textbooks under various conditions. We will start with the case when all system parameters are precisely known, and a feedback linearization based controller is constructed to give proper performance. Afterwards, we assume that most parameters in the robot model are not known but a regressor can be derived such that all uncertain parameters are collected into an unknown vector. Conventional adaptive strategies can thus be applied to give update laws to this unknown parameter vector, and closed loop stability can also be proved easily. However, implementation of this scheme requires the information of joint accelerations which is impractical in most industrial applications. What is worse is that the estimation in the inertia matrix might suffer the singularity problem. A well-known design proposed by Slotine and Li is then reviewed to get rid of the need for joint acceleration feedback and avoid the singularity problem. In the above designs, the robot dynamics should be linearly parameterized into a known regressor matrix multiplied by an unknown parameter vector. The regressor matrix is known to be tedious in its derivation for a robot with degree of freedom more than 4. The regressor matrix is not unique for a given robot, but depending on the selection of the parameter vector. The entries of this vector should be constants that are combinations of unknown system parameters. However, these parameters are mostly easier to be found than the derivation of the regressor matrix. For example, the weight, length, moment of inertia and gravity center of a link are frequently seen in the parameter vector and their values are very easy to measure in practice. It is not reasonable to construct a controller whose design needs to know an extremely complex regressor matrix but to update an easy-to-obtain parameter vector. Motivated by this reasoning, the regressor-free adaptive control approach is developed. The uncertain matrices and vectors in the robot model will be represented as finite combinations of basis functions. Update laws for the weighting matrices can be obtained by the Lyapunov-like design. The effect of the approximation error is investigated with rigorous mathematical justifications. The output error can thus be proved to be uniformly ultimately bounded. Finally, the trajectory of the output error is bounded by a weighted exponential function plus some constant. With proper adjustment of controller parameters, both the transient performance and the steady state error can be modified.

Compliant motion control of a rigid robot

Item 2, 4, 6 and 8 in Table 1.1 relate to the compliant motion control of robot manipulators whose dynamic model include the effect of the external force vector exerted by the environment. Many control strategies are available for rigid robots to give closed loop stability in the compliant motion applications. Among them, the impedance control employed in this book is the most widely used one which is a unified approach for controlling robot manipulators in both free space tracking and compliant motion phases. The impedance controller makes the robot system behave like a target impedance in the Cartesian space, and the target impedance is specified as a mass-spring-damper system. For rigid robots, we start with the case when the robot system and the environment are known and the impedance controller is designed. The regressor-based adaptive impedance controller is then derived for robot systems containing uncertainties. To avoid derivation of the regressor matrix, the regressor-free adaptive impedance controller using function approximation techniques is introduced. For the impedance controller of EDRRE, FJRE and EDFJRE, much more complex derivations will be involved due to the complexity in the system model. Unlike the rigid robots, the regressor-based designs of these robots need additional information such as the derivative of the regressor matrix, the joint accelerations and derivative of the external force. All of these are not generally available, and hence regressor-free designs are introduced to get rid of their necessity. The unified approach in the FAT-based regressor-free adaptive impedance controller designs for RRE, EDRRE, FJRE and EDFJRE can all give uniformly ultimately bounded performance to the output error and the transient performance can also be evaluated by using the bound for the output error signal.

Consideration of the actuator dynamics

The control problem of rigid robot manipulators has been well developed under the assumption that all actuator dynamics are neglected. However, it had been reported that the robot control problem should carefully consider the actuator dynamics to have good tracking performance, especially in the cases of high-velocity movement and highly varying loads. Therefore, in item 3, 4, 7 and 8 of Table 1.1, we include considerations of actuator dynamics in the system equations to investigate their effects in performance improvement. The input vector to a robot without consideration of the actuator dynamics contains torques to the joints, while the input vector to electrically-driven robots is with signals in voltages. This special cascade structure connecting the actuator and the robot dynamics enables us to employ backstepping-like designs to eliminate uncertainties entering the system in a mismatched fashion. The regressor-based adaptive designs are introduced first for items 3, 4, 7 and 8 followed by their regressor-free counterparts. Implementations of most regressor-based methods introduced here need the information of the derivative of the regressor matrix and joint accelerations, but the regressor-free designs do not. Besides, the uniformly ultimately bounded performance can be proved to be maintained when considering the actuator dynamics by using the regressor-free approach. Simulation cases for justifying performance improvements are designed with high speed tracking problems. All of these regressor-free schemes give good performance regardless of various system uncertainties.

Consideration of the joint flexibility

Many industrial robots use harmonic drives to reduce speed and amplify output torque. A cup-shape component in the harmonic drive provides elastic deformation to enable large speed reduction. Therefore, it is known that harmonic drives introduce significant torsional flexibility into the robot joints. To have a high performance robot control system, elastic coupling between actuators and links cannot be neglected. Modeling of these effects, however, produces an enormously complicated model. For simplicity, most researches regard the flexibility as an effect of the linear torsional spring connecting the shaft of the motor and the end about which the link is rotating. Two second-order differential equations should be used to describe the dynamic of a flexible joint: one for the motor shaft and one for the link. This implies that the number of degree-of-freedom is twice the number for a rigid robot, since the motion of the motor shaft is no longer simply related to the link angle by the gear ratio. The high system order and highly nonlinear coupling in the dynamics equation result in difficulties in the controller design. If the system model contains inaccuracies and uncertainties, the controller design problem becomes extremely difficult. In this book, we are going to design conventional regressor-based adaptive controllers for this system first and followed by a regressor-free control strategy. In addition, adaptive controllers for impedance control of flexible joint robot will also be derived. Furthermore, the actuator dynamics are to be considered so that in the most complex case a regressor-free adaptive impedance controller will be designed for an EDFJRE. When considering the joint flexibility, the realization of the regressor-based adaptive controller requires the knowledge of joint accelerations, the regressor matrix, and their derivatives. The regressor-free designs, however, do not need these additional information.

The regressor-free adaptive controller design

Calculation of the regressor matrix is a must in the traditional adaptive control of robot manipulators which is because the update laws are able to be designed only when the parameter vectors are unknown constants. Parameterization of the uncertainties into multiplications of the regressor matrix with the unknown parameter vector need to be done based on the system model. With proper definitions of the entries in the parameter vector, the regressor matrix can then be determined. Since these definitions are not unique, the regressor matrix for a given robot is not unique either. Some definitions will give relatively simple forms for the regressor matrix, while some will become very complex. When the degree of freedom of the robot is more than four, the derivation of the regressor matrix becomes very tedious. In general, the entries in the parameter vector are combinations of the quantities such as the link masses, dimensions of the links, and moments of inertia. These quantities are relatively easier to measure compared with the derivation of the regressor matrix. However, the traditional adaptive designs are only capable of updating these easy-to-measure parameters, but require the complex regressor matrix to be known. In addition, in every control cycle of the real-time implementation, the calculation of the regressor matrix is also time consuming which largely limits the computation hardware selections, especially in the embedded applications. In this book, a unified approach for the design of regressor-free adaptive controllers for robot manipulators is introduced which is feasible for robots with considerations of the actuator dynamics, joint flexibilities as well as the interaction with the environment. All of these designs will end up with the uniformly ultimately bounded closed loop performance via the proofs using the Lyapunov-like techniques.

The FAT-based design

Two main approaches are available for dealing with uncertainties in control systems. The robust strategies need to know the worst case of the system so that a fixed controller is able to be constructed to cover the uncertainties. In most cases, the worst case of the system is evaluated by proper modeling of the uncertainties either in the time domain or frequency domain. The variation bounds estimated from the uncertainty model are then used to design the robust terms in the controller. In some practical cases, however, these variation bounds are not available, and hence most robust strategies are infeasible. The other approach for dealing with system uncertainties is the adaptive method. Although intuitively we think that an adaptive controller should be able to give good performance to a system with time-varying uncertainties, conventional adaptive designs can actually be useful to systems with constant uncertainties. Therefore, to be feasible for the adaptive designs all time-varying parts in the system dynamics should be collected into a known regressor matrix, while the unknown constant parameters are put into a parameter vector. This process is called the linear parameterization of the uncertainties which is almost a must for adaptive designs. After the parameterization, proper update laws can then be derived to provide sufficient information to the certainty equivalence based adaptive controller such that the closed loop system can give good performance. However, there are some practical cases whose uncertainties are not able to be linearly parameterized (e.g., various friction effects), and some others are linearly parameterizable but the regressor matrices are too complex to derive (e.g., robot manipulators).

Now let us consider a case when the uncertainties are time-varying and their variation bounds are not available. Since they are time-varying, most traditional robust designs fail. Because their variation bounds are unknown, most conventional adaptive strategies are infeasible. In this book, we are going to call this kind of uncertainties the *general uncertainties*. For a system with general uncertainties, few control schemes are available to stabilize the closed loop system. Because the regressor-free adaptive controller design for robot manipulators should avoid the use of the regressor matrix, a new representation for the system uncertainties is needed. In this case, it is more practical to regard the uncertainties in the robot model to be general uncertainties, and the controller design problem is a challenge. Here, in this book, we employ the function approximation technique to represent the uncertainties into finite combinations of basis functions. This effectively transforms a general uncertainty into a known basis vector multiplied by a vector of unknown coefficients. Since these coefficients are constants, update laws can be derived by using the Lyapunov-like methods. Due to the fact that the mathematical background for the function approximation has well been established and the controller design portion follows the traditional adaptive strategies, the FAT-based adaptive method provides an effective tool in dealing with controller design problems involving the general uncertainties.

Organization of the book

Robot systems considered in this book are all listed in Tables 1.1 according to the complexity in their dynamics. For better presentation, however, they will be arranged into the chapters different from the order as shown in the table. In Chapter 2, the backgrounds for mathematics and control theories useful in this book are reviewed. Readers familiar to these fundamentals are suggested to go directly to the next chapter. Various concepts from the linear algebra and real analysis are briefly presented in this chapter. Some emphasis will be placed on the spaces where the function approximation techniques are valid. Various orthonormal functions are also listed with their effective ranges for the convenience in the selection of basis functions for the FAT-based designs. Then the Lyapunov stability theory and the Lyapunov-like methods are reviewed in detail followed by the introduction of the control theories such as the sliding control and model reference adaptive control. After these conventional robust and adaptive designs, the concept of general uncertainties is presented. Limitations in the sliding controller designs when the variation bounds for the uncertainties are not available are investigated. Likewise, the problem for the model reference adaptive control when the system contains time-varying parameters is illustrated. Finally, the FAT-based adaptive controller is designed for these systems with general uncertainties in detail.

Chapter 3 collects all dynamic equations for systems listed in Table 1.1. These equations will be used in later chapters for controller designs. Examples for these systems will also be presented, and they will also be used in the simulation studies later. Adaptive control strategies for the rigid robots are introduced in Chapter 4. Traditional regressor-based adaptive rules will be derived first followed by some investigation into the detail of the regressor matrix and the parameter vector. This justifies the necessity for the regressor-free adaptive designs. A FAT-based regressor-free adaptive controller is then derived for the rigid robot with consideration of the approximation errors. The rigorous proof for the closed loop stability is presented to give uniformly ultimately bounded performance. Next, the actuator dynamics is included into the system model and adaptive controllers are derived using regressor-based designs and regressor-free designs. Significant amount of simulation results are provided to justify the efficacy of the controllers when actuator dynamics are considered.

Chapter 5 considers the compliant motion control of rigid robot manipulators. The impedance controller is employed to enable the robot to interact with the environment compliantly while maintaining good performance in the free space tracking. The traditional impedance controller is reviewed first for the system with known dynamics. A regressor-based and a regressor-free adaptive controller are then derived. Finally, the actuator dynamics is considered to improve the control performance.

Chapter 6 includes joint flexibility into consideration such that the order of the system model is doubled compared with its rigid joint counterpart. Control of a known FJR is firstly reviewed. The regressor-based adaptive controller is then introduced followed by the derivation of the regressor-free controller. The actuator dynamics will be considered in the last section of this chapter. A 5th order differential equation should be used to describe a single link in this case which makes the controller design problem become extremely challenging.

The last chapter deals with the problem of the adaptive impedance control for FJR. We review the control of a known robot first to have some basic understanding of this problem. The regressor-based adaptive controller is then designed, but it requires some impractical knowledge in the real-time implementation. The regressor-free adaptive controller is derived without any requirements on additional information. Consideration of the actuator dynamics further complicated the problem, and the regressor-free adaptive design is still able to give good performance to the closed loop system.

Chapter 2

Preliminaries

2.1 Introduction

Some mathematical backgrounds are reviewed in this chapter. They can be found in most elementary mathematics books; therefore, most results are provided without proof. On the other hand, some preliminaries in control theories will also be presented in this chapter as the background for the theoretical development introduced in the later chapters. In Section 2.2, some notions of vector spaces are introduced. Some best approximation problems in the Hilbert space will be mentioned in Section 2.3. Various orthogonal functions are collected in Section 2.4 to facilitate the selection of basis functions in FAT applications. The vector and matrix analysis is reviewed in Section 2.5 which includes their differential calculus operations. Norms for functions, vectors and matrices are summarized in Section 2.6, and some normed spaces are also introduced. Section 2.7 illustrates the approximation representations of functions, vectors and matrices. The Lyapunov stability theory is reviewed in Section 2.8. The concept of sliding control is provided in Section 2.9 as an introduction to the robust design for a system containing uncertainties defined in compact sets. Section 2.10 gives the basics of the model reference adaptive control to linear time-invariant systems. To robustify the adaptive control loop, some modifications of the adaptive designs are also presented in Section 2.10. The concept of general uncertainties is clarified in Section 2.11. The limitations for traditional MRAC and robust designs will also be discussed. Finally, the FAT-based adaptive controller is introduced in Section 2.12 to cover the general uncertainties.

2.2 Vector Spaces

Vector spaces provide an appropriate framework for the study of approximation techniques. In this section we review some concepts and results useful in this book.

A nonempty set X is a (*real*) *vector space* if the following axioms are satisfied:

$$\mathbf{x} + \mathbf{y} \in X, \ \forall \mathbf{x}, \mathbf{y} \in X$$

$$\mathbf{x} + \mathbf{y} = \mathbf{y} + \mathbf{x}, \ \forall \mathbf{x}, \mathbf{y} \in X$$

$$(\mathbf{x} + \mathbf{y}) + \mathbf{z} = \mathbf{x} + (\mathbf{y} + \mathbf{z}), \ \forall \mathbf{x}, \mathbf{y}, \mathbf{z} \in X$$

There is a unique vector $\mathbf{0} \in X$ such that $\mathbf{x} + \mathbf{0} = \mathbf{x}, \ \forall \mathbf{x} \in X$

$\forall \mathbf{x} \in X$, there exists a unique vector $-\mathbf{x} \in X$ such that $\mathbf{x} + (-\mathbf{x}) = \mathbf{0}$

$$\alpha \mathbf{x} \in X, \ \forall \mathbf{x} \in X, \ \forall \alpha \in \mathfrak{R}$$

$$\alpha(\beta \mathbf{x}) = (\alpha\beta)\mathbf{x}, \ \forall \mathbf{x} \in X, \ \forall \alpha, \beta \in \mathfrak{R}$$

$$1\mathbf{x} = \mathbf{x}, \ \forall \mathbf{x} \in X$$

$$(\alpha + \beta)\mathbf{x} = \alpha \mathbf{x} + \beta \mathbf{x}, \ \forall \mathbf{x} \in X, \ \forall \alpha, \beta \in \mathfrak{R}$$

$$\alpha(\mathbf{x} + \mathbf{y}) = \alpha \mathbf{x} + \alpha \mathbf{y}, \ \forall \mathbf{x}, \mathbf{y} \in X, \ \forall \alpha \in \mathfrak{R}$$

For example, the set of all n-tuples of real numbers is a real vector space and is known as $\mathfrak{R}^n$. The set of all real-valued functions defined over an interval $[a,b] \in \mathfrak{R}$ is also a vector space. The set of all functions maps the interval $[a,b] \in \mathfrak{R}$ into $\mathfrak{R}^n$ can also be proved to be a vector space. In some literature, the real vector space is also known as the *real linear space*.

If $\mathbf{x}_1, ..., \mathbf{x}_k \in \mathfrak{R}^n$ and $c_1, ..., c_k \in \mathfrak{R}$, a vector of the form $c_1 \mathbf{x}_1 + ... + c_k \mathbf{x}_k$ is said to be a *linear combination* of the vectors $\mathbf{x}_1, ..., \mathbf{x}_k$. The set of vectors $\mathbf{x}_1, ..., \mathbf{x}_k \in \mathfrak{R}^n$ is said to be *independent* if the relation $c_1 \mathbf{x}_1 + ... + c_k \mathbf{x}_k = 0$ implies $c_i = 0, i = 1, ..., k$; otherwise, the set of vectors is *dependent*. If $S \subset \mathfrak{R}^n$ and if R is the set of all linear combinations of elements of S, then S spans R, or we may say that R is the *span* of S. An independent subset of a vector space $X \subset \mathfrak{R}^n$ which spans X is called a *basis* of X. A vector space is *n-dimensional* if it contains an independent set of n vectors, but every set of $n+1$ vectors is a dependent set.

2.2.1 Metric space

A set X of elements p, q, r,... is said to be a *metric space* if with any two points p and q there is associated a real number $d(p,q)$, the distance between p and q, such that

$$d(p,q) > 0 \ \text{ if } \ p \neq q \, ;$$

$$d(p,p) = 0 \, ;$$

$$d(p,q) = d(q,p) \, ;$$

$$d(p,q) \leq d(p,r) + d(r,q) \ \text{ for any } \ r \in X \, .$$

The distance function on a metric space can be thought of as the length of a vector; therefore, many useful concepts can be defined. A set $E \subset X$ is said to be *open* if for every $x \in E$, there is a ball $B(x,r) = \{ y \in X \mid d(y,x) < r \}$ such that $B(x,r) \subset E$ for some positive r. A set is *closed* if and only if its complement in X is open. A set E is *bounded* if $\exists r > 0$ such that $d(x,y) < r \ \forall x, y \in E$. Let $S \subset T$ be two subsets of X. S is said to be *dense* in T if for each element t in T and each $\varepsilon > 0$, there exists an element s in S such that $d(s,t) < \varepsilon$. Thus every element of T can be approximated to arbitrary precision by elements of S.

Let X and Y be metric spaces with distance functions d_X and d_Y, respectively. A function $f : X \rightarrow Y$ is said to be *continuous* at a point $x \in X$ if $f(x + \delta x) \rightarrow f(x)$ whenever $\delta x \rightarrow 0$. Or, we may say, f is continuous at x if given $\varepsilon > 0$, there exist $\delta > 0$ such that $d_X(x,y) < \delta \Rightarrow d_Y(f(x), f(y)) < \varepsilon$. A function f is continuous on $E \subset X$ if it is continuous at every points of E, and it is *uniformly continuous* on E if given $\varepsilon > 0$, there exist $\delta(\varepsilon) > 0$ such that $d_X(x,y) < \delta \Rightarrow d_Y(f(x), f(y)) < \varepsilon$ for all $x, y \in E$.

The distance function generates the notion of convergence: A sequence $\{x_i\}$ in a metric space X is said to be *convergent* to an element x if for each $\varepsilon > 0$ there exists an integer n such that $d(x,x_i) \leq \varepsilon$ whenever $i > n$. A set $E \subset X$ is *compact* if each sequence of points in E contains a subsequence which converges to a point in E. In particular, a compact subset of $\Re^n$ is necessarily closed and bounded. The sequence $\{x_i\}$ is a *Cauchy sequence* if for each $\varepsilon > 0$ there exists an integer n such that $p,q > n \Rightarrow d(x_p, x_q) \leq \varepsilon$. Clearly, every convergent sequence is a Cauchy sequence, but the converse is not true in general. A *complete* metric space X is a space where every Cauchy sequence converges to a point in X.

2.2.2 Normed vector space

Let X be a vector space, a real-valued function $\|\cdot\|$ defined on X is said to be a *norm* on X if it satisfies the following properties

$$\|\mathbf{x}\| > 0, \forall \mathbf{x} \in X, \mathbf{x} \neq \mathbf{0}$$

$$\|\mathbf{0}\| = 0$$

$$\|\alpha \mathbf{x}\| = |\alpha| \|\mathbf{x}\| \text{ for any scalar } \alpha$$

$$\|\mathbf{x} + \mathbf{y}\| \leq \|\mathbf{x}\| + \|\mathbf{y}\|, \forall \mathbf{x}, \mathbf{y} \in X$$

A *normed vector space* $(X, \|\cdot\|)$ is a metric space with the metric defined by $d(\mathbf{x}, \mathbf{y}) = \|\mathbf{x} - \mathbf{y}\|$, $\forall \mathbf{x}, \mathbf{y} \in X$. The concept of sequence convergence can be defined using the norm as the distance function. Hence, we are now ready to define convergence of series. The series $\sum_{i=1}^{\infty} \mathbf{x}_i$ is said to converge to $\mathbf{x} \in X$ if the sequence of partial sums converges to $\mathbf{x}$, i.e., if $\forall \varepsilon > 0, \exists n > 0$ such that

$$\left\| \sum_{i=1}^{m} \mathbf{x}_i - \mathbf{x} \right\| < \varepsilon \text{ whenever } m > n.$$

A complete normed vector space is called a *Banach space*. In a normed vector space, the length of any vector is defined by its norm. To define the angle between any two vectors, in particular the concept of orthogonality between vectors, we need the notion of the inner product space.

An *inner product* $\langle \mathbf{x}, \mathbf{y} \rangle$ on a real vector space X is a real-valued mapping of the pair $\mathbf{x}, \mathbf{y} \in X$ with the properties

$$\langle \mathbf{x}, \mathbf{y} \rangle = \langle \mathbf{y}, \mathbf{x} \rangle$$

$$\langle \alpha \mathbf{x}, \mathbf{y} \rangle = \alpha \langle \mathbf{x}, \mathbf{y} \rangle$$

$$\langle \mathbf{x} + \mathbf{y}, \mathbf{z} \rangle = \langle \mathbf{x}, \mathbf{z} \rangle + \langle \mathbf{y}, \mathbf{z} \rangle, \forall \mathbf{z} \in X$$

$$\langle \mathbf{x}, \mathbf{x} \rangle > 0 \; \forall \mathbf{x} \neq \mathbf{0}$$

A real vector space with an inner product defined is called a *real inner product space*. An inner product space is a normed vector space and hence a metric space with

$$d(\mathbf{x}, \mathbf{y}) = \|\mathbf{x} - \mathbf{y}\| = \sqrt{\langle \mathbf{x} - \mathbf{y}, \mathbf{x} - \mathbf{y} \rangle}$$

For any two vectors $\mathbf{x}$ and $\mathbf{y}$ in an inner product space, we have the *Schwarz inequality* in the form $|\langle \mathbf{x}, \mathbf{y} \rangle| \leq \|\mathbf{x}\| \|\mathbf{y}\|$. The equality holds if and only if $\mathbf{x}$ and $\mathbf{y}$ are dependent. Two vectors $\mathbf{x}, \mathbf{y}$ are *orthogonal* if $\langle \mathbf{x}, \mathbf{y} \rangle = 0$. Let $\{\mathbf{x}_i\}$ be a set of elements in an inner product space X. $\{\mathbf{x}_i\}$ is an *orthogonal set* if $\langle \mathbf{x}_i, \mathbf{x}_j \rangle = 0, \forall i \neq j$. If in addition every vector in the set has unit norm, the set is *orthonormal*.

A *Hilbert space* is a complete inner product space with the norm induced by its inner product. For example, $\mathfrak{R}^n$ is a Hilbert space with inner product

$$\langle \mathbf{x}, \mathbf{y} \rangle = \sum_{i=1}^{n} x_i y_i$$

Suppose functions $x(t)$ and $y(t)$ are defined in a domain D, then L_2 is a Hilbert space with the inner product $\langle x, y \rangle = \int_D x(t) y(t) dt$.

2.3 Best Approximation Problem in Hilbert Space

Let U be a set of vectors in a Hilbert space H. The *algebraic span* $S(U)$ is defined as the set of all finite linear combinations of vectors $\mathbf{x}_i \in U$ (Stakgold 1979). The set $\overline{S}(U)$ is the closure of $S(U)$ and is called the *closed span* of U. For example, if $U = \{1, x, x^2, ...\} \subset L_2$, then $S(U)$ is the set of all polynomials, whereas $\overline{S}(U) = L_2$. The set U is a *spanning set* of H if $\overline{S}(U)$ is dense in H. The Hilbert space H is *separable* if it contains a countable spanning set U. The space L_2 is separable since the countable set $\{1, x, x^2, ...\}$ is a spanning set. Any finite-dimensional Hilbert space is separable because its basis is a countable spanning set. Since an infinite-dimensional space cannot have a finite spanning set, a separable infinite-dimensional Hilbert space must contain a countably infinite set $U = \{\mathbf{x}_i\}$ so that each vector $\mathbf{x} \in H$ can be approximated to any desired accuracy by a linear combination of a finite number of elements of U. This can be rewritten as: given $\varepsilon > 0$ and $\mathbf{x} \in H$, there exist an integer n such that $\left\| \mathbf{x} - \sum_{i=1}^{n} c_i \mathbf{x}_i \right\| < \varepsilon$ where $c_i \in \Re, i = 1, \dots, n$. It can be proved that a separable Hilbert space H contains an orthonormal spanning set, and the spanning set is necessarily an orthonormal basis of H. An orthonormal basis is also known as a *complete* orthonormal set.

The best approximation problem in a separable Hilbert space is to approximate an arbitrary vector $\mathbf{x} \in H$ by a linear combination of the given independent set $U = \{\mathbf{x}_1, \dots, \mathbf{x}_n\}$. Since the set of linear combination of $\mathbf{x}_1, \dots, \mathbf{x}_n$ is an n-dimensional linear manifold M_n, an orthonormal basis $\{\mathbf{e}_1, \dots, \mathbf{e}_n\}$ for M_n can be constructed from U by using the Gram-Schmidt procedure. Therefore, the *approximation error* can be calculated as

$$\left\| \mathbf{x} - \sum_{i=1}^{n} c_i \mathbf{e}_i \right\|^2 = \|\mathbf{x}\|^2 + \sum_{i=1}^{n} \left| \langle \mathbf{x}, \mathbf{e}_i \rangle - c_i \right|^2 - \sum_{i=1}^{n} \left| \langle \mathbf{x}, \mathbf{e}_i \rangle \right|^2 \tag{1}$$

Hence, the minimum error can be obtained when $c_i = \langle \mathbf{x}, \mathbf{e}_i \rangle$. These c_i are known as the *Fourier coefficients* of $\mathbf{x}$ with respect to the orthonormal basis $\{\mathbf{e}_1, \dots, \mathbf{e}_n\}$. With these coefficients, the vector $\mathbf{x} \in H$ is approximated as $\sum_{i=1}^{n} \langle \mathbf{x}, \mathbf{e}_i \rangle \mathbf{e}_i$, and the approximation error becomes

$$\left\| \mathbf{x} - \sum_{i=1}^{n} \langle \mathbf{x}, \mathbf{e}_i \rangle \mathbf{e}_i \right\|^2 = \|\mathbf{x}\|^2 - \sum_{i=1}^{n} \left| \langle \mathbf{x}, \mathbf{e}_i \rangle \right|^2 \tag{2}$$

If an additional vector $\mathbf{e}_{n+1}$ is included into the orthonormal set, the vector $\mathbf{x} \in H$ is thus approximated by the series $\sum_{i=1}^{n+1} \langle \mathbf{x}, \mathbf{e}_i \rangle \mathbf{e}_i$, which is the same as the previous one except that one extra term $\langle \mathbf{x}, \mathbf{e}_{i+1} \rangle \mathbf{e}_{i+1}$ is added. This implies that previously calculated Fourier coefficients do not need to be recalculated. It can also be observed that the right hand side of (2) gets smaller when the orthonormal set is taken larger. Hence, the best approximation to $\mathbf{x}$ improves as we use more terms in the orthonormal set. As the number of terms used goes to infinity, the approximating series becomes the *Fourier series* $\sum_{i=1}^{\infty} \langle \mathbf{x}, \mathbf{e}_i \rangle \mathbf{e}_i$.

Hence, (2) implies

$$\|\mathbf{x}\|^2 = \sum_{i=1}^{\infty} |\langle \mathbf{x}, \mathbf{e}_i \rangle|^2 \tag{3}$$

which is known as the *Parseval's identity*. Convergence of the Fourier series can be proved by using the Riesz-Fischer theorem with the fact that the partial sum of the series $\sum_{i=1}^{\infty} |\langle \mathbf{x}, \mathbf{e}_i \rangle|^2$ is monotonically increasing and is bounded above by $\|\mathbf{x}\|^2$. Consequently, it is easy to prove that $\lim_{i \to \infty} \langle \mathbf{x}, \mathbf{e}_i \rangle = 0$, i.e., the coefficients of the Fourier series vanish as $i \to \infty$.

2.4 Orthogonal Functions

In the previous section we have reviewed the general framework for the best approximation problem in the Hilbert space. In this section, we would like to restrict the scope to the function approximation problem using orthogonal functions.

The set of real-valued functions $\{\phi_i(x)\}$ defined over some interval $[a,b]$ is said to form an *orthogonal set* on that interval if

$$\int_a^b \phi_i(x)\phi_j(x)dx \begin{cases} =0 & i \neq j \\ \neq 0 & i = j \end{cases} \tag{1}$$

An orthogonal set $\{\phi_i(x)\}$ on $[a,b]$ having the property $\int_a^b \phi_i^2(x)dx = 1$ for all i is called an *orthonormal set* on $[a,b]$. The set of real-valued functions $\{\phi_i(x)\}$ defined over some interval $[a,b]$ is orthogonal with respect to the weight function $p(x)$ on that interval if

$$\int_a^b p(x)\phi_i(x)\phi_j(x)dx \begin{cases} =0 & i \neq j \\ \neq 0 & i = j \end{cases} \tag{2}$$

Any set of functions orthogonal with respect to a weight function $p(x)$ can be converted into a set of functions orthogonal to 1 simply by multiplying each member of the set by $\sqrt{p(x)}$ if $p(x) > 0$ on that interval. For any set of orthonormal functions $\{\phi_i(x)\}$ on $[a,b]$, an arbitrary function $f(x)$ can be represented in terms of $\phi_i(x)$ by a series

$$f(x) = c_1\phi_1(x) + c_2\phi_2(x) + \cdots + c_n\phi_n(x) + \cdots \tag{3}$$

This series is called a *generalized Fourier series* of $f(x)$ and its coefficients are Fourier coefficients of $f(x)$ with respect to $\{\phi_i(x)\}$. Multiplying by $\phi_n(x)$ and integrating over the interval $[a,b]$ and using the orthogonality property, the series becomes

$$\int_a^b f(x)\phi_n(x)dx = c_n \int_a^b \phi_n^2(x)dx \tag{4}$$

Hence, the coefficient c_n can be obtained from the quotient

$$c_n = \frac{\int_a^b f(x)\phi_n(x)dx}{\int_a^b \phi_n^2(x)dx} \tag{5}$$

It should be noted that although the orthogonality property can be used to determine all coefficients in (3), it is not sufficient to conclude convergence of the series. To guarantee convergence of the approximating series, the orthogonal set should be complete. An orthogonal set $\{\phi_i(x)\}$ on $[a,b]$ is said to be *complete* if the relation $\int_a^b g(x)\phi_i(x)dx = 0$ can hold for all values of i only if $g(x)$ can have non-zero values in a measure zero set in $[a,b]$. Here, $g(x)$ is called a *null function* on $[a,b]$ satisfying $\int_a^b g^2(x)dx = 0$. It is easy to prove that if $\{\phi_i(x)\}$ is a complete orthonormal set on $[a,b]$ and the expansion $c_1\phi_1(x) + c_2\phi_2(x) + \cdots + c_n\phi_n(x) + \cdots$ of $f(x)$ converges and can be integrated term by term, then the sum of the series differs from $f(x)$ by at most a null function.

Examples of orthonormal functions

Since there are many areas of applications of orthonormal functions, a sizable body of literature can be easily found. In this section, we consider some of the orthonormal functions that are frequently encountered in engineering problems and useful in our applications.

1. Taylor polynomials

In the calculus courses, it is well known that given a function $f(x)$ and a point c in the domain of f, suppose the function is n-times differentiable at c, then we can construct a polynomial

$$P_n(x) = f(c) + f'(c)(x-c) + \frac{f''(c)}{2!}(x-c)^2$$
$$+ \cdots + \frac{f^{(n)}(c)}{n!}(x-c)^n \tag{6}$$

where $P_n(x)$ is called the nth-degree Taylor polynomial approximation of f at c. The Taylor polynomial is not well suited to approximate a function $f(x)$ over an interval $[a,b]$ if the approximation is to be uniformly accurate over the entire domain. Taylor polynomial approximation is known to yield very small error near a given point, but the error increases in a considerable amount as we move away from that point. The following orthogonal polynomials, however, can give a more uniform approximation error over the specified interval.

2. Chebyshev polynomials

The set of Chebyshev polynomials is orthogonal with respect to the weight function $(1-x^2)^{-\frac{1}{2}}$ on the interval $[-1,1]$. The first two polynomials are $T_0(x) = 1$ and $T_1(x) = x$, and the remaining polynomials can be determined by the recurrence relation

$$T_{n+1}(x) = 2xT_n(x) - T_{n-1}(x) \tag{7}$$

for all $n = 1, 2, \ldots$ For convenience, we list the first 7 polynomials below

$$T_0(x) = 1$$

$$T_1(x) = x$$

$$T_2(x) = 2x^2 - 1$$

$$T_3(x) = 4x^3 - 3x$$

$$T_4(x) = 8x^4 - 8x^2 + 1$$

$$T_5(x) = 16x^5 - 20x^3 + 5x$$

$$T_6(x) = 32x^6 - 48x^4 + 18x^2 - 1$$

$$T_7(x) = 64x^7 - 112x^5 + 56x^3 - 7x$$

3. Legendre polynomials

The set of Legendre polynomials is orthogonal with respect to the weight function $p(x) = 1$ on the interval $[-1,1]$. The first two polynomials are $L_0(x) = 1$ and $L_1(x) = x$, and the remaining polynomials can be determined by the recurrence relation

$$(n+1)L_{n+1}(x) = (2n+1)xL_n(x) - nL_{n-1}(x) \qquad (8)$$

for all $n = 1, 2, \ldots$ Here, we list the first 7 polynomials for convenience

$$L_0(x) = 1$$

$$L_1(x) = x$$

$$L_2(x) = \frac{1}{2}(3x^2 - 1)$$

$$L_3(x) = \frac{1}{2}(5x^3 - 3x)$$

$$L_4(x) = \frac{1}{8}(35x^4 - 30x^2 + 3)$$

$$L_5(x) = \frac{1}{8}(63x^5 - 70x^3 + 15x)$$

$$L_6(x) = \frac{1}{16}(231x^6 - 315x^4 + 105x^2 - 5)$$

$$L_7(x) = \frac{1}{16}(429x^7 - 693x^5 + 315x^3 - 35x)$$

4. Hermite polynomials

The set of Hermite polynomials is orthogonal with respect to the weight function $p(x) = e^{-x^2}$ on the interval $(-\infty, \infty)$. The first two polynomials are

$H_0(x) = 1$ and $H_1(x) = 2x$, and the remaining polynomials can be determined by the recurrence relation

$$H_{n+1}(x) = 2xH_n(x) - 2nH_{n-1}(x) \tag{9}$$

for all $n = 1, 2, \ldots$ Here, we list the first 7 polynomials as

$$H_0(x) = 1$$

$$H_1(x) = 2x$$

$$H_2(x) = 4x^2 - 2$$

$$H_3(x) = 8x^3 - 12x$$

$$H_4(x) = 16x^4 - 48x^2 + 12$$

$$H_5(x) = 32x^5 - 160x^3 + 120x$$

$$H_6(x) = 64x^6 - 480x^4 + 720x^2 - 120$$

$$H_7(x) = 128x^7 - 1344x^5 + 3360x^3 - 1680x$$

5. Laguerre polynomials

The set of Laguerre polynomials is orthogonal with respect to the weight function $p(x) = e^{-x}$ on the interval $[0, \infty)$. The first two polynomials are $L_0(x) = 1$ and $L_1(x) = -x + 1$, and the remaining polynomials can be determined by the recurrence relation

$$L_{n+1}(x) = (2n + 1 - x)L_n(x) - n^2 L_{n-1}(x) \tag{10}$$

for all $n = 1, 2, \ldots$ The following are the first 7 polynomials

$$L_0(x) = 1$$

$$L_1(x) = -x + 1$$

$$L_2(x) = x^2 - 4x + 2$$

$$L_3(x) = -x^3 + 9x^2 - 18x + 6$$

$$L_4(x) = x^4 - 16x^3 + 72x^2 - 96x + 24$$

$$L_5(x) = -x^5 + 25x^4 - 200x^3 + 600x^2 - 600x + 120$$

$$L_6(x) = x^6 - 36x^5 + 450x^4 - 2400x^3 + 5400x^2 - 4320x + 720$$

$$L_7(x) = -x^7 + 49x^6 - 882x^5 + 7350x^4 - 29400x^3$$
$$+ 52920x^2 - 35280x + 5040$$

6. Bessel polynomials

The set of Bessel polynomials is orthogonal with respect to the weight function $p(x) = x$ on the interval $[0,b]$ in the form

$$\int_0^b x J_n(k_i x) J_n(k_j x) dx = 0 \tag{11}$$

for all $i \neq j$. The Bessel polynomials can be calculated with

$$J_n(x) = x^n \sum_{m=0}^{\infty} \frac{(-1)^m x^{2m}}{2^{2m+n} m!(n+m)!} \tag{12}$$

In particular, for $n=0$, 1, the Bessel polynomials are

$$J_0(x) = 1 - \frac{x^2}{2^2} + \frac{x^4}{2^2 \cdot 4^2} - \frac{x^6}{2^2 \cdot 4^2 \cdot 6^2} + \cdots$$

$$J_1(x) = \frac{x}{2} - \frac{x^3}{2^2 \cdot 4} + \frac{x^5}{2^2 \cdot 4^2 \cdot 6} - \frac{x^7}{2^2 \cdot 4^2 \cdot 6^2 \cdot 8} + \cdots$$

These two series converge very rapidly, so that they are useful in computations. The recurrence relation below can also be used to find other Bessel polynomials based on $J_0(x)$ and $J_1(x)$ given above.

$$J_{n+1}(x) = -J_{n-1}(x) + \frac{2n}{x} J_n(x) \tag{13}$$

It should be noted that k_i, $i=1,2,\ldots$ in (11) are real numbers so that $J_n(k_ib) = 0$, i.e., they are distinct roots of $J_n = 0$. These roots for $n=0, 1$ are listed here for reference

$$J_0(x) = 0 \text{ for } x = 2.405, 5.520, 8.654, 11.792, 14.931, \ldots$$

$$J_1(x) = 0 \text{ for } x = 0, 3.832, 7.016, 10.173, 13.324, \ldots$$

Having the orthogonality property in (11), we can represent a given function $f(x)$ in a series of the form in $[0,b]$ with a given n

$$f(x) = \sum_{i=1}^{\infty} c_i J_n(k_i x) \tag{14}$$

This series is called a *Fourier-Bessel series* or simply a *Bessel series*.

7. Fourier series

A bounded period function $f(x)$ can be expanded in the form

$$f(x) = \frac{a_0}{2} + \sum_{n=1}^{\infty} \left[a_n \cos \frac{n\pi x}{T} + b_n \sin \frac{n\pi x}{T} \right] \tag{15}$$

if in any one period it has at most a finite number of local extreme values and a finite number of discontinuities. (15) is called the *Fourier series* of function $f(x)$. The constants a_0, a_n and b_n, $n=1,2,3,\ldots$ are called *Fourier coefficients*, and the value $2T$ is the period of $f(x)$. It can be proved that the Fourier series converges to $f(x)$ at all points where $f(x)$ is continuous and converges to the average of the right- and left-hand limits of $f(x)$ at each point where it is discontinuous.

Table 2.1 summarizes the orthonormal functions introduced in this section. When using these functions in approximation applications, it is very important that the valid range for the functions to be orthonormal is ensured.

Table 2.1 Some useful orthonormal functions

Polynomial	Valid Interval	Forms
Taylor	$[a,b]$	$P_n(x) = f(c) + f'(c)(x-c) + \dfrac{f''(c)}{2!}(x-c)^2$ $+ \cdots + \dfrac{f^{(n)}(c)}{n!}(x-c)^n$
Chebyshev	$[-1,1]$	$T_0(x) = 1$ $T_1(x) = x$ $T_{n+1}(x) = 2xT_n(x) - T_{n-1}(x)$
Legendre	$[-1,1]$	$L_0(x) = 1$ $L_1(x) = x$ $(n+1)L_{n+1}(x) = (2n+1)xL_n(x) - nL_{n-1}(x)$
Hermite	$(-\infty, \infty)$	$H_0(x) = 1$ $H_1(x) = 2x$ $H_{n+1}(x) = 2xH_n(x) - 2nH_{n-1}(x)$
Laguerre	$[0, \infty)$	$L_0(x) = 1$ $L_1(x) = -x + 1$ $L_{n+1}(x) = (2n+1-x)L_n(x) - n^2 L_{n-1}(x)$
Bessel	$[0,b]$	$J_0(x) = 1 - \dfrac{x^2}{2^2} + \dfrac{x^4}{2^2 \cdot 4^2} - \dfrac{x^6}{2^2 \cdot 4^2 \cdot 6^2} + \cdots$ $J_1(x) = \dfrac{x}{2} - \dfrac{x^3}{2^2 \cdot 4} + \dfrac{x^5}{2^2 \cdot 4^2 \cdot 6} - \cdots$ $J_{n+1}(x) = -J_{n-1}(x) + \dfrac{2n}{x} J_n(x)$
Fourier series	One period	$f(x) = \dfrac{a_0}{2} + \displaystyle\sum_{n=1}^{\infty} \left[a_n \cos \dfrac{n\pi x}{T} + b_n \sin \dfrac{n\pi x}{T} \right]$

2.5 Vector and Matrix Analysis

2.5.1 Properties of matrices

Let $\mathbf{A} \in \Re^{n \times m}$ be a matrix with n rows and m columns, and $a_{ij} \in \Re$ be the (i, j)th element of $\mathbf{A}$. Matrix $\mathbf{A}$ can also be represented as $[a_{ij}]$. If the rows and columns are interchanged, then the resulting $m \times n$ matrix is called the *transpose* of $\mathbf{A}$, and is denoted by $\mathbf{A}^T$. The transpose operation has the following properties:

$$(\mathbf{A}^T)^T = \mathbf{A} \tag{1a}$$

$$(\mathbf{A} + \mathbf{B})^T = \mathbf{A}^T + \mathbf{B}^T \quad \forall \mathbf{B} \in \mathfrak{R}^{n \times m} \tag{1b}$$

$$(\mathbf{AB})^T = \mathbf{B}^T \mathbf{A}^T \quad \forall \mathbf{B} \in \mathfrak{R}^{m \times n} \tag{1c}$$

A matrix $\mathbf{A} \in \mathfrak{R}^{n \times n}$ is *symmetric* if $\mathbf{A} = \mathbf{A}^T$, and is *skew-symmetric* if $\mathbf{A} = -\mathbf{A}^T$. For any matrix $\mathbf{A} \in \mathfrak{R}^{n \times n}$, $\mathbf{A} + \mathbf{A}^T$ is symmetric and $\mathbf{A} - \mathbf{A}^T$ is skew-symmetric. If $\mathbf{A}$ is a $m \times n$ matrix, then $\mathbf{A}^T \mathbf{A}$ is symmetric. A matrix $\mathbf{A} \in \mathfrak{R}^{n \times n}$ is *diagonal* if $a_{ij} = 0, \forall i \neq j$. A diagonal matrix $\mathbf{A}$ can be written as $diag(a_{11}, ..., a_{nn})$. An *identity matrix* is a diagonal matrix with $a_{11} = \cdots = a_{nn} = 1$.

A matrix $\mathbf{A} \in \mathfrak{R}^{n \times n}$ is *nonsingular* if $\exists \mathbf{B} \in \mathfrak{R}^{n \times n}$ such that $\mathbf{AB} = \mathbf{BA} = \mathbf{I}$ where $\mathbf{I}$ is an $n \times n$ identity matrix. If $\mathbf{B}$ exists, then it is known as the *inverse* of $\mathbf{A}$ and is denoted by $\mathbf{A}^{-1}$. The inverse operation has the following properties:

$$(\mathbf{A}^{-1})^{-1} = \mathbf{A} \tag{2a}$$

$$(\mathbf{A}^T)^{-1} = (\mathbf{A}^{-1})^T \tag{2b}$$

$$(\alpha \mathbf{A})^{-1} = \frac{1}{\alpha} \mathbf{A}^{-1} \quad \forall \alpha \in \mathfrak{R}, \alpha \neq 0 \tag{2c}$$

$$(\mathbf{AB})^{-1} = \mathbf{B}^{-1} \mathbf{A}^{-1} \quad \forall \mathbf{B} \in \mathfrak{R}^{n \times n} \text{ with valid inverse} \tag{2d}$$

A matrix $\mathbf{A} \in \mathfrak{R}^{n \times n}$ is said to be *positive semi-definite* (denoted by $\mathbf{A} \geq \mathbf{0}$) if $\mathbf{x}^T \mathbf{A} \mathbf{x} \geq 0 \quad \forall \mathbf{x} \in \mathfrak{R}^n$. It is *positive definite* (denoted by $\mathbf{A} > \mathbf{0}$) if $\mathbf{x}^T \mathbf{A} \mathbf{x} > 0$ $\forall \mathbf{x} \in \mathfrak{R}^n, \mathbf{x} \neq \mathbf{0}$. It is *negative (semi-)definite* if $-\mathbf{A}$ is positive (semi-)definite. A time-varying matrix $\mathbf{A}(t)$ is *uniformly positive definite* if $\exists \alpha > 0$ such that $\mathbf{A}(t) \geq \alpha \mathbf{I}$.

Let $\mathbf{A} \in \mathfrak{R}^{n \times n}$, then the *trace* of $\mathbf{A}$ is defined as

$$Tr(\mathbf{A}) = \sum_{i=1}^{n} a_{ii} \tag{3}$$

where a_{ii} is the ith diagonal element of $\mathbf{A}$. The trace operation has the following properties:

$$Tr(\mathbf{A}) = Tr(\mathbf{A}^T) \tag{4a}$$

$$Tr(\alpha \mathbf{A}) = \alpha Tr(\mathbf{A}) \quad \forall \alpha \in \mathfrak{R} \tag{4b}$$

$$Tr(\mathbf{A} + \mathbf{B}) = Tr(\mathbf{A}) + Tr(\mathbf{B}) \quad \forall \mathbf{B} \in \mathfrak{R}^{n \times n} \tag{4c}$$

$$Tr(\mathbf{AB}) = Tr(\mathbf{BA}) = Tr(\mathbf{A}^T \mathbf{B}^T) \quad \forall \mathbf{A} \in \mathfrak{R}^{n \times m}, \mathbf{B} \in \mathfrak{R}^{m \times n} \tag{4d}$$

$$\mathbf{x}^T \mathbf{y} = Tr(\mathbf{x}\mathbf{y}^T) = Tr(\mathbf{y}\mathbf{x}^T) = \mathbf{y}^T \mathbf{x} \quad \forall \mathbf{x}, \mathbf{y} \in \mathfrak{R}^n \tag{4e}$$

2.5.2 Differential calculus of vectors and matrices

Suppose $f(\mathbf{x}): \mathfrak{R}^n \to \mathfrak{R}$ is a differentiable function, then the *gradient* vector is defined as

$$\frac{\partial}{\partial \mathbf{x}} f(\mathbf{x}) = \left[\frac{\partial f}{\partial x_1} \quad \cdots \quad \frac{\partial f}{\partial x_n} \right]^T \tag{5}$$

and if f is twice differentiable, then the *Hessian matrix* can be defined as

$$\frac{\partial^2}{\partial \mathbf{x}^2} f(\mathbf{x}) = \begin{bmatrix} \dfrac{\partial^2 f}{\partial x_1^2} & \cdots & \dfrac{\partial^2 f}{\partial x_1 x_n} \\ \vdots & \ddots & \vdots \\ \dfrac{\partial^2 f}{\partial x_n x_1} & \cdots & \dfrac{\partial^2 f}{\partial x_n^2} \end{bmatrix} \tag{6}$$

Let $a_{ij}(x)$ be the (i,j)th element of matrix $\mathbf{A}(x) \in \mathfrak{R}^{n \times m}$ with $x \in \mathfrak{R}$, then the derivative of $\mathbf{A}$ with respect to x is computed as

$$\frac{d}{dx} \mathbf{A}(x) = \begin{bmatrix} \dfrac{da_{11}(x)}{dx} & \cdots & \dfrac{da_{1m}(x)}{dx} \\ \vdots & \ddots & \vdots \\ \dfrac{da_{n1}(x)}{dx} & \cdots & \dfrac{da_{nm}(x)}{dx} \end{bmatrix} \tag{7}$$

Let $f(.): \Re^{n \times m} \to \Re$ be a real-valued mapping, then

$$\frac{\partial}{\partial \mathbf{A}} f(\mathbf{A}) = \begin{bmatrix} \dfrac{\partial f}{\partial a_{11}} & \cdots & \dfrac{\partial f}{\partial a_{1m}} \\ \vdots & \ddots & \vdots \\ \dfrac{\partial f}{\partial a_{n1}} & \cdots & \dfrac{\partial f}{\partial a_{nm}} \end{bmatrix} \tag{8}$$

The following basic properties of matrix calculus are useful in this book.

$$\frac{d}{dx}(\mathbf{A} + \mathbf{B}) = \frac{d\mathbf{A}}{dx} + \frac{d\mathbf{B}}{dx} \quad \forall \mathbf{A}, \mathbf{B} \in \Re^{n \times m} \tag{9a}$$

$$\frac{d}{dx}(\mathbf{AB}) = \frac{d\mathbf{A}}{dx}\mathbf{B} + \mathbf{A}\frac{d\mathbf{B}}{dx} \quad \forall \mathbf{A} \in \Re^{n \times m}, \mathbf{B} \in \Re^{m \times n} \tag{9b}$$

$$\frac{d}{dx}\mathbf{A}^{-1} = -\mathbf{A}^{-1}\frac{d\mathbf{A}}{dx}\mathbf{A}^{-1} \quad \forall \mathbf{A} \in \Re^{n \times n} \tag{9c}$$

$$\frac{\partial}{\partial \mathbf{x}}(\mathbf{x}^T \mathbf{y}) = \mathbf{y} \quad \forall \mathbf{x}, \mathbf{y} \in \Re^n \tag{9d}$$

$$\frac{\partial}{\partial \mathbf{y}}(\mathbf{x}^T \mathbf{y}) = \mathbf{x} \quad \forall \mathbf{x}, \mathbf{y} \in \Re^n \tag{9e}$$

$$\frac{\partial}{\partial \mathbf{x}}(\mathbf{A}\mathbf{x}) = \mathbf{A}^T \quad \forall \mathbf{A} \in \Re^{n \times m}, \mathbf{x} \in \Re^m \tag{9f}$$

$$\frac{\partial}{\partial \mathbf{A}}(\mathbf{x}^T \mathbf{A}\mathbf{x}) = \mathbf{x}\mathbf{x}^T \quad \forall \mathbf{A} \in \Re^{n \times n}, \mathbf{x} \in \Re^n \tag{9g}$$

$$\frac{\partial}{\partial \mathbf{A}}(\mathbf{x}^T \mathbf{A}^T \mathbf{A}\mathbf{x}) = 2\mathbf{A}\mathbf{x}\mathbf{x}^T \quad \forall \mathbf{A} \in \Re^{n \times n}, \mathbf{x} \in \Re^n \tag{9h}$$

$$\frac{\partial}{\partial \mathbf{x}}(\mathbf{x}^T \mathbf{A}\mathbf{x}) = \mathbf{A}\mathbf{x} + \mathbf{A}^T \mathbf{x} = 2\mathbf{A}\mathbf{x} \quad \forall \mathbf{A} \in \Re^{n \times n}, \mathbf{x} \in \Re^n \tag{9i}$$

$$\frac{\partial}{\partial \mathbf{y}}(\mathbf{y}^T \mathbf{A}\mathbf{x}) = \mathbf{A}\mathbf{x} \quad \forall \mathbf{A} \in \Re^{n \times m}, \mathbf{x} \in \Re^m, \mathbf{y} \in \Re^n \tag{9j}$$

$$\frac{\partial}{\partial \mathbf{x}}(\mathbf{y}^T \mathbf{A}\mathbf{x}) = \mathbf{A}^T \mathbf{y} \quad \forall \mathbf{A} \in \mathfrak{R}^{n \times m}, \mathbf{x} \in \mathfrak{R}^m, \mathbf{y} \in \mathfrak{R}^n \tag{9k}$$

$$\frac{\partial}{\partial \mathbf{A}}Tr(\mathbf{A}\mathbf{B}) = \frac{\partial}{\partial \mathbf{A}}Tr(\mathbf{A}^T \mathbf{B}^T) = \frac{\partial}{\partial \mathbf{A}}Tr(\mathbf{B}^T \mathbf{A}^T)$$

$$= \frac{\partial}{\partial \mathbf{A}}Tr(\mathbf{B}\mathbf{A}) = \mathbf{B}^T \quad \forall \mathbf{A} \in \mathfrak{R}^{n \times m}, \mathbf{B} \in \mathfrak{R}^{m \times n} \tag{9l}$$

$$\frac{\partial}{\partial \mathbf{A}}Tr(\mathbf{B}\mathbf{A}\mathbf{C}) = \frac{\partial}{\partial \mathbf{A}}Tr(\mathbf{B}^T \mathbf{A}^T \mathbf{C}^T) = \frac{\partial}{\partial \mathbf{A}}Tr(\mathbf{C}^T \mathbf{A}^T \mathbf{B}^T)$$

$$= \frac{\partial}{\partial \mathbf{A}}Tr(\mathbf{A}\mathbf{C}\mathbf{B}) = \frac{\partial}{\partial \mathbf{A}}Tr(\mathbf{C}\mathbf{B}\mathbf{A})$$

$$= \frac{\partial}{\partial \mathbf{A}}Tr(\mathbf{A}^T \mathbf{B}^T \mathbf{C}^T) = \mathbf{B}^T \mathbf{C}^T \quad \forall \mathbf{A}, \mathbf{B}, \mathbf{C} \in \mathfrak{R}^{n \times n} \tag{9m}$$

2.6 Various Norms

2.6.1 Vector norms

Let $\mathbf{x} \in \mathfrak{R}^n$, then the p-norm of $\mathbf{x}$ can be defined as

$$\|\mathbf{x}\|_p = \left(\sum_{i=1}^n |x_i|^p \right)^{\frac{1}{p}} \tag{1}$$

for $1 \le p \le \infty$. Three most commonly used vector norms are

$$l_1 \text{ norm on } \mathfrak{R}^n: \quad \|\mathbf{x}\|_1 = \sum_{i=1}^n |x_i| \tag{2a}$$

$$l_2 \text{ norm on } \mathfrak{R}^n: \quad \|\mathbf{x}\|_2 = \sqrt{\sum_{i=1}^n |x_i|^2} \tag{2b}$$

$$l_\infty \text{ norm on } \mathfrak{R}^n: \quad \|\mathbf{x}\|_\infty = \max_{1 \le i \le n} |x_i| \tag{2c}$$

The l_2 norm on $\mathfrak{R}^n$ is known as the *Euclidean norm* on $\mathfrak{R}^n$, which is also an inner product norm. All p-norms are equivalent in the sense that if

$\|\cdot\|_{p_1}$ and $\|\cdot\|_{p_2}$ are two different norms, then there exist $\alpha_1, \alpha_2 \in \Re_+$ such that $\alpha_1 \|\mathbf{x}\|_{p_1} \leq \|\mathbf{x}\|_{p_2} \leq \alpha_2 \|\mathbf{x}\|_{p_1}$ for all $\mathbf{x} \in \Re^n$. For $\mathbf{A} \in \Re^{n \times n}$, any nonzero vector $\mathbf{x}_i$ satisfying $\mathbf{A}\mathbf{x}_i = \lambda_i \mathbf{x}_i$ is called an *eigenvcetor* associated with an *eigenvalue* λ_i of $\mathbf{A}$. Let $\lambda_{\min}(\mathbf{A})$ and $\lambda_{\max}(\mathbf{A})$ be the maximum and minimum eigenvalues of $\mathbf{A} \in \Re^{n \times n}$, respectively, then we have the useful inequality

$$\lambda_{\min}(\mathbf{A})\|\mathbf{x}\|^2 \leq \mathbf{x}^T \mathbf{A}\mathbf{x} \leq \lambda_{\max}(\mathbf{A})\|\mathbf{x}\|^2. \tag{3}$$

2.6.2 Matrix norms

Let $\mathbf{A} \in \Re^{n \times n}$, then the *matrix norm* induced by the *p*-norm of vector $\mathbf{x} \in \Re^n$ is defined as

$$\|\mathbf{A}\|_p = \sup_{\mathbf{x} \neq 0} \frac{\|\mathbf{A}\mathbf{x}\|_p}{\|\mathbf{x}\|_p} = \sup_{\|\mathbf{x}\|_p = 1} \|\mathbf{A}\mathbf{x}\|_p \tag{4}$$

for $1 \leq p \leq \infty$. In particular, when $p = 1, 2, \infty$, we have

$$\|\mathbf{A}\|_1 = \max_{1 \leq j \leq n} \sum_{i=1}^{n} |a_{ij}| \tag{5a}$$

$$\|\mathbf{A}\|_2 = \sqrt{\lambda_{\max}(\mathbf{A}^T \mathbf{A})} \tag{5b}$$

$$\|\mathbf{A}\|_\infty = \max_{1 \leq i \leq n} \sum_{i=1}^{n} |a_{ij}| \tag{5c}$$

The following relations are useful in this book.

$$\|\mathbf{A}\mathbf{B}\| \leq \|\mathbf{A}\|\|\mathbf{B}\| \tag{6a}$$

$$\|\mathbf{A} + \mathbf{B}\| \leq \|\mathbf{A}\| + \|\mathbf{B}\| \tag{6b}$$

$$\|\mathbf{A} - \mathbf{B}\| \geq \|\mathbf{A}\| - \|\mathbf{B}\| \tag{6c}$$

2.6.3 Function norms and normed function spaces

A real-valued function defined on $\Re^+$ is *measurable* if and only if it is the limit of a sequence of piecewise constant functions over $\Re^+$ except for some measure zero sets. Let $f(t):\Re_+ \to \Re$ be a measurable function, then its p-norm is defined as

$$\|f\|_p = \left[\int_0^\infty |f(t)|^p \, dt \right]^{\frac{1}{p}} \text{ for } p \in [1,\infty) \tag{7a}$$

$$\|f\|_\infty = \sup_{t\in[0,\infty)} |f(t)| \text{ for } p = \infty \tag{7b}$$

The normed function spaces with $p = 1, 2, \infty$ are defined as

$$L_1 = \left\{ f(t):\Re_+ \to \Re \,\Big|\, \|f\|_1 = \int_0^\infty |f(t)| \, dt < \infty \right\} \tag{8a}$$

$$L_2 = \left\{ f(t):\Re_+ \to \Re \,\Big|\, \|f\|_2 = \sqrt{\int_0^\infty |f(t)|^2 \, dt} < \infty \right\} \tag{8b}$$

$$L_\infty = \left\{ f(t):\Re_+ \to \Re \,\Big|\, \|f\|_\infty = \sup_{t\in[0,\infty)} |f(t)| < \infty \right\} \tag{8c}$$

Let $\mathbf{f}:\Re_+ \to \Re^n$ with $\mathbf{f}(t) = [f_1(t) \cdots f_n(t)]^T$ be a measurable vector function, then the corresponding p-norm spaces are defined as

$$L_1^n = \left\{ \mathbf{f}(t):\Re_+ \to \Re^n \,\Big|\, \|\mathbf{f}\|_1 = \int_0^\infty \sum_{i=1}^n |f_i(t)| dt < \infty \right\} \tag{9a}$$

$$L_2^n = \left\{ \mathbf{f}(t):\Re_+ \to \Re^n \,\Big|\, \|\mathbf{f}(t)\|_2 = \sqrt{\int_0^\infty \sum_{i=1}^n |f_i(t)|^2 dt} < \infty \right\} \tag{9b}$$

$$L_\infty^n = \left\{ \mathbf{f}(t):\Re_+ \to \Re^n \,\Big|\, \|\mathbf{f}(t)\|_\infty = \max_{1\le i\le n} |f_i(t)| < \infty \right\} \tag{9c}$$

2.7 Representations for Approximation

In this section some representations for approximation of scalar functions, vectors and matrices using finite-term orthonormal functions are reviews. Let us consider a set of real-valued functions $\{z_i(t)\}$ that are orthonormal in $[t_1, t_2]$ such that

$$\int_{t_1}^{t_2} z_i(t)z_j(t)dt = \begin{cases} 0, & i \neq j \\ 1, & i = j \end{cases} \tag{1}$$

With the definition of the inner product $<f, g> = \int_{t_1}^{t_2} f(t)g(t)dt$ and its corresponding norm $\|f\| = \sqrt{<f, f>}$, the space of functions for which $\|f\|$ exists and is finite is a Hilbert space. If $\{z_i(t)\}$ is an orthonormal basis in the sense of (1) then every $f(t)$ with $\|f\|$ finite can be expanded in the form

$$f(t) = \sum_{i=1}^{\infty} w_i z_i(t) \tag{2}$$

where $w_i = <f, z_i>$ is the Fourier coefficient, and the series converges in the sense of mean square as

$$\lim_{n \to \infty} \int_{t_1}^{t_2} \left| f(t) - \sum_{i=1}^{k} w_i z_i(t) \right|^2 dt = 0. \tag{3}$$

This implies that any function $f(t)$ in the current Hilbert space can be approximated to arbitrarily prescribed accuracy by finite linear combinations of the orthonormal basis $\{z_i(t)\}$ as

$$f(t) \approx \sum_{i=1}^{k} w_i z_i(t). \tag{4a}$$

An excellent property of (4a) is its linear parameterization of the time-varying function $f(t)$ into a basis function vector $\mathbf{z}(t) = [z_1(t) \cdots z_k(t)]^T$ and a time-invariant coefficient vector $\mathbf{w} = [w_1 \cdots w_k]^T$, i.e.,

$$f(t) \approx \mathbf{w}^T \mathbf{z}(t) \tag{4b}$$

We would like to abuse the notation by writing the approximation as

$$f(t) = \mathbf{w}^T \mathbf{z}(t) \tag{4c}$$

provided a sufficient number of the basis functions are used. In this book, equation (4) is used to represent time-varying parameters in the system dynamic equation. The time-varying vector $\mathbf{z}(t)$ is known while $\mathbf{w}$ is an unknown constant vector. With this approximation, the estimation of the unknown time-varying function $f(t)$ is reduced to the estimation of a vector of unknown constants $\mathbf{w}$.

In the following, three representations are introduced for approximating a matrix $\mathbf{M}(t) \in \Re^{p \times q}$. By letting $q=1$, the same technique can be used to approximate vectors.

Representation 1: We may use the technique in (4) to represent individual matrix elements. Let $\mathbf{w}_{ij}, \mathbf{z}_{ij} \in \Re^k$ for all i, j, then matrix $\mathbf{M}$ is represented to be

$$\mathbf{M} = \begin{bmatrix} m_{11} & m_{12} & \cdots & m_{1q} \\ m_{21} & m_{22} & \cdots & m_{2q} \\ \vdots & \vdots & \ddots & \vdots \\ m_{p1} & m_{p2} & \cdots & m_{pq} \end{bmatrix} = \begin{bmatrix} \mathbf{w}_{11}^T \mathbf{z}_{11} & \mathbf{w}_{12}^T \mathbf{z}_{12} & \cdots & \mathbf{w}_{1q}^T \mathbf{z}_{1q} \\ \mathbf{w}_{21}^T \mathbf{z}_{21} & \mathbf{w}_{22}^T \mathbf{z}_{22} & \cdots & \mathbf{w}_{2q}^T \mathbf{z}_{2q} \\ \vdots & \vdots & \ddots & \vdots \\ \mathbf{w}_{p1}^T \mathbf{z}_{p1} & \mathbf{w}_{p2}^T \mathbf{z}_{p2} & \cdots & \mathbf{w}_{pq}^T \mathbf{z}_{pq} \end{bmatrix} \tag{5}$$

An operation $\otimes$ can be defined to separate the above representation into two parts as

$$\begin{bmatrix} \mathbf{w}_{11}^T \mathbf{z}_{11} & \mathbf{w}_{12}^T \mathbf{z}_{12} & \cdots & \mathbf{w}_{1q}^T \mathbf{z}_{1q} \\ \mathbf{w}_{21}^T \mathbf{z}_{21} & \mathbf{w}_{22}^T \mathbf{z}_{22} & \cdots & \mathbf{w}_{2q}^T \mathbf{z}_{2q} \\ \vdots & \vdots & \ddots & \vdots \\ \mathbf{w}_{p1}^T \mathbf{z}_{p1} & \mathbf{w}_{p2}^T \mathbf{z}_{p2} & \cdots & \mathbf{w}_{pq}^T \mathbf{z}_{pq} \end{bmatrix}$$

$$= \begin{bmatrix} \mathbf{w}_{11}^T & \mathbf{w}_{12}^T & \cdots & \mathbf{w}_{1q}^T \\ \mathbf{w}_{21}^T & \mathbf{w}_{22}^T & \cdots & \mathbf{w}_{2q}^T \\ \vdots & \vdots & \ddots & \vdots \\ \mathbf{w}_{p1}^T & \mathbf{w}_{p2}^T & \cdots & \mathbf{w}_{pq}^T \end{bmatrix} \otimes \begin{bmatrix} \mathbf{z}_{11} & \mathbf{z}_{12} & \cdots & \mathbf{z}_{1q} \\ \mathbf{z}_{21} & \mathbf{z}_{22} & \cdots & \mathbf{z}_{2q} \\ \vdots & \vdots & \ddots & \vdots \\ \mathbf{z}_{p1} & \mathbf{z}_{p2} & \cdots & \mathbf{z}_{pq} \end{bmatrix}$$

Or, we may write the above relation in the following form

$$\mathbf{M} = \mathbf{W}^T \otimes \mathbf{Z} \tag{6}$$

where $\mathbf{W}$ is a matrix containing all $\mathbf{w}_{ij}$ and $\mathbf{Z}$ is a matrix of all $\mathbf{z}_{ij}$. Since this is not a conventional operation of matrices, dimensions of all involved matrices do not follow the rule for matrix multiplication. Here, $\mathbf{W}^T$ is a $p \times kq$ matrix and $\mathbf{Z}$ is a $kp \times q$ matrix, but the dimension of $\mathbf{M}$ after the operation is still $p \times q$. This notation can be used to facilitate the derivation of update laws.

Representation 2: Let us assume that all matrix elements are approximated using the same number, say β, of orthonormal functions, and then the matrix $\mathbf{M}(t) \in \Re^{p \times q}$ can be represented in the conventional form for matrix multiplications

$$\mathbf{M} = \mathbf{W}^T \mathbf{Z} \tag{7}$$

where $\mathbf{M}, \mathbf{Z} \in \Re^{pq\beta \times p}$ are in the form

$$\mathbf{W}^T = \begin{bmatrix} \mathbf{w}_{11}^T & 0 & \cdots & 0 & | & \mathbf{w}_{12}^T & 0 & \cdots & 0 & | & \cdots & | & \mathbf{w}_{1q}^T & 0 & \cdots & 0 \\ 0 & \mathbf{w}_{21}^T & \cdots & 0 & | & 0 & \mathbf{w}_{22}^T & \cdots & 0 & | & \cdots & | & 0 & \mathbf{w}_{2q}^T & \cdots & 0 \\ \vdots & \vdots & \ddots & \vdots & | & \vdots & \vdots & \ddots & \vdots & | & \cdots & | & \vdots & \vdots & \ddots & \vdots \\ 0 & 0 & \cdots & \mathbf{w}_{p1}^T & | & 0 & 0 & \cdots & \mathbf{w}_{p2}^T & | & \cdots & | & 0 & 0 & \cdots & \mathbf{w}_{pq}^T \end{bmatrix}$$

$$\mathbf{Z}^T = \begin{bmatrix} \mathbf{z}_{11}^T & \mathbf{z}_{21}^T & \cdots & \mathbf{z}_{p1}^T & | & 0 & 0 & \cdots & 0 & | & \cdots & | & 0 & 0 & \cdots & 0 \\ 0 & 0 & \cdots & 0 & | & \mathbf{z}_{12}^T & \mathbf{z}_{22}^T & \cdots & \mathbf{z}_{p2}^T & | & \cdots & | & 0 & 0 & \cdots & 0 \\ \vdots & \vdots & \ddots & \vdots & | & \vdots & \vdots & \ddots & \vdots & | & \cdots & | & \vdots & \vdots & \ddots & \vdots \\ 0 & 0 & \cdots & 0 & | & 0 & 0 & \cdots & 0 & | & \cdots & | & \mathbf{z}_{1q}^T & \mathbf{z}_{2q}^T & \cdots & \mathbf{z}_{pq}^T \end{bmatrix}$$

The matrix elements $\mathbf{w}_{ij}$ and $\mathbf{z}_{ij}$ are $\beta \times 1$ vectors. It can be easily check that

$$\mathbf{M} = \mathbf{W}^T \mathbf{Z} = \begin{bmatrix} \mathbf{w}_{11}^T \mathbf{z}_{11} & \mathbf{w}_{12}^T \mathbf{z}_{12} & \cdots & \mathbf{w}_{1q}^T \mathbf{z}_{1q} \\ \mathbf{w}_{21}^T \mathbf{z}_{21} & \mathbf{w}_{22}^T \mathbf{z}_{22} & \cdots & \mathbf{w}_{2q}^T \mathbf{z}_{2q} \\ \vdots & \vdots & \ddots & \vdots \\ \mathbf{w}_{p1}^T \mathbf{z}_{p1} & \mathbf{w}_{p2}^T \mathbf{z}_{p2} & \cdots & \mathbf{w}_{pq}^T \mathbf{z}_{pq} \end{bmatrix} \tag{8}$$

In this representation, we use the usual matrix operation to represent $\mathbf{M}$, but the sizes of $\mathbf{W}$ and $\mathbf{Z}$ are apparently much larger than those in the representation 1.

Since this representation is compatible to all conventional matrix operations, it is used in this book for representing functions, vectors and matrices.

Representation 3: In the above representations, all matrix elements are approximated by the same number of orthonormal functions. In many applications, however, it may be desirable to use different number of orthonormal functions for different matrix elements. Suppose the component form of a vector field $\mathbf{f}(\mathbf{x})$ is written as

$$\mathbf{f}(\mathbf{x}) = [f_1(\mathbf{x}) \quad f_2(\mathbf{x}) \quad \cdots \quad f_m(\mathbf{x})]^T \tag{9}$$

and we may approximate the real-valued function $f_i(\mathbf{x})$, $i = 1, \ldots, m$ as

$$f_i(\mathbf{x}) = \mathbf{w}_{f_i}^T \mathbf{z}_{f_i} \tag{10}$$

where $\mathbf{w}_{f_i}, \mathbf{z}_{f_i} \in \Re^{p_i \times 1}$ and p_i is the number of terms of the basis functions selected to approximate f_i. Hence, (9) can be written as

$$\mathbf{f}(\mathbf{x}) = [\mathbf{w}_{f_1}^T \mathbf{z}_{f_1} \quad \mathbf{w}_{f_2}^T \mathbf{z}_{f_2} \quad \cdots \quad \mathbf{w}_{f_m}^T \mathbf{z}_{f_m}]^T$$

$$= \begin{bmatrix} w_{11}z_{11} + w_{12}z_{12} + \cdots + w_{1p_1}z_{1p_1} \\ w_{21}z_{21} + w_{22}z_{22} + \cdots + w_{2p_2}z_{2p_2} \\ \vdots \\ w_{m1}z_{m1} + w_{m2}z_{m2} + \cdots + w_{mp_m}z_{mp_m} \end{bmatrix} \tag{11}$$

Define $p_{\max} = \max_{i=1,\cdots,m} p_i$ and let $w_{ij} = 0$ for all $i = 1, \ldots, m$ and $j > p_i$, then (11) can be further written in the following form

$$\mathbf{f}(\mathbf{x}) = \begin{bmatrix} w_{11} & 0 & \cdots & 0 \\ 0 & w_{21} & \cdots & 0 \\ \vdots & \vdots & \ddots & \vdots \\ 0 & 0 & \cdots & w_{m1} \end{bmatrix} \begin{bmatrix} z_{11} \\ z_{21} \\ \vdots \\ z_{m1} \end{bmatrix} + \cdots + \begin{bmatrix} w_{1p_{\max}} & 0 & \cdots & 0 \\ 0 & w_{2p_{\max}} & \cdots & 0 \\ \vdots & \vdots & \ddots & \vdots \\ 0 & 0 & \cdots & w_{mp_{\max}} \end{bmatrix} \begin{bmatrix} z_{1p_{\max}} \\ z_{2p_{\max}} \\ \vdots \\ z_{mp_{\max}} \end{bmatrix} \tag{12}$$

Define

$$\mathbf{W}_i = \begin{bmatrix} w_{1i} & 0 & \cdots & 0 \\ 0 & w_{2i} & \cdots & 0 \\ \vdots & \vdots & \ddots & \vdots \\ 0 & 0 & \cdots & w_{mi} \end{bmatrix}, \quad \mathbf{z}_i = [z_{1i} \quad z_{2i} \quad \cdots \quad z_{mi}]^T, \tag{13}$$

where $i = 1, \dots, p_{max}$, and then (8) can be expressed in the form

$$\mathbf{f}(\mathbf{x}) = \sum_{i=1}^{p_{max}} \mathbf{W}_i \mathbf{z}_i \qquad (14)$$

For approximating the matrix $\mathbf{M}(t) \in \mathfrak{R}^{p \times q}$, we may rewrite it into a row vector as $\mathbf{M} = [\mathbf{m}_1 \cdots \mathbf{m}_q]$ where $\mathbf{m}_i \in \mathfrak{R}^p$. Therefore, we may approximate $\mathbf{m}_i$ using the technique above as

$$\mathbf{M} = \left[\sum_{i=1}^{p_{1max}} \mathbf{W}_{1i} \mathbf{z}_{1i} \quad \cdots \quad \sum_{i=1}^{p_{qmax}} \mathbf{W}_{qi} \mathbf{z}_{qi} \right] \qquad (15)$$

2.8 Lyapunov Stability Theory

The Lyapunov stability theory is widely used in the analysis and design of control systems. To ensure closed loop stability and boundedness of internal signals, all controllers derived in this book will be based on the rigorous mathematical proof via the Lyapunov or Lyapunov-like theories. The concept of stability in the sense of Lyapunov will be introduced first in this section followed by Lyapunov stability theorems for autonomous and non-autonomous systems. The invariant set theorem will be reviewed to facilitate the proof for asymptotically stability of autonomous systems when only negative semi-definite of the time derivative of the Lyapunov function can be concluded. On the other hand, a Lypunov-like technique summarized in Barbalat's lemma will also be reviewed. It is going to be used for almost every controller designed in this book.

2.8.1 Concepts of stability

Let us consider a nonlinear dynamic system described by the differential equation

$$\dot{\mathbf{x}} = \mathbf{f}(\mathbf{x}, t) \qquad (1)$$

where $\mathbf{x} \in \mathfrak{R}^n$ and $\mathbf{f} : \mathfrak{R}^n \times \mathfrak{R}_+ \to \mathfrak{R}^n$. If the function $\mathbf{f}$ dose not explicitly depend on time t, i.e., the system is in the form

$$\dot{\mathbf{x}} = \mathbf{f}(\mathbf{x}) \tag{2}$$

then the system is called an *autonomous system*; otherwise, a *non-autonomous system*. The state trajectory of an autonomous system is independent of the initial time, but that of a non-autonomous system is generally not. Therefore, in studying the behavior of a non-autonomous system, we have to consider the initial time explicitly. The system (1) is *linear* if $\mathbf{f}(\mathbf{x},t) = \mathbf{A}(t)\mathbf{x}$ for some mapping $\mathbf{A}(\cdot): \mathfrak{R}_+ \to \mathfrak{R}^{n \times n}$. If the matrix $\mathbf{A}$ is a function of time, the system is *linear time-varying*; otherwise, *linear time-invariant*.

Stability is the most important property of a control system. The concept of stability of a dynamic system is usually related to the ability to remain in a state regardless of small perturbations. This leads to the definition of the concept of the equilibrium state or equilibrium point. A state $\mathbf{x}_e$ is said to be an *equilibrium point* of (1), if $\mathbf{f}(\mathbf{x}_e,t) = \mathbf{0}$ for all $t > 0$. For simplicity, we often transform the system equations in such a way that the equilibrium point is the origin of the state space.

The equilibrium point $\mathbf{x}_e = \mathbf{0}$ of the autonomous system (2) is said to be (i) *stable*, if $\forall R > 0, \exists r > 0$ such that $\|\mathbf{x}(0)\| < r \implies \|\mathbf{x}(t)\| < R$, $\forall t \geq 0$; (ii) *asymptotically stable*, if it is stable and if $\exists r_1 > 0$ such that $\|\mathbf{x}(0)\| < r_1$ implies that $\mathbf{x}(t) \to 0$ as $t \to \infty$; (iii) *exponentially stable*, if $\exists \alpha, \lambda > 0$, such that $\|\mathbf{x}(t)\| \leq \alpha \|\mathbf{x}(0)\| e^{-\lambda t}$ for all $t > 0$ in some neighborhood N of the origin; (iv) *globally asymptotically (or exponentially) stable*, if the property holds for any initial condition.

The equilibrium point $\mathbf{x}_e = \mathbf{0}$ of the non-autonomous system (1) is said to be (i) *stable at* t_0, if $\forall R > 0, \exists r(R,t_0) > 0$ such that $\|\mathbf{x}(t_0)\| < r(R,t_0) \implies \|\mathbf{x}(t)\| < R$, $\forall t \geq t_0$; otherwise the equilibrium point is *unstable* (ii) *asymptotically stable at* t_0 if it is stable and $\exists r_1(t_0) > 0$ such that $\|\mathbf{x}(t_0)\| < r_1(t_0)$ implies that $\|\mathbf{x}(t)\| \to 0$ as $t \to \infty$; (iii) *uniformly stable* if $\forall R > 0, \exists r(R) > 0$ such that $\|\mathbf{x}(t_0)\| < r(R) \implies \|\mathbf{x}(t)\| < R$, $\forall t \geq t_0$; (iv) *uniformly asymptotically stable* if it is uniformly stable and $\exists r_1 > 0$ such that $\|\mathbf{x}(t_0)\| < r_1$ implies that $\|\mathbf{x}(t)\| \to 0$ as $t \to \infty$; (v) *exponentially stable at* t_0, if $\exists \alpha, \lambda > 0$, such that $\|\mathbf{x}(t)\| \leq \alpha \|\mathbf{x}(t_0)\| e^{-\lambda(t-t_0)}$ for all $t > t_0$ in some ball around the origin; (vi) *globally (uniformly) asymptotically (or exponentially) stable*, if the property holds for any initial conditions. It is noted that exponential stability always implies uniform asymptotic stability. Likewise, uniform asymptotic stability always implies asymptotic stability, but the converse is not generally true.

2.8.2 Lyapunov stability theorem

A continuous function $\alpha(r): \Re \to \Re$ is said to belong to *class K* if

$$\alpha(0) = 0$$

$$\alpha(r) > 0 \ \forall r > 0$$

$$\alpha(r_1) \geq \alpha(r_2) \ \forall r_1 > r_2 .$$

A continuous function $V(\mathbf{x},t): \Re^n \times \Re_+ \to \Re$ is *locally positive definite* if there exists a class K function $\alpha(\cdot)$ such that $V(\mathbf{x},t) \geq \alpha(\|\mathbf{x}\|)$ for all $t \geq 0$ in the neighborhood N of the origin of $\Re^n$. It is *positive definite* if $N = \Re^n$. A continuous function $V(\mathbf{x},t): \Re^n \times \Re_+ \to \Re$ is *locally decrescent* if there exists a class K function $\beta(\cdot)$ such that $V(\mathbf{x},t) \leq \beta(\|\mathbf{x}\|)$ for all $t \geq 0$ in the neighborhood N of the origin of $\Re^n$. It is *decrescent* if $N = \Re^n$. A continuous function $V(\mathbf{x},t): \Re^n \times \Re_+ \to \Re$ is *radially unbounded* if $V(\mathbf{x},t) \to \infty$ uniformly in time as $\|\mathbf{x}\| \to \infty$. If function $V(\mathbf{x},t)$ is locally positive definite and has continuous partial derivatives, and if its time derivative along the trajectory of (1) is negative semi-definite then it is called a *Lyapunov function* for system (1).

Lyapunov stability theorem for autonomous systems

Given the autonomous system (2) with an equilibrium point at the origin, and let N be a neighborhood of the origin, then the origin is (i) *stable* if there exists a scalar function $V(\mathbf{x}) > 0 \ \forall \mathbf{x} \in N$ such that $\dot{V}(\mathbf{x}) \leq 0$; (ii) *asymptotic stable* if $V(\mathbf{x}) > 0$ and $\dot{V}(\mathbf{x}) < 0$; (iii) *globally asymptotically stable* if $V(\mathbf{x}) > 0$, $\dot{V}(\mathbf{x}) < 0$ and $V(\mathbf{x})$ is radially unbounded.

LaSalle's theorem (Invariant set theorem)

Given the autonomous system (2) suppose $V(\mathbf{x}) > 0$ and $\dot{V}(\mathbf{x}) \leq 0$ along the system trajectory. Then (2) is asymptotically stable if $\dot{V}$ does not vanish identically along any trajectory of (2) other than the trivial solution $\mathbf{x} = \mathbf{0}$. The result is global if the properties hold for the entire state space and $V(\mathbf{x})$ is radially unbounded.

Lyapunov stability theorem for non-autonomous systems

Given the non-autonomous system (1) with an equilibrium point at the origin, and let N be a neighborhood of the origin, then the origin is (i) *stable* if

$\forall \mathbf{x} \in N$, there exists a scalar function $V(\mathbf{x}, t)$ such that $V(\mathbf{x}, t) > 0$ and $\dot{V}(\mathbf{x}, t) \leq 0$; (ii) *uniformly stable* if $V(\mathbf{x}, t) > 0$ and decrescent and $\dot{V}(\mathbf{x}, t) \leq 0$; (iii) *asymptotically stable* if $V(\mathbf{x}, t) > 0$ and $\dot{V}(\mathbf{x}, t) < 0$; (iv) *globally asymptotically stable* if $\forall \mathbf{x} \in \Re^n$, there exists a scalar function $V(\mathbf{x}, t)$ such that $V(\mathbf{x}, t) > 0$ and $\dot{V}(\mathbf{x}, t) < 0$ and $V(\mathbf{x}, t)$ is radially unbounded; (v) *uniformly asymptotically stable* if $\forall \mathbf{x} \in N$, there exists a scalar function $V(\mathbf{x}, t)$ such that $V(\mathbf{x}, t) > 0$ and decrescent and $\dot{V}(\mathbf{x}, t) < 0$; (vi) *globally uniformly asymptotically stable* if $\forall \mathbf{x} \in \Re^n$, there exists a scalar function $V(\mathbf{x}, t)$ such that $V(\mathbf{x}, t) > 0$ and decrescent and is radially unbounded and $\dot{V}(\mathbf{x}, t) < 0$; (vii) *exponentially stable* if there exits $\alpha, \beta, \gamma > 0$ such that $\forall \mathbf{x} \in N$, $\alpha \|\mathbf{x}\|^2 \leq V(\mathbf{x}, t) \leq \beta \|\mathbf{x}\|^2$ and $\dot{V}(\mathbf{x}, t) \leq -\gamma \|\mathbf{x}\|^2$; (viii) *globally exponentially stable* if it is exponentially stable and $V(\mathbf{x}, t)$ is radially unbounded.

Barbalat's lemma

La Salle's theorem is very useful in the stability analysis of autonomous systems when asymptotic stability is desired but only with negative semi-definite result for the time derivative of the Lyapunov function. Unfortunately, La Salle's theorem does not apply to non-autonomous systems. Therefore, to conclude asymptotic stability of a non-autonomous system with $\dot{V} \leq 0$, we need to find a new approach. A simple and powerful tool called Barbalat's lemma can be used to partially remedy this situation. Let $f(t)$ be a differentiable function, then Barbalat's lemma states that if $\lim_{t \to \infty} f(t) = k < \infty$ and $\dot{f}(t)$ is uniformly continuous, then $\lim_{t \to \infty} \dot{f}(t) = 0$. It can be proved that a differentiable function is uniformly continuous if its derivative is bounded. Hence, the lemma can be rewritten as: if $\lim_{t \to \infty} f(t) = k < \infty$ and $\ddot{f}(t)$ exists and is bounded, then $\dot{f} \to 0$ as $t \to \infty$. In the Lyapunov stability analysis, Barbalat's lemma can be applied in the fashion similar to La Salle's theorem: If $V(\mathbf{x}, t)$ is lower bounded, $\dot{V} \leq 0$, and $\ddot{V}$ is bounded, then $\dot{V} \to 0$ as $t \to \infty$. It should be noted that the Lyapunov function is only required to be lower bounded in stead of positive definite. In addition, we can only conclude convergence of $\dot{V}$, not the states. In this book, we would like to use the other form of Barbalat's lemma to prove closed loop stability. If we can prove that a time function e is bounded and square integrable, and its time derivative is also bounded, then e is going to converge to zero asymptotically. It can be restated as: if $e \in L_\infty \cap L_2$ and $\dot{e} \in L_\infty$, then $e \to 0$ as $t \to \infty$.

2.9 Sliding Control

A practical control system should be designed to ensure system stability and performance to be invariant under perturbations from internal parameter variation, unmodeled dynamics excitation and external disturbances. For nonlinear systems, the sliding control is perhaps the most popular approach to achieve the robust performance requirement. In the sliding control, a sliding surface is designed so that the system trajectory is force to converge to the surface by some worst-case control efforts. Once on the surface, the system dynamics is reduced to a stable linear time invariant system which is irrelevant to the perturbations no matter from internal or external sources. Convergence of the output error is then easily achieved. In this section, we are going to review the sliding controller design including two smoothing techniques to eliminate the chattering activity in the control effort.

Let us consider a non-autonomous system

$$x^{(n)} = f(\mathbf{x},t) + g(\mathbf{x},t)u + d(t) \tag{1}$$

where $\mathbf{x} = [x \; \dot{x} \cdots x^{(n-1)}]^T \in \mathfrak{R}^n$ is the state vector, $x \in \mathfrak{R}$ the output of interest, and $u(t) \in \mathfrak{R}$ the control input. The function $f(\mathbf{x},t) \in \mathfrak{R}$ and the disturbance $d(t) \in \mathfrak{R}$ are both unknown functions of time, but bounds of their variations should be available. The control gain function $g(\mathbf{x},t) \in \mathfrak{R}$ is assumed to be non-singular for all admissible $\mathbf{x}$ and for all time t. In the following derivation, we would like to design a sliding controller with the knowledge of $g(\mathbf{x},t)$ first, and then a controller is constructed with unknown $g(\mathbf{x},t)$. Let us assume that $f(\mathbf{x},t)$ and $d(t)$ can be modeled as

$$f = f_m + \Delta f \tag{2a}$$

$$d = d_m + \Delta d \tag{2b}$$

where f_m and d_m are known nominal values of f and d, respectively. The uncertain terms Δf and Δd are assumed to be bounded by some known functions $\alpha(\mathbf{x},t) > 0$ and $\beta(\mathbf{x},t) > 0$, respectively, as

$$|\Delta f| \le \alpha(\mathbf{x},t) \tag{3a}$$

$$|\Delta d| \le \beta(\mathbf{x},t) \tag{3b}$$

Since the system contains uncertainties, the inversion-based controller

$$u = \frac{1}{g(\mathbf{x},t)}[-f(\mathbf{x},t) - d(t) + v] \tag{4}$$

is not realizable. We would like to design a tracking controller so that the output x tracks the desired trajectory x_d asymptotically regardless of the presence of uncertainties. Let us define a *sliding surface* $s(\mathbf{x},t) = 0$ as a desired error dynamics, where $s(\mathbf{x},t)$ is a linear stable differential operator acting on the tracking error $e = x - x_d$ as

$$s = (\frac{d}{dt} + \lambda)^{n-1} e \tag{5}$$

where $\lambda > 0$ determines the behavior of the error dynamics. Selection of the sliding surface is not unique, but the one in (5) is preferable simply because it is linear and it will result in a relative degree one dynamics, i.e. u appears when we differentiate s once. One way to achieve output error convergence is to find a control u such that the state trajectory converges to the sliding surface. Once on the surface, the system behaves like a stable linear system $(\frac{d}{dt} + \lambda)^{n-1} e = 0$; therefore, asymptotic convergence of the tracking error can be obtained. Now, the problem is how to drive the system trajectory to the sliding surface. With $s(\mathbf{x},t) = 0$ as the boundary, the state space can be decomposed into two parts: the one with $s > 0$ and the other with $s < 0$. Intuitively, to make the sliding surface attractive, we can design a control u so that s will decrease in the $s > 0$ region, and it will increase in the $s < 0$ region. This condition is called the *sliding condition* which can be written in a compact form

$$s\dot{s} < 0 \tag{6}$$

Using (5), the sliding surface for system (1) is of the form

$$s = (\frac{d}{dt} + \lambda)^{n-1} e$$
$$= c_1 e + c_2 \dot{e} + \cdots + c_{n-1} e^{(n-2)} + c_n e^{(n-1)} \tag{7}$$

where $c_k = \dfrac{(n-1)!\lambda^{n-k}}{(n-k)!(k-1)!}$, $k=1,...,n-1$, and $c_n=1$. Let us differentiate s with respect to time once to make u appear

$$\dot{s} = c_1\dot{e} + \cdots + c_{n-1}e^{(n-1)} + c_n e^{(n)}$$
$$= c_1\dot{e} + \cdots + c_{n-1}e^{(n-1)} + x^{(n)} - x_d^{(n)}$$
$$= c_1\dot{e} + \cdots + c_{n-1}e^{(n-1)} + f + gu + d - x_d^{(n)} \tag{8}$$

Let us select the control u as

$$u = \frac{1}{g}[-c_1\dot{e} - \cdots - c_{n-1}e^{(n-1)} - f_m - d_m + x_d^{(n)} - \eta_1 \operatorname{sgn}(s)] \tag{9}$$

where $\eta_1 > 0$ is a design parameter to be determined, so that (8) becomes

$$\dot{s} = \Delta f + \Delta d - \eta_1 \operatorname{sgn}(s) \tag{10}$$

To satisfy the sliding condition (6), let us multiply both sides of (10) with s as

$$s\dot{s} = (\Delta f + \Delta d)s - \eta_1|s|$$
$$\leq (\alpha + \beta)|s| - \eta_1|s| \tag{11}$$

By picking $\eta_1 = \alpha + \beta + \eta$ with $\eta > 0$, the above inequality becomes

$$s\dot{s} \leq -\eta|s| \tag{12}$$

Therefore, with the controller (9), the sliding surface (7) is attractive, and the tracking error converges asymptotically regardless of the uncertainties in f and d, once the sliding surface is reached. Now, let us consider the case when $g(\mathbf{x}, t)$ is not available, but we do know that it is non-singular for all admissible state and time t, and its variation bound is known with $0 < g_{\min} \leq g \leq g_{\max}$. In stead of the additive uncertainty model we used for f and d, a multiplicative model is chosen for g as

$$g = g_m \Delta g \tag{13}$$

where Δg satisfies the relation

$$0 \le \gamma_{min} \equiv \frac{g_{min}}{g_m} \le \Delta g \le \frac{g_{max}}{g_m} \equiv \gamma_{max} . \tag{14}$$

In this case, the controller is chosen as

$$u = \frac{1}{g_m}[-c_1\dot{e} - \cdots - c_{n-1}e^{(n-1)} - f_m - d_m + x_d^{(n)} - \eta_1 \, \mathrm{sgn}(s)] \tag{15}$$

Substituting (15) into (8), we have

$$\dot{s} = (1 - \Delta g)[c_1\dot{e} + \cdots + c_{n-1}e^{(n-1)} + f_m + d_m - x_d^{(n)}]$$
$$+ \Delta f + \Delta d - \Delta g \eta_1 \, \mathrm{sgn}(s) \tag{16}$$

Multiplying both sides with s, equation (16) becomes

$$s\dot{s} = (1 - \Delta g)[c_1\dot{e} + \cdots + c_{n-1}e^{(n-1)} + f_m + d_m - x_d^{(n)}]s$$
$$+ (\Delta f + \Delta d)s - \Delta g \eta_1 |s|$$
$$\le (1 - \gamma_{min})\left|c_1\dot{e} + \cdots + c_{n-1}e^{(n-1)} + f_m + d_m - x_d^{(n)}\right||s|$$
$$+ (\alpha + \beta)|s| - \gamma_{min}\eta_1 |s| \tag{17}$$

The parameter η_1 can thus be selected as

$$\eta_1 = \frac{1}{\gamma_{min}}[(1 - \gamma_{min})\left|c_1\dot{e} + \cdots + c_{n-1}e^{(n-1)} + f_m + d_m - x_d^{(n)}\right|$$
$$+ (\alpha + \beta) + \eta] \tag{18}$$

where η is a positive number. Therefore, we can also have the result in (12).

Smoothed sliding control law

Both controllers (9) and (15) contain the switching function sgn(s). In practical implementation, the switching induced from this function will sometimes result in control chattering. Consequently, the tracking performance degrades, and the high-frequency unmodeled dynamics may be excited. In some cases, the switching controller has to be modified with a continuous approximation. One approach is to use the saturation function $\text{sat}(\sigma)$ defined below instead of the signum function $\text{sgn}(s)$.

$$\text{sat}(\sigma) = \begin{cases} \sigma & \text{if } |\sigma| \leq \phi \\ \text{sgn}(\sigma) & \text{if } |\sigma| > \phi \end{cases} \qquad (19)$$

where $\phi > 0$ is called the *boundary layer* of the sliding surface. When s is outside the boundary layer, i.e., $|s| > \phi$, the sliding controller with $\text{sgn}(s)$ is exactly the same as the one with $\text{sat}(\dfrac{s}{\phi})$. Hence, the boundary layer is also attractive. When s is inside the boundary layer, equation (10) becomes

$$\dot{s} + \eta_1 \frac{s}{\phi} = \Delta f + \Delta d \qquad (20)$$

This implies that the signal s is the output of a stable first-order filter whose input is the bounded model error $\Delta f + \Delta d$. Thus, the chattering behavior can indeed be eliminated with proper selection of the filer bandwidth and as long as the high-frequency unmodeled dynamics is not excited. One drawback of this smoothed sliding controller is the degradation of the tracking accuracy. At best we can say that once the signal s converges to the boundary layer, the output tracking error is bounded by the value ϕ.

Instead of the saturation function, we may also use $\dfrac{s}{\phi}$ to have a smoothed version of the sliding controller. This selection is very easy in implementation, because the robust term is linear in the signal s. For example, controller (9) can be smoothed in the form

$$u = \frac{1}{g}[-c_1\dot{e} - \cdots - c_{n-1}e^{(n-1)} - f_m - d_m + x_d^{(n)} - \eta_1 \frac{s}{\phi}] \qquad (21)$$

To justify its effectiveness, the following analysis is performed. When s is outside the boundary layer, (10) can be rewritten in the form

$$\dot{s} = \Delta f + \Delta d - \eta_1 \frac{s}{\phi} \qquad (22)$$

With the selection of $\eta_1 = \alpha + \beta + \eta$, the sliding condition can be checked as

$$s\dot{s} = (\Delta f + \Delta d)s - \eta_1 \frac{s^2}{\phi}$$

$$\leq (\alpha + \beta)|s| - (\alpha + \beta + \eta)\frac{s^2}{\phi}$$

$$= (\alpha + \beta)|s|\left[1 - \frac{|s|}{\phi}\right] - \eta\frac{s^2}{\phi} \tag{23}$$

Since $|s| > \phi$, i.e., when outside the boundary layer, we may have the result $s\dot{s} \leq -\eta\frac{s^2}{\phi}$. Hence, the boundary layer is still attractive. When s is inside the boundary layer, equation (20) can be obtained; therefore, effective chattering elimination can be achieved. Let us now consider the case when $g(\mathbf{x},t)$ is unknown and the sliding controller is smoothed with $\frac{s}{\phi}$ as

$$u = \frac{1}{g_m}[\bar{u} + x_d^{(n)} - \eta_1\frac{s}{\phi}] \tag{24}$$

where $\bar{u} = -c_1\dot{e} - \cdots - c_{n-1}e^{(n-1)} - f_m - d_m$. By selecting η_1 according to (18), the sliding condition is checked with

$$s\dot{s} \leq [(1 - \beta_{\min})|\bar{u}| + \alpha + \beta]|s|\left[1 - \frac{|s|}{\phi}\right] - \eta\frac{s^2}{\phi} \tag{25}$$

If s is outside the boundary layer, (25) implies $s\dot{s} \leq -\eta\frac{s^2}{\phi}$, i.e., all trajectories will eventually converge to the boundary layer even though the system contains uncertainties. Since inside the boundary layer there is also an equivalent first order filter dynamics, the chattering activity can be effectively eliminated. The output tracking error, however, can only be concluded to be uniformly bounded. There is one more drawback for the smoothing using $\frac{s}{\phi}$, i.e., the initial control effort may become enormously large if there is a significant difference between the desired trajectory and the initial state. To overcome this problem, desired trajectory and initial conditions should be carefully selected.

2.10 Model Reference Adaptive Control (MRAC)

Adaptive control and robust control are two main approaches for controlling systems containing uncertainties and disturbances. The sliding control introduced in the previous section is one of the robust designs widely used in the literature. In this section, the well-known MRAC is reviewed as an example in the traditional adaptive approach. For an adaptive controller to be feasible the system structure is assumed to be known and a set of unknown constant system parameters (or equivalently the corresponding controller parameters) are to be estimated so that the closed loop stability is ensured via a certainty equivalence based controller. In this section, the MRAC for a linear time-invariant scalar system is introduced first, followed by the design for the vector case. The persistent excitation condition is investigated for the convergence of estimated parameters. Two modifications to the update law are introduced to robustify the adaptive loop when the system contains unmodeled dynamics or external disturbances.

2.10.1 MRAC of LTI scalar systems

Consider a linear time-invariant system described by the differential equation

$$\dot{x}_p = a_p x_p + b_p u \tag{1}$$

where $x_p \in \Re$ is the state of the plant and $u \in \Re$ the control input. The parameters a_p and b_p are unknown constants, but $\text{sgn}(b_p)$ is available. The pair (a_p, b_p) is controllable. The problem is to design a control u and an update law so that all signals in the closed loop plant are bounded and the system output x_p tracks the output x_m of the reference model

$$\dot{x}_m = a_m x_m + b_m r \tag{2}$$

asymptotically, where a_m and b_m are known constants with $a_m < 0$, and r is a bounded reference signal. If plant parameters a_p and b_p are available, the model reference control (MRC) rule can be designed as

$$u = ax_p + br \tag{3}$$

where $a = \dfrac{a_m - a_p}{b_p}$ and $b = \dfrac{b_m}{b_p}$ are perfect gains for transforming dynamics in (1) into (2). Since the values of a_p and b_p are not given, we may not select these perfect gains to complete the MRC design in (3) and the model reference adaptive control (MRAC) rule is constructed instead

$$u = \hat{a}(t)x_p + \hat{b}(t)r \tag{4}$$

where $\hat{a}$ and $\hat{b}$ are estimates of a and b, respectively, and proper update laws are to be selected to give $\hat{a} \to a$ and $\hat{b} \to b$. Define the output tracking error as

$$e = x_p - x_m \tag{5}$$

then the error dynamics can be computed as following

$$\dot{e} = a_m e + b_p(\hat{a} - a)x_p + b_p(\hat{b} - b)r \tag{6}$$

Define the parameter errors $\tilde{a} = \hat{a} - a$ and $\tilde{b} = \hat{b} - b$, then equation (6) is further written as

$$\dot{e} = a_m e + b_p \tilde{a} x_p + b_p \tilde{b} r \tag{7}$$

This is the dynamics of the output error e, which is a stable linear system driven by the parameter errors. Therefore, if update laws are found to have convergence of these parameter errors, convergence of the output error is then ensured. To find these update laws for $\hat{a}$ and $\hat{b}$, let us define a Lyapunov function candidate

$$V(e, \tilde{a}, \tilde{b}) = \frac{1}{2}e^2 + \frac{1}{2}\left|b_p\right|(\tilde{a}^2 + \tilde{b}^2) \tag{8}$$

Taking the time derivative of V along the trajectory of (7), we have

$$
\begin{aligned}
\dot{V} &= e\dot{e} + \left|b_p\right|(\tilde{a}\dot{\hat{a}} + \tilde{b}\dot{\hat{b}}) \\
&= e(a_m e + b_p \tilde{a} x_p + b_p \tilde{b} r) + \left|b_p\right|(\tilde{a}\dot{\hat{a}} + \tilde{b}\dot{\hat{b}}) \\
&= a_m e^2 + \tilde{a}\left|b_p\right|[\mathrm{sgn}(b_p)ex_p + \dot{\hat{a}}] + \tilde{b}\left|b_p\right|[\mathrm{sgn}(b_p)er + \dot{\hat{b}}]
\end{aligned} \tag{9}
$$

If the update laws are selected as

$$\dot{\hat{a}} = -\text{sgn}(b_p)ex_p \tag{10a}$$

$$\dot{\hat{b}} = -\text{sgn}(b_p)er \tag{10b}$$

then (9) becomes

$$\dot{V} = a_m e^2 \le 0 \tag{11}$$

This implies that $e, \tilde{a}, \tilde{b} \in L_\infty$. From the simple derivation

$$\int_0^\infty e^2 dt = -a_m^{-1} \int_0^\infty \dot{V} dt = a_m^{-1}(V_0 - V_\infty) < \infty$$

we know that $e \in L_2$. The result $\dot{e} \in L_\infty$ can easily be concluded from (7). Therefore, if follows from Barbalat's lemma that the output error $e(t)$ converges to zero asymptotically as $t \to \infty$. In summary, the controller (4) together with update laws in (10) make the system (1) track the reference model (2) asymptotically with boundedness of all internal signals. It can be observed in (10) that the update laws are driven by the tracking error e. Once e gets close to zero, the estimated parameters converges to some values. We cannot predict the exact values these parameters will converge to from the above derivation. Let us consider the situation when the system gets into the steady state, i.e., when $t \to \infty$. The error dynamics (7) becomes

$$\tilde{a}x_p + \tilde{b}r = 0 \tag{12}$$

If r is a constant, then x_p in (12) can be found as $x_p = x_m = kr$, where $k = -\dfrac{b_m}{a_m}$ is the d.c. gain of the reference model (2). Equation (12) further implies

$$k\tilde{a} + \tilde{b} = 0 \tag{13}$$

which is exactly a straight line in the parameter error space. Therefore, for a constant reference input r, both the estimated parameters do not necessarily converge to zero. To investigate the problem of parameter convergence, we need the concept of persistent excitation. A signal $\mathbf{v} \in \mathfrak{R}^n$ is said to satisfy the *persistent excitation* (PE) condition if $\exists \alpha, T > 0$ such that

$$\int_{t}^{t+T} \mathbf{v}\mathbf{v}^{T} dt \geq \alpha \mathbf{I} \tag{14}$$

Define $\mathbf{v} = [x_p \quad r]^{T}$ and $\tilde{\theta} = [\tilde{a} \quad \tilde{b}]^{T}$, and equation (12) is able to be represented into the vector form

$$\mathbf{v}^{T}\tilde{\theta} = [x_p \quad r]\begin{bmatrix} \tilde{a} \\ \tilde{b} \end{bmatrix} = 0 \tag{15}$$

Since $\mathbf{v}\mathbf{v}^{T}\tilde{\theta} = 0$, its integration in $[t, t+T]$ is

$$\int_{t}^{t+T} \mathbf{v}\mathbf{v}^{T}\tilde{\theta} dt = 0 \tag{16}$$

When $t \to \infty$, update laws in (10) imply $\dot{\tilde{\theta}} \to 0$; therefore, (16) becomes

$$\int_{t}^{t+T} \mathbf{v}\mathbf{v}^{T} dt\tilde{\theta} = 0 \tag{17}$$

Hence, if $\mathbf{v}$ is PE, equation (17) implies $\tilde{\theta} = 0$, i.e., parameter convergence when $t \to \infty$. A more general treatment of the PE condition will be presented in Section 2.10.3.

2.10.2 MRAC of LTI systems: vector case

Consider a linear time-invariant system

$$\dot{\mathbf{x}}_p = \mathbf{A}_p\mathbf{x}_p + \mathbf{B}_p\mathbf{u} \tag{18}$$

where $\mathbf{x}_p \in \Re^{n}$ is the state vector, $\mathbf{u} \in \Re^{m}$ is the control vector, and $\mathbf{A}_p \in \Re^{n \times n}$ and $\mathbf{B}_p \in \Re^{n \times m}$ are unknown constant matrices. The pair $(\mathbf{A}_p, \mathbf{B}_p)$ is controllable. The problem is to find a control $\mathbf{u}$ so that all signals in the closed loop system are bounded and the system states asymptotically track the states of the reference model

$$\dot{\mathbf{x}}_m = \mathbf{A}_m\mathbf{x}_m + \mathbf{B}_m\mathbf{r} \tag{19}$$

where $\mathbf{A}_m \in \Re^{n \times n}$ and $\mathbf{B}_m \in \Re^{n \times m}$ are known and $\mathbf{r} \in \Re^m$ is a bounded reference input vector. All eigenvalues of $\mathbf{A}_m$ are assumed to be with strictly negative real parts. A control law for this problem can be designed in the form

$$\mathbf{u} = \mathbf{B}\mathbf{A}\mathbf{x}_p + \mathbf{B}\mathbf{r} \tag{20}$$

where $\mathbf{A}$ and $\mathbf{B}$ are feedback gain matrices. Substituting (20) into (18), the closed loop system is derived as

$$\dot{\mathbf{x}}_p = (\mathbf{A}_p + \mathbf{B}_p \mathbf{B}\mathbf{A})\mathbf{x}_p + \mathbf{B}_p \mathbf{B}\mathbf{r} \tag{21}$$

Hence, if $\mathbf{A}$ and $\mathbf{B}$ are chosen so that

$$\mathbf{A}_p + \mathbf{B}_p \mathbf{B}\mathbf{A} = \mathbf{A}_m \tag{22a}$$

$$\mathbf{B}_p \mathbf{B} = \mathbf{B}_m \tag{22b}$$

then the behavior of system (18) is identical to that of the reference model. The control in (20) is called the *MRC rule* if $\mathbf{A}$ and $\mathbf{B}$ satisfy (22). Since the values of $\mathbf{A}_p$ and $\mathbf{B}_p$ are not given, the MRC rule is not realizable. Now, let us replace the values $\mathbf{A}$ and $\mathbf{B}$ in (20) with their estimates $\hat{\mathbf{A}}(t)$ and $\hat{\mathbf{B}}(t)$, respectively, to have the MRAC rule

$$\mathbf{u} = \hat{\mathbf{B}}(t)\hat{\mathbf{A}}(t)\mathbf{x}_p + \hat{\mathbf{B}}(t)\mathbf{r} \tag{23}$$

We would like to design proper update laws to have $\hat{\mathbf{A}} \to \mathbf{A}$ and $\hat{\mathbf{B}} \to \mathbf{B}$. Define the tracking error

$$\mathbf{e} = \mathbf{x}_p - \mathbf{x}_m \tag{24}$$

then the error dynamics can be computed as following

$$\begin{aligned}
\dot{\mathbf{e}} &= \mathbf{A}_m(\mathbf{x}_p - \mathbf{x}_m) + \mathbf{B}_m(\hat{\mathbf{A}} - \mathbf{A})\mathbf{x}_p \\
&\quad + (\mathbf{B}_p\hat{\mathbf{B}} - \mathbf{B}_m)\hat{\mathbf{A}}\mathbf{x}_p + (\mathbf{B}_p\hat{\mathbf{B}} - \mathbf{B}_m)\mathbf{r} \\
&= \mathbf{A}_m\mathbf{e} + \mathbf{B}_m\tilde{\mathbf{A}}\mathbf{x}_p + (\mathbf{B}_p\hat{\mathbf{B}} - \mathbf{B}_m)(\hat{\mathbf{A}}\mathbf{x}_p + \mathbf{r})
\end{aligned} \tag{25}$$

where $\tilde{\mathbf{A}} = \hat{\mathbf{A}} - \mathbf{A}$. Using the relations in (22), the above equation is reduced to

$$\dot{\mathbf{e}} = \mathbf{A}_m\mathbf{e} + \mathbf{B}_m\tilde{\mathbf{A}}\mathbf{x}_p + \mathbf{B}_m(\mathbf{B}^{-1}\hat{\mathbf{B}} - \mathbf{I})\hat{\mathbf{B}}^{-1}(\hat{\mathbf{B}}\hat{\mathbf{A}}\mathbf{x}_p + \hat{\mathbf{B}}\mathbf{r})$$
$$= \mathbf{A}_m\mathbf{e} + \mathbf{B}_m\tilde{\mathbf{A}}\mathbf{x}_p + \mathbf{B}_m(\mathbf{B}^{-1} - \hat{\mathbf{B}}^{-1})\mathbf{u} \qquad (26)$$

Define $\tilde{\mathbf{B}}_1 = \mathbf{B}^{-1} - \hat{\mathbf{B}}^{-1}$, then equation (26) becomes

$$\dot{\mathbf{e}} = \mathbf{A}_m\mathbf{e} + \mathbf{B}_m\tilde{\mathbf{A}}\mathbf{x}_p + \mathbf{B}_m\tilde{\mathbf{B}}_1\mathbf{u} \qquad (27)$$

To find update laws for $\hat{\mathbf{A}}$ and $\hat{\mathbf{B}}$, let us define a Lyapunov function candidate

$$V(\mathbf{e}, \tilde{\mathbf{A}}, \tilde{\mathbf{B}}_1) = \mathbf{e}^T\mathbf{P}\mathbf{e} + Tr(\tilde{\mathbf{A}}^T\tilde{\mathbf{A}} + \tilde{\mathbf{B}}_1^T\tilde{\mathbf{B}}_1) \qquad (28)$$

where $\mathbf{P} \in \mathfrak{R}^{n \times n}$ is a positive definite matrix satisfying $\mathbf{A}_m^T\mathbf{P} + \mathbf{P}\mathbf{A}_m = -\mathbf{Q}$ for some positive definite matrix $\mathbf{Q} \in \mathfrak{R}^{n \times n}$. Taking time derivative of V along the trajectory of (27) and selecting the update laws

$$\dot{\hat{\mathbf{A}}} = -\mathbf{B}_m^T\mathbf{P}\mathbf{e}\mathbf{x}_p^T \qquad (29a)$$

$$\dot{\hat{\mathbf{B}}}_1 = -\mathbf{B}_m^T\mathbf{P}\mathbf{e}\mathbf{u}^T \qquad (29b)$$

we have the result $\dot{V} = -\mathbf{e}^T\mathbf{Q}\mathbf{e} \leq 0$. By using the identity (2.5-9c), update law (29b) can be transformed to

$$\dot{\hat{\mathbf{B}}} = -\hat{\mathbf{B}}\mathbf{B}_m^T\mathbf{P}\mathbf{e}\mathbf{u}^T\hat{\mathbf{B}} \qquad (29c)$$

Therefore, we have proved that the origin $(\mathbf{e}, \tilde{\mathbf{A}}, \tilde{\mathbf{B}}_1) = 0$ is uniformly stable using controller (23) with update laws in (29). It should be noted that the control scheme can only guarantee uniform stability of the origin in the $\{\mathbf{e}, \tilde{\mathbf{A}}, \tilde{\mathbf{B}}\}$ space. Since the Lyapunov function in (28) is not radially unbounded, the global behavior cannot be concluded. If the reference input is PE, convergence of the estimated parameters can further be proved.

2.10.3 Persistent excitation

We have introduced MRAC laws for linear time-invariant systems to have asymptotic tracking error convergence performance. However, we can only obtain uniformly boundedness of parameter errors in those systems. In this section, we are going to investigate the problem of persistency of excitation of signals in the closed loop system, which relates to the convergence of the parameter vector. Consider a special linear system (Marino and Tomei 1996)

$$\dot{\mathbf{x}} = \mathbf{A}\mathbf{x} + \Omega^T(t)\mathbf{z} \tag{30a}$$

$$\dot{\mathbf{z}} = -\Omega(t)\mathbf{P}\mathbf{x} \tag{30b}$$

where $\mathbf{x} \in \mathfrak{R}^n$ and $\mathbf{z} \in \mathfrak{R}^p$. $\mathbf{A} \in \mathfrak{R}^{n \times n}$ is a Hurwitz matrix, and $\mathbf{P} \in \mathfrak{R}^{n \times n}$ is a positive definite matrix satisfying $\mathbf{A}^T\mathbf{P} + \mathbf{P}\mathbf{A} = -\mathbf{Q}$ for some positive definite $n \times n$ matrix $\mathbf{Q}$. The $p \times n$ real matrix Ω has the property that $\|\Omega(t)\|$ and $\|\dot{\Omega}(t)\|$ are uniformly bounded. System (30) is frequently encountered in the parameter convergence analysis of adaptive systems. Equation (30a) usually corresponds to the tracking error dynamics and (30b) is the update law. These can be confirmed with MRAC systems introduced in the previous section. To have parameter convergence in those systems, it is equivalent to require convergence of vector $\mathbf{z}$ in (30b). Here, we would like to prove that as long as the persistent excitation condition is satisfied by the signal matrix Ω, the equilibrium point $(\mathbf{x}, \mathbf{z}) = \mathbf{0}$ is globally exponentially stable.

We first show that the tracking error vector $\mathbf{x}$ converges to zero asymptotically as $t \to \infty$. Define a Lyapunov function candidate

$$V(\mathbf{x}, \mathbf{z}) = \mathbf{x}^T\mathbf{P}\mathbf{x} + \mathbf{z}^T\mathbf{z} \tag{31}$$

Along the trajectory of (30), the time derivative of V can be computed to have

$$\dot{V} = \mathbf{x}^T(\mathbf{A}^T\mathbf{P} + \mathbf{P}\mathbf{A})\mathbf{x} = -\mathbf{x}^T\mathbf{Q}\mathbf{x} \le 0 \tag{32}$$

Hence, the origin of (30) is uniformly stable, and $\mathbf{x} \in L_\infty^n$, $\mathbf{z} \in L_\infty^p$. From the computation

$$\int_0^\infty \mathbf{x}^T\mathbf{Q}\mathbf{x}dt = -\int_0^\infty \dot{V}dt = V_0 - V_\infty < \infty$$

we have $\mathbf{x} \in L_2^n$. Boundedness of $\dot{\mathbf{x}}$ can be obtained by observing (30a). Therefore, by Barbalat's lemma, we have proved $\mathbf{x} \to \mathbf{0}$ as $t \to \infty$.

To prove asymptotic convergence of $\mathbf{z}$, we need to prove that $\forall \varepsilon > 0$, $\exists T_\varepsilon > 0$ such that $\|\mathbf{z}(t)\| < \varepsilon$, $\forall t \geq T_\varepsilon$. Since when $t \geq T_\varepsilon$, V in (31) satisfies $V(T_\varepsilon) \geq V(t)$, or equivalently

$$\mathbf{x}^T(T_\varepsilon)\mathbf{P}\mathbf{x}(T_\varepsilon) + \mathbf{z}^T(T_\varepsilon)\mathbf{z}(T_\varepsilon) \geq \mathbf{x}^T(t)\mathbf{P}\mathbf{x}(t) + \mathbf{z}^T(t)\mathbf{z}(t)$$
$$\geq \mathbf{z}^T(t)\mathbf{z}(t) \tag{33}$$

Using relation (2.6-3), inequality (33) becomes

$$\lambda_{\max}(\mathbf{P})\|\mathbf{x}(T_\varepsilon)\|^2 + \|\mathbf{z}(T_\varepsilon)\|^2 \geq \|\mathbf{z}(t)\|^2$$

This further implies

$$\|\mathbf{z}(t)\| \leq \sqrt{\lambda_{\max}(\mathbf{P})\|\mathbf{x}(T_\varepsilon)\|^2 + \|\mathbf{z}(T_\varepsilon)\|^2} \tag{34}$$

Since we have proved $\mathbf{x} \to \mathbf{0}$ as $t \to \infty$, this implies that $\forall \varepsilon > 0$, $\exists t_\varepsilon > 0$ such that $\forall t \geq t_\varepsilon$,

$$\|\mathbf{x}(t)\| \leq \frac{\varepsilon}{\sqrt{2\lambda_{\max}(\mathbf{P})}}. \tag{35}$$

Plug (35) into (34), yields

$$\|\mathbf{z}(t)\| \leq \sqrt{\frac{\varepsilon^2}{2} + \|\mathbf{z}(T_\varepsilon)\|^2} \tag{36}$$

If we may claim that $\forall \varepsilon > 0$, $T > 0$ and for any initial conditions of $\mathbf{x}$ and $\mathbf{z}$, $\exists t > T$ such that $\|\mathbf{z}(t)\| < \varepsilon$, then we may use this property to say that $\forall \varepsilon > 0$, $\exists T_\varepsilon > t_\varepsilon$ such that $\|\mathbf{z}(T_\varepsilon)\| < \dfrac{\varepsilon}{\sqrt{2}}$. Under this condition, (36) can be further written as $\|\mathbf{z}(t)\| \leq \sqrt{\dfrac{\varepsilon^2}{2} + \dfrac{\varepsilon^2}{2}} = \varepsilon$ for all $t \geq T_\varepsilon$. This completes the proof of $\mathbf{z} \to \mathbf{0}$ as $t \to \infty$, if the claim is justified. Since all properties hold for all $\mathbf{x}$ and $\mathbf{z}$, and are uniform respect to the initial time, the equilibrium point $(\mathbf{x}, \mathbf{z}) = \mathbf{0}$ is globally uniformly stable. Since the system is linear, it is also globally exponentially stable.

Now let us prove the above claim by contradiction, i.e., to prove that $\forall \varepsilon > 0$, we cannot find $t_1 > 0$ such that $\|\mathbf{z}(t)\| > \varepsilon$ for all $t \geq t_1$. Suppose there exists a $t_1 > 0$ such that $\|\mathbf{z}(t)\| > \varepsilon$ for all $t \geq t_1$. The PE condition says that there exist T, $k > 0$, such that

$$\int_t^{t+T} \Omega(\tau)\Omega^T(\tau)d\tau \geq k\mathbf{I} > 0 \quad \forall t \geq t_0 . \tag{37}$$

For all $\mathbf{w} \in \mathfrak{R}^p$, $\|\mathbf{w}\| = 1$, inequality (37) implies

$$\int_t^{t+T} \mathbf{w}^T \Omega(\tau)\mathbf{P}\Omega^T(\tau)\mathbf{w}d\tau \geq k\lambda_{\min}(\mathbf{P}) \quad \forall t \geq t_0 \tag{38}$$

Let $\mathbf{w} = \dfrac{\mathbf{z}}{\|\mathbf{z}\|}$, then (38) becomes

$$\int_t^{t+T} \mathbf{z}^T \Omega(\tau)\mathbf{P}\Omega^T(\tau)\mathbf{z}d\tau \geq k\lambda_{\min}(\mathbf{P})\|\mathbf{z}\|^2 \quad \forall t \geq t_0$$

$$\geq k\lambda_{\min}(\mathbf{P})\varepsilon^2 \quad \forall t \geq t_1 \tag{39}$$

Consider the bounded function for some $T > 0$

$$\phi_T(\mathbf{z}(t),t) = \frac{1}{2}[\mathbf{z}^T(t+T)\mathbf{z}(t+T) - \mathbf{z}^T(t)\mathbf{z}(t)] \tag{40}$$

Its time derivative is computed as following

$$\dot{\phi}_T = \mathbf{z}^T(t+T)\dot{\mathbf{z}}(t+T) - \mathbf{z}^T(t)\dot{\mathbf{z}}(t)$$

$$= \int_t^{t+T} \frac{d}{d\tau}[\mathbf{z}^T(\tau)\dot{\mathbf{z}}(\tau)]d\tau \tag{41}$$

Using (30) and (39), equation (41) can be derived as

$$\dot{\phi}_T = \int_t^{t+T} [\mathbf{x}^T\mathbf{P}\Omega^T\Omega\mathbf{P}\mathbf{x} - \mathbf{z}^T\dot{\Omega}\mathbf{P}\mathbf{x} - \mathbf{z}^T\Omega\mathbf{P}A\mathbf{x}]d\tau$$

$$- \int_t^{t+T} \mathbf{z}^T\Omega\mathbf{P}\Omega^T\mathbf{z}d\tau$$

$$\leq \int_t^{t+T} [\mathbf{x}^T\mathbf{P}\Omega^T\Omega\mathbf{P}\mathbf{x} - \mathbf{z}^T\dot{\Omega}\mathbf{P}\mathbf{x} - \mathbf{z}^T\Omega\mathbf{P}A\mathbf{x}]d\tau$$

$$- k\lambda_{\min}(\mathbf{P})\varepsilon^2 \quad \forall t \geq t_1 \tag{42}$$

Since $\mathbf{x}$, $\mathbf{z}$, Ω and $\dot{\Omega}$ are all uniformly bounded, there exist $\overline{\Omega}$, $\overline{\mathbf{x}}$ and $\overline{\mathbf{z}}$ such that $\overline{\Omega} = \max\{\sup_t \|\Omega(t)\|, \sup_t \|\dot{\Omega}(t)\|\}$, $\overline{\mathbf{x}} = \sup_t \|\mathbf{x}(t)\|$ and $\overline{\mathbf{z}} = \sup_t \|\mathbf{z}(t)\|$. Then the integration term in (42) can be represented in the form

$$\int_t^{t+T} [\mathbf{x}^T \mathbf{P}\Omega^T \Omega \mathbf{P}\mathbf{x} - \mathbf{z}^T \dot{\Omega} \mathbf{P}\mathbf{x} - \mathbf{z}^T \Omega \mathbf{P}\mathbf{A}\mathbf{x}] d\tau$$

$$\leq [\lambda_{\max}^2(\mathbf{P})\overline{\Omega}^2\overline{\mathbf{x}} + \lambda_{\max}(\mathbf{P})\overline{\Omega}\overline{\mathbf{z}} + \lambda_{\max}(\mathbf{P})\overline{\Omega}\overline{\mathbf{z}}\|\mathbf{A}\|] \int_t^{t+T} \|\mathbf{x}(\tau)\| d\tau \quad (43)$$

Since we have proved that $\|\mathbf{x}\| \to 0$ as $t \to \infty$, then $\exists t_2 > 0$ such that $\forall t \geq t_2$

$$\int_t^{t+T} [\mathbf{x}^T \mathbf{P}\Omega^T \Omega \mathbf{P}\mathbf{x} - \mathbf{z}^T \dot{\Omega} \mathbf{P}\mathbf{x} - \mathbf{z}^T \Omega \mathbf{P}\mathbf{A}\mathbf{x}] d\tau \leq \frac{1}{2} k\lambda_{\min}(\mathbf{P})\varepsilon^2 \quad (44)$$

Plug (44) into (42) for all $t \geq \max\{t_1, t_2\}$, we have

$$\dot{\phi}_T \leq \frac{1}{2} k\lambda_{\min}(\mathbf{P})\varepsilon^2 - k\lambda_{\min}(\mathbf{P})\varepsilon^2 = -\frac{1}{2} k\lambda_{\min}(\mathbf{P})\varepsilon^2 < 0$$

Hence, ϕ_T is an unbounded function, which is a contradiction to the assumption in (40). Therefore, we have proved the claim.

2.10.4 Robust adaptive control

The adaptive controllers presented in Section 2.10.1 and 2.10.2 are developed for LTI systems without external disturbances or unmodeled dynamics. For practical control systems, uncertain parameters may vary with time and the system may contain some non-parametric uncertainties. Rohrs et. al. (1985) showed that in the presence of a small amount of measurement noise and high-frequency unmodeled dynamics, an adaptive control system presents slow parameter drift behavior and the system output suddenly diverges sharply after a finite interval of time. For practical implementation, an adaptive control system should be designed to withstand all kinds of non-parametric uncertainties. Some modifications of the adaptive laws have been developed to deal with these problems. In the following, a technique called dead-zone is introduced followed by the review of the well-known σ-modification.

Dead-Zone

Consider the uncertain linear time-invariant system

$$\dot{x}_p = a_p x_p + b_p u + d(t) \tag{45}$$

where a_p is unknown but $b_p \neq 0$ is available. The disturbance $d(t)$ is assumed to be bounded by some $\delta > 0$. A reference model is designed as

$$\dot{x}_m = a_m x_m + b_m r \tag{46}$$

where $a_m < 0$ and $b_m = b_p$. Let $a = \dfrac{a_m - a_p}{b_p}$ and $b = 1$ be ideal feedback gains for the MRAC law (4). Since a_p is not given, a practical feedback law is designed to be

$$u = \hat{a}(t) x_p + br \tag{47}$$

where $\hat{a}(t)$ is the estimate of a. Let $e = x_p - x_m$ and $\tilde{a} = \hat{a} - a$, then the error dynamics is computed as

$$\dot{e} = a_m e + b_p \tilde{a} x_p + d(t) \tag{48}$$

The time derivative of the Lyapunov function

$$V(e, \tilde{a}) = \frac{1}{2}(e^2 + \tilde{a}^2)$$

along the trajectory of (48) is

$$\dot{V} = a_m e^2 + \tilde{a}(b_p e x_p + \dot{\hat{a}}) + ed \tag{49}$$

With the selection of the update law

$$\dot{\hat{a}} = -b_p e x_p \tag{50}$$

(49) becomes

$$\begin{aligned}
\dot{V} &= a_m e^2 + ed \\
&= (d - |a_m|e)e \\
&\leq (\delta - |a_m||e|)|e|
\end{aligned} \tag{51}$$

If $\delta - |a_m||e| < 0$, i.e., $|e| > \dfrac{\delta}{|a_m|}$, then (51) implies that V is non-increasing.

Let D be the set where V will grow unbounded, i.e. $D = \left\{ (e, \tilde{a}) \Big| |e| \leq \dfrac{\delta}{|a_m|} \right\}$;

therefore, the modified update law

$$\dot{\hat{a}} = \begin{cases} -ex_p & \text{if } (e, \tilde{a}) \in D^c \\ 0 & \text{if } (e, \tilde{a}) \in D \end{cases} \tag{52}$$

assures boundedness of all signals in the system. The notation D^c denotes the complement of D. The modified update law (52) implies that when the error e is within the dead-zone D, the update law is inactive to avoid possible parameter drift. It should be noted that, however, the asymptotic convergence of the error signal e is no longer valid after the dead-zone modification even when the disturbance is removed.

σ-modification

In applying the dead-zone modification, the upper bound of the disturbance signal is required to be given. Here, a technique called σ-modification is introduced which does not need the information of disturbance bounds.

Instead of (52), the update law (50) is modified as

$$\dot{\hat{a}} = -b_p ex_p - \sigma \hat{a} \tag{53}$$

where σ is a small positive constant. Then (51) becomes

$$\begin{aligned} \dot{V} &= a_m e^2 - \sigma \tilde{a}\hat{a} + ed \\ &\leq -|a_m|e^2 - \sigma \tilde{a}^2 - \sigma \tilde{a}a + |e|\delta \end{aligned} \tag{54}$$

for some unknown $\delta > 0$. Rewrite the two terms in (54) involving e as

$$-|a_m|e^2 + |e|\delta = -\frac{1}{2}\left[\sqrt{|a_m|}|e| - \frac{\delta}{\sqrt{|a_m|}} \right]^2 - \frac{1}{2}|a_m|e^2 + \frac{1}{2}\frac{\delta^2}{|a_m|}$$

$$\leq -\frac{1}{2}|a_m|e^2 + \frac{1}{2}\frac{\delta^2}{|a_m|} \tag{55}$$

Likewise, the rest two terms in (54) are derived as

$$-\sigma \tilde{a}^2 - \sigma \tilde{a}a \le -\sigma \tilde{a}^2 + \sigma |\tilde{a}||a|$$

$$\le -\frac{1}{2}\sigma \tilde{a}^2 + \frac{1}{2}\sigma |a|^2 \tag{56}$$

Substituting (55) and (56) into (54), we have

$$\dot{V} \le -\frac{1}{2}|a_m|e^2 + \frac{1}{2}\frac{\delta^2}{|a_m|} - \frac{1}{2}\sigma \tilde{a}^2 + \frac{1}{2}\sigma |a|^2 \tag{57}$$

Adding and subtracting αV for some $\alpha > 0$, (57) becomes

$$\dot{V} \le -\alpha V + \frac{1}{2}\frac{\delta^2}{|a_m|} + \frac{1}{2}\sigma |a|^2 + \frac{1}{2}[\alpha - |a_m|]e^2 + \frac{1}{2}[\alpha - \sigma]\tilde{a}^2 \tag{58}$$

Picking $\alpha < \min\{|a_m|, \sigma\}$, we obtain

$$\dot{V} \le -\alpha V + \frac{1}{2}\frac{\delta^2}{|a_m|} + \frac{1}{2}\sigma |a|^2 \tag{59}$$

Therefore, $\dot{V} \le 0$, if

$$V \ge \frac{1}{2\alpha}\frac{\delta^2}{|a_m|} + \frac{1}{2\alpha}\sigma |a|^2. \tag{60}$$

This implies that signals in the closed loop system are uniformly bounded. Hence, the additional term $\sigma \hat{a}$ in the update law makes the adaptive control system robust to bounded external disturbances, although bounds of these disturbances are not given. One drawback of this method is that the origin of the system (48) and (53) is no longer an equilibrium point, i.e., the error signal e will not converge to zero even when the disturbance is removed.

2.11 General Uncertainties

We have seen in Section 2.9 that, to derive a sliding controller, the variation bounds of the parametric uncertainties should be give. Availability for the knowledge of the uncertainty variation bounds is a must for almost all robust control strategies. This is because the robust controllers need to cover system uncertainties even for the worst case. One the other hand, we know from Section 2.10 that for the adaptive controller to be feasible the unknown parameters should be time-invariant. This is also almost true for most adaptive control schemes. Let us now consider the case when a system contains time-varying uncertainties whose variation bounds are not known. Since it is time-varying, traditional adaptive design is not feasible. Because the variation bounds are not given, the robust strategies fail. We would like to call this kind of uncertainties the *general uncertainties*. It is challenging to design controllers for systems containing general uncertainties.

In Section 2.11.1, we are going to have some investigation on the difficulties for the design of adaptive controllers when the system has time-varying parameters. In Section 2.11.2, we will look at the problem in designing robust controllers for systems containing uncertain parameters without knowing their bounds.

2.11.1 MRAC of LTV systems

In the conventional design of adaptive control systems such as the one introduced in Section 2.10, there is a common assumption that the unknown parameters to be updated should be time-invariant. This can be understood by considering the scalar linear time-varying system

$$\dot{x}_p = a_p(t)x_p + b_p u \tag{1}$$

where a_p is a time-varying unknown parameter and $b_p > 0$ is known. A controller is to be constructed such that the system behaves like the dynamics of the reference model

$$\dot{x}_m = a_m x_m + b_m r \tag{2}$$

where $a_m < 0$ and $b_m = b_p$. Let $a(t) = \dfrac{a_m - a_p(t)}{b_p}$ and $b = 1$ be ideal feedback gains for the MRC law $u = a(t)x_p + br$. Since $a_p(t)$ is not given, a practical feedback law based on the MRAC is designed

$$u = \hat{a}(t)x_p + br \tag{3}$$

where $\hat{a}(t)$ is an adjustable parameter of the controller. Let $e = x_p - x_m$ and $\tilde{a}(t) = \hat{a}(t) - a(t)$, then the error dynamics is computed to be

$$\dot{e} = a_m e + b_p \tilde{a} x_p \tag{4}$$

Take the time derivative of the Lyapunov function candidate

$$V(e, \tilde{a}) = \frac{1}{2} e^2 + \frac{1}{2} b_p \tilde{a}^2 \tag{5}$$

along the trajectory of (4), we have

$$\dot{V} = a_m e^2 + b_p \tilde{a}(e x_p + \dot{\hat{a}} - \dot{a}) \tag{6}$$

If we choose the update law similar to the one in (2.10-10a)

$$\dot{\hat{a}} = -e x_p \tag{7}$$

then (6) becomes

$$\dot{V} = a_m e^2 - b_p \tilde{a} \dot{a} \tag{8}$$

Since $\tilde{a}$ and $\dot{a}$ are not available, the definiteness of $\dot{V}$ can not be determined. Therefore, we are not able to conclude anything about the properties of the signals in the closed loop system. From here, we know that the assumption for the unknown parameters to be time-invariant is very important for the feasibility of the design of adaptive controllers. It is equivalent to say that the traditional MRAC fails in controlling systems with time-varying uncertainties.

2.11.2 Sliding control for systems with unknown variation bounds

Consider a first order uncertain nonlinear system

$$\dot{x} = f(x,t) + g(x,t)u \tag{9}$$

where $f(x,t)$ is a bounded uncertainty and $g(x,t)$ is a known nonsingular function. The uncertainty $f(x,t)$ is modeled as the summation of the known nominal value f_m and the unknown variation Δf.

$$f(x,t) = f_m + \Delta f \tag{10}$$

Since Δf is a bounded function with unknown bounds, there is a positive constant α satisfying

$$|\Delta f| \le \alpha \tag{11}$$

Let us select the sliding variable $s = x - x_d$, where x_d is the desired trajectory. The dynamics of the sliding variable is computed as

$$\dot{s} = f + gu - x_d \tag{12}$$

By selecting the sliding control law as

$$u = \frac{1}{g}[-f_m + x_d - \eta_1 \operatorname{sgn}(s)] \tag{13}$$

equation (12) becomes

$$\dot{s} = \Delta f - \eta_1 \operatorname{sgn}(s) \tag{14}$$

Multiplying s to the both sides to have

$$
\begin{aligned}
s\dot{s} &= \Delta f s - \eta_1 |s| \\
&\le (\alpha - \eta_1)|s|
\end{aligned}
\tag{15}
$$

Since α is not given, we may not select η_1 similar to the one we have in (2.9-11). Therefore, the sliding condition can not be satisfied and the sliding control fails in this case.

Bound estimation for uncertain parameter

An intuitive attempt in circumvent the difficulty here is to estimate α by using conventional adaptive strategies. Since α itself is a constant, it might be possible to design a proper update law $\dot{\hat{\alpha}}$ for its estimate $\hat{\alpha}$.

Let η be a positive number, then we may pick $\eta_1 = \hat{\alpha} + \eta$ so that equation (15) becomes

$$s\dot{s} \le \tilde{\alpha}|s| - \eta|s| \tag{16}$$

where $\tilde{\alpha} = \alpha - \hat{\alpha}$. Consider a Lyapunov function candidate

$$V = \frac{1}{2}s^2 + \frac{1}{2}\tilde{\alpha}^2 \qquad (17)$$

Its time derivative can be found as

$$\dot{V} = s\dot{s} - \tilde{\alpha}\dot{\hat{\alpha}}$$
$$\leq -\eta|s| + \tilde{\alpha}(|s| - \dot{\hat{\alpha}}) \qquad (18)$$

By selecting the update law as

$$\dot{\hat{\alpha}} = |s| \qquad (19)$$

equation (18) becomes

$$\dot{V} \leq -\eta|s| \qquad (20)$$

It seems that the estimation of the uncertainty bound can result in closed loop stability. However, in practical applications, the error signal s will never be zero, and the update law (19) implies an unbounded $\hat{\alpha}$. Therefore, the concept in estimation of the uncertainty bound is not realizable.

2.12 FAT-Based Adaptive Controller Design

In practical realization of control systems, the mathematical model inevitably contains uncertainties. If the variation bounds of these uncertainties are available, traditional robust control strategies such as the Lyapunov redesign and sliding control are applicable. If their bounds are not given, but we know that these uncertainties are time-invariant, various adaptive control schemes are useful. It is possible that system uncertainties are time-varying without knowing their bounds (general uncertainties); therefore, the above tools are not feasible. In this book, we would like to use the FAT-based designs to overcome the given problem. The basic idea of the FAT is to represent the general uncertainties by using a set of known basis functions weighted by a set of unknown coefficients (Huang and Kuo 2001, Huang and Chen 2004b, Chen and Huang 2004). Since these coefficients are constants, the Lyapunov designs can thus be applied to derive proper update laws to ensure closed loop stability. This approach has been successfully applied to the control of many systems, such as robot manipulators (Chien and Huang 2004, 2006a, 2006b, 2007a, 2007b, Huang and Chen 2004a, Huang et. al. 2006, Huang and Liao 2006), active suspensions (Chen and Huang 2005a, 2005b, 2006), pneumatic servo (Tsai and Huang 2008a, 2008b), vibration control (Chang and Shaw 2007), DC motors (Liang et. al. 2008) and jet engine control (Tyan and Lee 2005).

In Section 2.11.1, we know that the traditional MRAC is unable to give proper performance to LTV systems. In the first part of this section, we would like to present the FAT-based MRAC to the same LTV system as in Section 2.11.1 without considering the approximation error. The asymptotical convergence can be obtained if a sufficient number of basis functions are used. In the second part of this section, we will investigate the effect of the approximation error in detail. By considering the approximation error in the adaptive loop, the output error can be proved to be uniformly ultimately bounded. The bound for the transient response of the output error can also be estimated as a weighted exponential function plus some constant offset.

FAT-based MRAC for LTV systems

Let us consider the linear time-varying system (2.11-1) again

$$\dot{x}_p = a_p(t)x_p + b_p u \tag{1}$$

We have proved in Section 2.11.1 that traditional MRAC is infeasible to give stable closed loop system due to the fact that a_p is time-varying. Let us apply the MRAC rule in (2.11-3) once again so that the error dynamics becomes

$$\dot{e} = a_m e + b_p(a - \hat{a})x_p \tag{2}$$

where $a(t) = \dfrac{a_m - a_p(t)}{b_p}$ is the perfect gain in the MRAC rule. Since it is time-varying, traditional MRAC design will end up with (2.11-8), and no conclusions for closed loop system stability can be obtained. Here, let us represent a and $\hat{a}$ using function approximation techniques shown in (2.7-4) as

$$a = \mathbf{w}^T \mathbf{z} + \varepsilon$$
$$\hat{a} = \hat{\mathbf{w}}^T \mathbf{z} \tag{3}$$

where $\mathbf{w} \in \Re^{n_a}$ is a vector of weightings, $\hat{\mathbf{w}} \in \Re^{n_a}$ is its estimate and $\mathbf{z} \in \Re^{n_a}$ is a vector of basis functions. The positive integer n_a is the number of terms we selected to perform the function approximation. In this case, we would like to assume that sufficient terms are employed so that the approximation error ε is ignorable. Later in this section, we are going to investigate the effect of the approximation error in detail. Define $\tilde{\mathbf{w}} = \mathbf{w} - \hat{\mathbf{w}}$, and then equation (2) can be represented into the form

$$\dot{e} = a_m e + b_p \tilde{\mathbf{w}}^T \mathbf{z} x_p \tag{4}$$

A new Lyapunov-like function candidate is given as

$$V(e, \tilde{\mathbf{w}}) = \frac{1}{2} e^2 + \frac{1}{2} b_p \tilde{\mathbf{w}}^T \tilde{\mathbf{w}} \tag{5}$$

Its time derivative along the trajectory of (4) is computed to be

$$\dot{V} = e\dot{e} - b_p \tilde{\mathbf{w}}^T \dot{\hat{\mathbf{w}}}$$
$$= a_m e^2 + b_p \tilde{\mathbf{w}}^T (\mathbf{z} x_p e - \dot{\hat{\mathbf{w}}}) \tag{6}$$

By selecting the update law

$$\dot{\hat{\mathbf{w}}} = \mathbf{z} x_p e \tag{7}$$

we may have

$$\dot{V} = a_m e^2 \leq 0 \tag{8}$$

This implies that both e and $\tilde{\mathbf{w}}$ are uniformly bounded. The output error e can also be concluded to be square integrable from (8). In addition, the boundedness of $\dot{e}$ can easily be observed from (4). Hence, it follows from Barbalat's lemma that e will converge to zero asymptotically.

Consideration of approximation error

Let us consider a more general non-autonomous system in the standard form

$$\dot{x}_1 = x_2$$
$$\dot{x}_2 = x_3$$
$$\vdots \qquad\qquad\qquad (9)$$
$$\dot{x}_{n-1} = x_n$$
$$\dot{x}_n = f(\mathbf{x},t) + g(\mathbf{x},t)u$$

where $\mathbf{x} = [x_1 \quad x_2 \quad \cdots \quad x_n]^T \in \Omega$, and Ω is a compact subset of $\Re^n$. $f(\mathbf{x},t)$ is an unknown function with unknown variation bound. The uncertain function $g(\mathbf{x},t)$ is assumed to be bounded by $0 < g_{\min}(\mathbf{x},t) \le g(\mathbf{x},t) \le g_{\max}(\mathbf{x},t)$ for some known functions $g_{\min}$ and $g_{\max}$ for all $\mathbf{x} \in \Omega$ and $t \in [t_0, \infty)$. Let $g_m = \sqrt{g_{\min} g_{\max}}$ be the nominal function, and then we may represent g in the form $g = g_m(\mathbf{x},t)\Delta g(\mathbf{x},t)$ where Δg is the multiplicative uncertainty satisfying

$$0 < \delta_{\min} \equiv \frac{g_{\min}}{g_m} \le \Delta g \le \frac{g_{\max}}{g_m} \equiv \delta_{\max}$$

We would like to design a controller such that the system state $\mathbf{x}$ tracks the desired trajectory $\mathbf{x}_d \in \Omega_d$, where Ω_d is a compact subset of Ω. Define the tracking error vector as $\mathbf{e} = \mathbf{x} - \mathbf{x}_d = [x_1 - x_{1d} \quad x_2 - x_{2d} \quad \cdots \quad x_n - x_{nd}]^T$. The control law can be selected as

$$u = \frac{1}{g_m}(-\hat{f} + v - u_r) \qquad\qquad (10)$$

where $\hat{f}$ is an estimate of f, u_r is a robust term to cover the uncertainties in g, and $v = \dot{x}_{nd} - \sum_{i=0}^{n-1} k_i e_{i+1}$ is to complete the desired dynamics. The coefficients k_i are selected so that the matrix

$$\mathbf{A} = \begin{bmatrix} 0 & 1 & 0 & \cdots & 0 \\ 0 & 0 & 1 & \cdots & 0 \\ \vdots & \vdots & \vdots & \ddots & \vdots \\ 0 & 0 & 0 & \cdots & 1 \\ -k_0 & -k_1 & -k_2 & \cdots & -k_{n-1} \end{bmatrix} \in \Re^{n \times n}$$

is Hurwitz. With the controller (10), the last line of (9) becomes

$$\dot{x}_n = f + \frac{g}{g_m}(-\hat{f} + v - u_r)$$
$$= (f - \hat{f}) + (1 - \Delta g)(\hat{f} - v) + v - \Delta g u_r$$

This can further be written into the form

$$\dot{e}_n + \sum_{i=0}^{n-1} k_i e_{i+1} = (f - \hat{f}) + (1 - \Delta g)(\hat{f} - v) - \Delta g u_r$$

Its state space representation is thus

$$\dot{\mathbf{e}} = \mathbf{A}\mathbf{e} + \mathbf{b}[(f - \hat{f}) + (1 - \Delta g)(\hat{f} - v) - \Delta g u_r] \qquad (11)$$

where $\mathbf{b} = [0 \quad 0 \quad \cdots \quad 1]^T \in \Re^n$. Since f is a general uncertainty, we may not use traditional adaptive strategies to have stable closed loop system. Let us apply the function approximation techniques to represent f and its estimate as

$$f = \mathbf{w}^T \mathbf{z} + \varepsilon$$
$$\hat{f} = \hat{\mathbf{w}}^T \mathbf{z}$$

Then (11) becomes

$$\dot{\mathbf{e}} = \mathbf{A}\mathbf{e} + \mathbf{b}[\tilde{\mathbf{w}}^T \mathbf{z} + \varepsilon + (1 - \Delta g)(\hat{f} - v) - \Delta g u_r] \qquad (12)$$

where $\tilde{\mathbf{w}} = \mathbf{w} - \hat{\mathbf{w}}$. To find the update law, let us consider the Lyapunov-like function candidate

$$V = \mathbf{e}^T \mathbf{P} \mathbf{e} + \tilde{\mathbf{w}}^T \mathbf{\Gamma} \tilde{\mathbf{w}} \qquad (13)$$

where $\mathbf{P}$ and $\mathbf{\Gamma}$ are positive definite matrices. In addition, $\mathbf{P}$ satisfies the Lyapunov equation $\mathbf{A}^T \mathbf{P} + \mathbf{P} \mathbf{A} = -\mathbf{Q}$ where $\mathbf{Q}$ is some positive definite matrix. Taking the time derivative of (13) along the trajectory of (12), we have

$$\dot{V} = \mathbf{e}^T(\mathbf{A}^T\mathbf{P} + \mathbf{P}\mathbf{A})\mathbf{e} + 2[(1 - \Delta g)(\hat{f} - v) - \Delta g u_r]\mathbf{b}^T\mathbf{P}\mathbf{e}$$
$$+ 2\varepsilon\mathbf{b}^T\mathbf{P}\mathbf{e} + 2\tilde{\mathbf{w}}^T(\mathbf{z}\mathbf{b}^T\mathbf{P}\mathbf{e} - \mathbf{\Gamma}\dot{\hat{\mathbf{w}}})$$
$$\leq -\mathbf{e}^T\mathbf{Q}\mathbf{e} + 2(1 + \delta_{\max})\left|\hat{f} - v\right|\left|\mathbf{b}^T\mathbf{P}\mathbf{e}\right| - 2\delta_{\min}u_r\mathbf{b}^T\mathbf{P}\mathbf{e}$$
$$+ 2\varepsilon\mathbf{b}^T\mathbf{P}\mathbf{e} + 2\tilde{\mathbf{w}}^T(\mathbf{z}\mathbf{b}^T\mathbf{P}\mathbf{e} - \mathbf{\Gamma}\dot{\hat{\mathbf{w}}})$$

We may thus select

$$u_r = \frac{1 + \delta_{\max}}{\delta_{\min}}\left|\hat{f} - v\right|\operatorname{sgn}(\mathbf{b}^T\mathbf{P}\mathbf{e}) \tag{14a}$$

$$\dot{\hat{\mathbf{w}}} = \mathbf{\Gamma}^{-1}(\mathbf{z}\mathbf{b}^T\mathbf{P}\mathbf{e} - \sigma\hat{\mathbf{w}}), \quad \sigma > 0 \tag{14b}$$

The signum function in (14a) might induce chattering control activity which would excite un-modeled system dynamics. Some modifications can be used to smooth out the control law. The most intuitive way is to replace the signum function with the saturation function as

$$u_r = \frac{1 + \delta_{\max}}{\delta_{\min}}\left|\hat{f} - v\right|\operatorname{sat}(\mathbf{b}^T\mathbf{P}\mathbf{e}) \tag{14c}$$

One drawback for this modification is the reduction in the output tracking accuracy. It is also noted that the σ-modification term in (14b) is to robustify the adaptive loop. With (14), the time derivative of V becomes

$$\dot{V} \leq -\mathbf{e}^T\mathbf{Q}\mathbf{e} + 2\varepsilon\mathbf{b}^T\mathbf{P}\mathbf{e} + 2\sigma\tilde{\mathbf{w}}^T\hat{\mathbf{w}}$$
$$= -\mathbf{e}^T\mathbf{Q}\mathbf{e} + 2\varepsilon\mathbf{b}^T\mathbf{P}\mathbf{e} + 2\sigma\tilde{\mathbf{w}}^T(\mathbf{w} - \tilde{\mathbf{w}})$$
$$\leq \underbrace{-\lambda_{\min}(\mathbf{Q})\|\mathbf{e}\|^2 + 2\lambda_{\max}(\mathbf{P})|\varepsilon|\|\mathbf{e}\|}_{(a)} + 2\sigma\underbrace{[\tilde{\mathbf{w}}^T\mathbf{w} - \|\tilde{\mathbf{w}}\|^2]}_{(b)} \tag{15}$$

Let us derive part (a) in (15) using straightforward manipulations as

$$-\lambda_{\min}(\mathbf{Q})\|\mathbf{e}\|^2 + 2\lambda_{\max}(\mathbf{P})|\varepsilon|\|\mathbf{e}\|$$

$$= -\frac{1}{2}\left(\sqrt{\lambda_{\min}(\mathbf{Q})}\|\mathbf{e}\| - \frac{2\lambda_{\max}(\mathbf{P})|\varepsilon|}{\sqrt{\lambda_{\min}(\mathbf{Q})}}\right)^2$$

$$-\frac{1}{2}\left(\lambda_{\min}(\mathbf{Q})\|\mathbf{e}\|^2 - \frac{4\lambda_{\max}^2(\mathbf{P})}{\lambda_{\min}(\mathbf{Q})}\varepsilon^2\right)$$

$$\leq -\frac{1}{2}\lambda_{\min}(\mathbf{Q})\|\mathbf{e}\|^2 + \frac{2\lambda_{\max}^2(\mathbf{P})}{\lambda_{\min}(\mathbf{Q})}\varepsilon^2$$

Likewise, part (b) can also be written as

$$\tilde{\mathbf{w}}^T\mathbf{w} - \|\tilde{\mathbf{w}}\|^2 \leq \|\tilde{\mathbf{w}}\|\|\mathbf{w}\| - \|\tilde{\mathbf{w}}\|^2$$

$$= -\frac{1}{2}(\|\tilde{\mathbf{w}}\| - \|\mathbf{w}\|)^2 - \frac{1}{2}(\|\tilde{\mathbf{w}}\|^2 - \|\mathbf{w}\|^2)$$

$$\leq -\frac{1}{2}(\|\tilde{\mathbf{w}}\|^2 - \|\mathbf{w}\|^2)$$

Therefore (15) becomes

$$\dot{V} \leq -\frac{1}{2}\lambda_{\min}(\mathbf{Q})\|\mathbf{e}\|^2 + \frac{2\lambda_{\max}^2(\mathbf{P})}{\lambda_{\min}(\mathbf{Q})}\varepsilon^2 - \sigma\|\tilde{\mathbf{w}}\|^2 + \sigma\|\mathbf{w}\|^2$$

$$= \underbrace{-\frac{1}{2}\lambda_{\min}(\mathbf{Q})\|\mathbf{e}\|^2 - \sigma\|\tilde{\mathbf{w}}\|^2}_{(c)} + \sigma\|\mathbf{w}\|^2 + \frac{2\lambda_{\max}^2(\mathbf{P})}{\lambda_{\min}(\mathbf{Q})}\varepsilon^2 \qquad (16)$$

We would like to relate (c) to V by considering

$$V = \mathbf{e}^T\mathbf{P}\mathbf{e} + \tilde{\mathbf{w}}^T\mathbf{\Gamma}\tilde{\mathbf{w}} \leq \lambda_{\max}(\mathbf{P})\|\mathbf{e}\|^2 + \lambda_{\max}(\mathbf{\Gamma})\|\tilde{\mathbf{w}}\|^2 \qquad (17)$$

Now (16) can be further derived as

$$\dot{V} \le -\alpha V + \left[\alpha \lambda_{max}(\mathbf{P}) - \frac{\lambda_{min}(\mathbf{Q})}{2} \right] \|\mathbf{e}\|^2$$

$$+ [\alpha \lambda_{max}(\mathbf{\Gamma}) - \sigma] \|\tilde{\mathbf{w}}\|^2 + \sigma \|\mathbf{w}\|^2 + \frac{2\lambda_{max}^2(\mathbf{P})}{\lambda_{min}(\mathbf{Q})} \varepsilon^2$$

Pick $\alpha \le \min \left\{ \dfrac{\lambda_{min}(\mathbf{Q})}{2\lambda_{max}(\mathbf{P})}, \dfrac{\sigma}{\lambda_{max}(\mathbf{\Gamma})} \right\}$, then we have

$$\dot{V} \le -\alpha V + \sigma \|\mathbf{w}\|^2 + \frac{2\lambda_{max}^2(\mathbf{P})}{\lambda_{min}(\mathbf{Q})} \varepsilon^2 \tag{18}$$

Hence, $\dot{V} < 0$ whenever

$$(\mathbf{e}, \tilde{\mathbf{w}}) \in E \equiv \left\{ (\mathbf{e}, \tilde{\mathbf{w}}) \,\middle|\, V > \frac{1}{\alpha} \left[\sigma \|\mathbf{w}\|^2 + \frac{2\lambda_{max}^2(\mathbf{P})}{\lambda_{min}(\mathbf{Q})} \sup_{\tau \ge t_0} \varepsilon^2(\tau) \right] \right\}.$$

This implies that $(\mathbf{e}, \tilde{\mathbf{w}})$ is uniformly ultimately bounded. Note that the size of the set E is adjustable by proper selection of α, σ, $\mathbf{P}$, and $\mathbf{Q}$. Smaller size of E implies more accurate in output tracking. However, this parameter adjustment is not always unlimited, because it might induce controller saturation in implementation.

The above derivation only demonstrates the boundedness of the closed loop system, but in practical applications the transient performance is also of great importance. For further development, we may solve the differential inequality in (18) to have the upper bound for V

$$V \le e^{-\alpha(t-t_0)} V(t_0) + \frac{\sigma}{\alpha} \|\mathbf{w}\|^2 + \frac{2\lambda_{max}^2(\mathbf{P})}{\alpha \lambda_{min}(\mathbf{Q})} \sup_{t_0 \le \tau \le t} \varepsilon^2(\tau) \tag{19}$$

By using the definition in (13), we may also find an upper bound of V as

$$V = \mathbf{e}^T \mathbf{P} \mathbf{e} + \tilde{\mathbf{w}}^T \mathbf{\Gamma} \tilde{\mathbf{w}} \ge \lambda_{min}(\mathbf{P}) \|\mathbf{e}\|^2 + \lambda_{min}(\mathbf{\Gamma}) \|\tilde{\mathbf{w}}\|^2$$

This gives the upper bound for the tracking error

$$\|\mathbf{e}\|^2 \le \frac{1}{\lambda_{min}(\mathbf{P})} [V - \lambda_{min}(\mathbf{\Gamma}) \|\tilde{\mathbf{w}}\|^2] \le \frac{1}{\lambda_{min}(\mathbf{P})} V$$

Taking the square root and using (19), we have

$$\|\mathbf{e}\| \leq \sqrt{\frac{V(t_0)}{\lambda_{min}(\mathbf{P})}} e^{-\frac{\alpha(t-t_0)}{2}} + \sqrt{\frac{\sigma}{\alpha\lambda_{min}(\mathbf{P})}} \|\mathbf{w}\|$$

$$+ \sqrt{\frac{2\lambda_{max}^2(\mathbf{P})}{\alpha\lambda_{min}(\mathbf{P})\lambda_{min}(\mathbf{Q})}} \sup_{t_0 \leq \tau \leq t} |\varepsilon(\tau)| \tag{20}$$

Hence, we have proved that the tracking error is bounded by a weighted exponential function plus a constant. This also implies that by adjusting controller parameters, we may improve output error convergence rate. However, it might also induce controller saturation problem in practice.

The case when the bound for the approximation error is known

If the bound for ε is known, i.e. there exists some $\beta > 0$ such that $|\varepsilon| \leq \beta$ for all $t \geq t_0$, then u_r in (14a) can be modified as

$$u_r = \frac{1+\delta_{max}}{\delta_{min}} \left|\hat{f} - v\right| \text{sgn}(\mathbf{b}^T\mathbf{Pe}) + \frac{\beta}{\delta_{min}} \text{sgn}(\mathbf{b}^T\mathbf{Pe})$$

If the control law and the update law are still selected as (10) and (14b) with $\sigma = 0$, then we may have

$$\dot{V} \leq -\mathbf{e}^T\mathbf{Qe} + 2|\varepsilon|\left|\mathbf{b}^T\mathbf{Pe}\right| - 2\beta\left|\mathbf{b}^T\mathbf{Pe}\right|$$

$$\leq -\mathbf{e}^T\mathbf{Qe} \leq 0$$

Therefore, we may also have asymptotical convergence of the output error by using Barbalat's lemma.

Chapter 3

Dynamic Equations for Robot Manipulators

3.1 Introduction

In this chapter, we review the mathematical models for the robot manipulators considered in this book. For their detailed derivation, please refer to any robotics textbooks. In Section 3.2, a set of $2n$ coupled nonlinear ordinary differential equations are used to describe the dynamics of an n-link rigid robot. When interacting with the environment, the external force is included into the dynamics equation which is presented in Section 3.3. The actuator dynamics is considered in Section 3.4, and a motor model is coupled to each joint dynamics, resulting in a set of $3n$ differential equations in its dynamics. In Section 3.5, the dynamics for an electrically driven rigid robot interacting with the environment is presented. It is composed of $3n$ differential equations with the inclusion of the external force. Section 3.6 takes the joint flexibility into account where a set of $4n$ differential equations are used to represent the dynamics of an n-link flexible-joint robot. To investigate the effect when interacting with the environment, the external force is added into the dynamics equation in Section 3.7. With consideration of the motor dynamics in each joint of the flexible joint robot, its model becomes a set of $5n$ differential equations in Section 3.8. Finally, we include the external force in Section 3.9 to have the most complex dynamics considered in this book, i.e., the dynamics for an electrically driven flexible joint robot interacting with the environment.

3.2 Rigid Robot (RR)

An n-link rigid robot manipulator without considering friction or other disturbances can be described by

$$\mathbf{D}(\mathbf{q})\ddot{\mathbf{q}} + \mathbf{C}(\mathbf{q},\dot{\mathbf{q}})\dot{\mathbf{q}} + \mathbf{g}(\mathbf{q}) = \boldsymbol{\tau} \qquad (1)$$

71

where $\mathbf{q} \in \Re^n$ is a vector of generalized coordinates, $\mathbf{D}(\mathbf{q})$ is the $n \times n$ inertia matrix, $\mathbf{C}(\mathbf{q},\dot{\mathbf{q}})\dot{\mathbf{q}}$ is the n-vector of centrifugal and Coriolis forces, $\mathbf{g}(\mathbf{q})$ is the gravitational force vector, and τ is the control torque vector. Although equation (1) presents a highly nonlinear and coupled dynamics, several good properties can be summarized as (Ge et. al. 1998)

Property 1: $\mathbf{D}(\mathbf{q}) = \mathbf{D}^T(\mathbf{q}) > 0$ and there exist positive constants α_1 and α_2, $\alpha_1 \le \alpha_2$ such that $\alpha_1 \mathbf{I}_n \le \mathbf{D}(\mathbf{q}) \le \alpha_2 \mathbf{I}_n$ for all $\mathbf{q} \in \Re^n$.

Property 2: $\dot{\mathbf{D}}(\mathbf{q}) - 2\mathbf{C}(\mathbf{q},\dot{\mathbf{q}})$ is skew-symmetric.

Property 3: The left-hand side of (1) can be linearly parameterized as the multiplication of a known regressor matrix $\mathbf{Y}(\mathbf{q},\dot{\mathbf{q}},\ddot{\mathbf{q}}) \in \Re^{n \times r}$ with a parameter vector $\mathbf{p} \in \Re^r$, i.e.

$$\mathbf{D}(\mathbf{q})\ddot{\mathbf{q}} + \mathbf{C}(\mathbf{q},\dot{\mathbf{q}})\dot{\mathbf{q}} + \mathbf{g}(\mathbf{q}) = \mathbf{Y}(\mathbf{q},\dot{\mathbf{q}},\ddot{\mathbf{q}})\mathbf{p}. \tag{2}$$

Example 3.1: A 2-D planar robot model

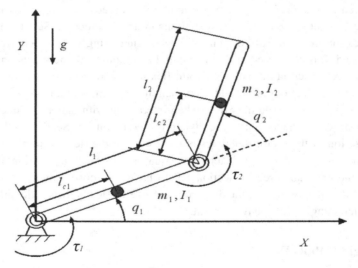

Figure 3.1 A 2-D planar robot

Consider a planar robot with two rigid links and two rigid revolute joints shown in Figure 3.1. Its governing equation can be represented by (Slotine and Li 1991)

$$\begin{bmatrix} d_{11} & d_{12} \\ d_{21} & d_{22} \end{bmatrix} \begin{bmatrix} \ddot{q}_1 \\ \ddot{q}_2 \end{bmatrix} + \begin{bmatrix} c_{11} & c_{12} \\ c_{21} & c_{22} \end{bmatrix} \begin{bmatrix} \dot{q}_1 \\ \dot{q}_2 \end{bmatrix} + \begin{bmatrix} g_1 \\ g_2 \end{bmatrix} = \begin{bmatrix} \tau_1 \\ \tau_2 \end{bmatrix} \tag{3}$$

where $d_{11} = m_1 l_{c1}^2 + I_1 + m_2 (l_1^2 + l_{c2}^2 + 2l_1 l_{c2} \cos q_2) + I_2$

$\quad d_{12} = d_{21} = m_2 l_1 l_{c2} \cos q_2 + m_2 l_{c2}^2 + I_2$

$\quad d_{22} = m_2 l_{c2}^2 + I_2$

$\quad c_{11} = -m_2 l_1 l_{c2} \sin q_2 \dot{q}_2$

$\quad c_{12} = -m_2 l_1 l_{c2} \sin q_2 (\dot{q}_1 + \dot{q}_2)$

$\quad c_{21} = -m_2 l_1 l_{c2} \sin q_2 \dot{q}_1$

$\quad c_{22} = 0$

$\quad g_1 = m_1 l_{c1} g \cos q_1 + m_2 g [l_{c2} \cos(q_1 + q_2) + l_1 \cos q_1]$

$\quad g_2 = m_2 l_{c2} g \cos(q_1 + q_2)$

Property 1 and 2 can be confirmed easily by direct derivation, while property 3 will further be investigated in Chapter 4.

3.3 Rigid Robot Interacting with Environment (RRE)

Suppose the robot manipulator will interact with a frictionless constraint surface, then equation (3.2-1) is modified to

$$\mathbf{D}(\mathbf{q})\ddot{\mathbf{q}} + \mathbf{C}(\mathbf{q}, \dot{\mathbf{q}})\dot{\mathbf{q}} + \mathbf{g}(\mathbf{q}) = \boldsymbol{\tau} - \mathbf{J}_a^T(\mathbf{q})\mathbf{F}_{ext} \tag{1}$$

where $\mathbf{J}_a(\mathbf{q}) \in \mathfrak{R}^{n \times n}$ is the Jacobian matrix which is assumed to be nonsingular, and $\mathbf{F}_{ext} \in \mathfrak{R}^n$ is the external force vector at the end-effector. During the free space tracking phase, there will be no external force and equation (1) degenerates to (3.2-1). To facilitate controller derivation, it is sometimes more convenient to represent (1) in the Cartesian space as

$$\mathbf{D}_x(\mathbf{x})\ddot{\mathbf{x}} + \mathbf{C}_x(\mathbf{x}, \dot{\mathbf{x}})\dot{\mathbf{x}} + \mathbf{g}_x(\mathbf{x}) = \mathbf{J}_a^{-T}(\mathbf{q})\boldsymbol{\tau} - \mathbf{F}_{ext} \tag{2}$$

where $\mathbf{x} \in \mathfrak{R}^n$ is the coordinate in the Cartesian space, and other symbols are defined as

$$\mathbf{D}_x(\mathbf{x}) = \mathbf{J}_a^{-T}(\mathbf{q})\mathbf{D}(\mathbf{q})\mathbf{J}_a^{-1}(\mathbf{q})$$

$$\mathbf{C}_x(\mathbf{x},\dot{\mathbf{x}}) = \mathbf{J}_a^{-T}(\mathbf{q})[\mathbf{C}(\mathbf{q},\dot{\mathbf{q}}) - \mathbf{D}(\mathbf{q})\mathbf{J}_a^{-1}(\mathbf{q})\dot{\mathbf{J}}_a(\mathbf{q})]\mathbf{J}_a^{-1}(\mathbf{q}) \qquad (3)$$

$$\mathbf{g}_x(\mathbf{x}) = \mathbf{J}_a^{-T}(\mathbf{q})\mathbf{g}(\mathbf{q})$$

Example 3.2: A 2-D planar robot interacting with the environment

Let us consider the 2-D robot in Example 3.1 again, but with a constraint surface as shown in Figure 3.2. Its equation of motion in the Cartesian space is represented as

$$\begin{bmatrix} d_{x11} & d_{x12} \\ d_{x21} & d_{x22} \end{bmatrix}\begin{bmatrix} \ddot{x} \\ \ddot{y} \end{bmatrix} + \begin{bmatrix} c_{x11} & c_{x12} \\ c_{x21} & c_{x22} \end{bmatrix}\begin{bmatrix} \dot{x} \\ \dot{y} \end{bmatrix} + \begin{bmatrix} g_{x1} \\ g_{x2} \end{bmatrix}$$
$$= \begin{bmatrix} J_{a11} & J_{a12} \\ J_{a21} & J_{a22} \end{bmatrix}^{-T}\begin{bmatrix} \tau_1 \\ \tau_2 \end{bmatrix} + \begin{bmatrix} f_{ext} \\ 0 \end{bmatrix} \qquad (4)$$

where

$$\begin{bmatrix} J_{a11} & J_{a12} \\ J_{a21} & J_{a22} \end{bmatrix} = \begin{bmatrix} -l_1\sin(q_1) - l_2\sin(q_1+q_2) & -l_2\sin(q_1+q_2) \\ l_1\cos(q_1) + l_2\cos(q_1+q_2) & l_2\cos(q_1+q_2) \end{bmatrix}$$

$$\begin{bmatrix} d_{x11} & d_{x12} \\ d_{x21} & d_{x22} \end{bmatrix} = \begin{bmatrix} J_{a11} & J_{a12} \\ J_{a21} & J_{a22} \end{bmatrix}^{-T}\begin{bmatrix} d_{11} & d_{12} \\ d_{21} & d_{22} \end{bmatrix}\begin{bmatrix} J_{a11} & J_{a12} \\ J_{a21} & J_{a22} \end{bmatrix}^{-1}$$

$$\begin{bmatrix} c_{x11} & c_{x12} \\ c_{x21} & c_{x22} \end{bmatrix} = \begin{bmatrix} J_{a11} & J_{a12} \\ J_{a21} & J_{a22} \end{bmatrix}^{-T}\left(\begin{bmatrix} c_{11} & c_{12} \\ c_{22} & c_{22} \end{bmatrix}\right.$$
$$\left. - \begin{bmatrix} d_{11} & d_{12} \\ d_{21} & d_{22} \end{bmatrix}\begin{bmatrix} J_{a11} & J_{a12} \\ J_{a21} & J_{a22} \end{bmatrix}^{-1}\begin{bmatrix} \dot{J}_{a11} & \dot{J}_{a12} \\ \dot{J}_{a21} & \dot{J}_{a22} \end{bmatrix}\right)\begin{bmatrix} J_{a11} & J_{a12} \\ J_{a21} & J_{a22} \end{bmatrix}^{-1}$$

$$\begin{bmatrix} g_{x1} \\ g_{x2} \end{bmatrix} = \begin{bmatrix} J_{a11} & J_{a12} \\ J_{a21} & J_{a22} \end{bmatrix}^{-T}\begin{bmatrix} g_1 \\ g_2 \end{bmatrix}$$

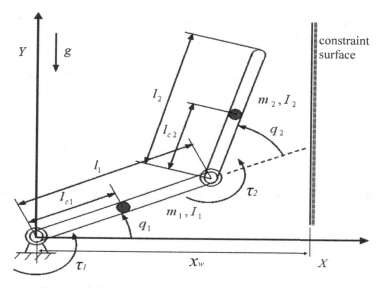

Figure 3.2 A 2-D planar robot interacting with a constraint surface

3.4 Electrically-Driven Rigid Robot (EDRR)

Let us consider the electrically-driven rigid robot. The dynamic equation of a rigid robot with consideration of motor dynamics is described as

$$\mathbf{D(q)\ddot{q} + C(q,\dot{q})\dot{q} + g(q) = Hi} \tag{1a}$$

$$\mathbf{L\dot{i} + Ri + K_b\dot{q} = u} \tag{1b}$$

where $\mathbf{i} \in \mathfrak{R}^n$ is the vector of motor armature currents, $\mathbf{u} \in \mathfrak{R}^n$ is the control input voltage, $\mathbf{H} \in \mathfrak{R}^{n \times n}$ is an invertible constant diagonal matrix characterizing the electro-mechanical conversion between the current vector and the torque vector, $\mathbf{L} \in \mathfrak{R}^{n \times n}$ is a constant diagonal matrix of electrical inductance, $\mathbf{R} \in \mathfrak{R}^{n \times n}$ represents the electrical resistance matrix, and $\mathbf{K}_b \in \mathfrak{R}^{n \times n}$ is a constant matrix for the motor back-emf effect. Inclusion of the actuator dynamics greatly increases the system order which gives large impact on the complexity of the controller design; especially, when the system contains uncertainties and disturbances.

Example 3.3: With the consideration of the motor dynamics, the 2-D robot introduced in Example 3.1 is now rewritten as

$$
\begin{bmatrix} d_{11} & d_{12} \\ d_{21} & d_{22} \end{bmatrix} \begin{bmatrix} \ddot{q}_1 \\ \ddot{q}_2 \end{bmatrix} + \begin{bmatrix} c_{11} & c_{12} \\ c_{21} & c_{22} \end{bmatrix} \begin{bmatrix} \dot{q}_1 \\ \dot{q}_2 \end{bmatrix} + \begin{bmatrix} g_1 \\ g_2 \end{bmatrix} = \begin{bmatrix} h_1 & 0 \\ 0 & h_2 \end{bmatrix} \begin{bmatrix} i_1 \\ i_2 \end{bmatrix} \tag{2a}
$$

$$
\begin{bmatrix} L_1 & 0 \\ 0 & L_2 \end{bmatrix} \begin{bmatrix} \dot{i}_1 \\ \dot{i}_2 \end{bmatrix} + \begin{bmatrix} r_1 & 0 \\ 0 & r_2 \end{bmatrix} \begin{bmatrix} i_1 \\ i_2 \end{bmatrix} + \begin{bmatrix} k_{b_1} & 0 \\ 0 & k_{b_2} \end{bmatrix} \begin{bmatrix} \dot{q}_1 \\ \dot{q}_2 \end{bmatrix} = \begin{bmatrix} u_1 \\ u_2 \end{bmatrix} \tag{2b}
$$

3.5 Electrically-Driven Rigid Robot Interacting with Environment (EDRRE)

The dynamics of a rigid-link electrically-driven robot interacting with the environment can be described by

$$
\mathbf{D}(\mathbf{q})\ddot{\mathbf{q}} + \mathbf{C}(\mathbf{q},\dot{\mathbf{q}})\dot{\mathbf{q}} + \mathbf{g}(\mathbf{q}) = \mathbf{H}\mathbf{i} - \mathbf{J}_a^T(\mathbf{q})\mathbf{F}_{ext} \tag{1a}
$$

$$
\mathbf{L}\dot{\mathbf{i}} + \mathbf{R}\mathbf{i} + \mathbf{K}_b\dot{\mathbf{q}} = \mathbf{u} \tag{1b}
$$

We may also use equation (3.3-3) to transform (1) to the Cartesian space

$$
\mathbf{D}_x(\mathbf{x})\ddot{\mathbf{x}} + \mathbf{C}_x(\mathbf{x},\dot{\mathbf{x}})\dot{\mathbf{x}} + \mathbf{g}_x(\mathbf{x}) = \mathbf{J}_a^{-T}(\mathbf{q})\mathbf{H}\mathbf{i} - \mathbf{F}_{ext} \tag{2a}
$$

$$
\mathbf{L}\dot{\mathbf{i}} + \mathbf{R}\mathbf{i} + \mathbf{K}_b\dot{\mathbf{q}} = \mathbf{u} \tag{2b}
$$

Example 3.4: The robot in Example 3.3 can be modified to include the effect of the interaction force with the environment in the Cartesian space as

$$
\begin{bmatrix} d_{x11} & d_{x12} \\ d_{x21} & d_{x22} \end{bmatrix} \begin{bmatrix} \ddot{x} \\ \ddot{y} \end{bmatrix} + \begin{bmatrix} c_{x11} & c_{x12} \\ c_{x21} & c_{x22} \end{bmatrix} \begin{bmatrix} \dot{x} \\ \dot{y} \end{bmatrix} + \begin{bmatrix} g_{x1} \\ g_{x2} \end{bmatrix} =
$$
$$
\begin{bmatrix} J_{a11} & J_{a12} \\ J_{a21} & J_{a22} \end{bmatrix}^{-T} \begin{bmatrix} h_1 & 0 \\ 0 & h_2 \end{bmatrix} \begin{bmatrix} i_1 \\ i_2 \end{bmatrix} + \begin{bmatrix} f_{ext} \\ 0 \end{bmatrix} \tag{3a}
$$

$$
\begin{bmatrix} L_1 & 0 \\ 0 & L_2 \end{bmatrix} \begin{bmatrix} \dot{i}_1 \\ \dot{i}_2 \end{bmatrix} + \begin{bmatrix} r_1 & 0 \\ 0 & r_2 \end{bmatrix} \begin{bmatrix} i_1 \\ i_2 \end{bmatrix} + \begin{bmatrix} k_{b_1} & 0 \\ 0 & k_{b_2} \end{bmatrix} \begin{bmatrix} \dot{q}_1 \\ \dot{q}_2 \end{bmatrix} = \begin{bmatrix} u_1 \\ u_2 \end{bmatrix} \tag{3b}
$$

3.6 Flexible Joint Robot (FJR)

The transmission mechanism in many industrial robots contains flexible components, such as the harmonic drives. Consideration of the joint flexibility in the controller design is one of the approaches to increase the control performance. For a robot with n links, we need to use $2n$ generalized coordinates to describe its whole dynamic behavior when taking the joint flexibility into account. Therefore, the modeling of the flexible joint robot is far more complex than that of the rigid robot.

The dynamics of an n-rigid link flexible-joint robot can be described by

$$\mathbf{D(q)\ddot{q} + C(q,\dot{q})\dot{q} + g(q) = K(\theta - q)} \tag{1a}$$

$$\mathbf{J\ddot{\theta} + B\dot{\theta} + K(\theta - q) = \tau}_a \tag{1b}$$

where $\mathbf{q} \in \Re^n$ is the vector of link angles, $\boldsymbol{\theta} \in \Re^n$ is the vector of actuator angles, $\boldsymbol{\tau}_a \in \Re^n$ is the vector of actuator input torques. $\mathbf{J}$, $\mathbf{B}$ and $\mathbf{K}$ are $n \times n$ constant diagonal matrices of actuator inertias, damping and joint stiffness, respectively.

Example 3.5: A single-link flexible-joint robot

Let us consider a single-link flexible-joint robot as shown in Figure 3.3. It can rotate in a vertical plane with the assumptions that its joint can only deform in the direction of joint rotation, the link is rigid, and the viscous damping is neglected. Its dynamic equation is in the form of (1) and can be represented in the state space as

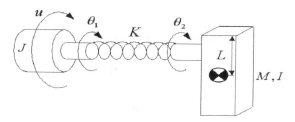

Figure 3.3 A single-link flexible-joint robot

$$\dot{x}_1 = x_2$$

$$\dot{x}_2 = -\frac{MgL}{I} \sin x_1 - \frac{K}{I}(x_1 - x_3)$$

$$\dot{x}_3 = x_4 \tag{2}$$

$$\dot{x}_4 = \frac{K}{J}(x_1 - x_3) + \frac{1}{J}u$$

where $x_i \in \mathfrak{R}$, $i = 1, \ldots, 4$ are state variables, and the output $y = x_1$ is the link angular displacement, i.e., θ_2 in the figure. The link inertial I, the rotor inertia J, the stiffness K, the link mass M, the gravity constant g, and the center of mass L are positive numbers. The control u is the torque delivered by the motor.

Example 3.6: A 2-D rigid-link flexible-joint robot

The equation of motion for the robot introduced in Example 3.1 with consideration of joint flexibility can be represented as

$$\begin{bmatrix} d_{11} & d_{12} \\ d_{21} & d_{22} \end{bmatrix} \begin{bmatrix} \ddot{q}_1 \\ \ddot{q}_2 \end{bmatrix} + \begin{bmatrix} c_{11} & c_{12} \\ c_{21} & c_{22} \end{bmatrix} \begin{bmatrix} \dot{q}_1 \\ \dot{q}_2 \end{bmatrix} + \begin{bmatrix} g_1 \\ g_2 \end{bmatrix} = \begin{bmatrix} k_1 & 0 \\ 0 & k_2 \end{bmatrix} \begin{bmatrix} \theta_1 - q_1 \\ \theta_2 - q_2 \end{bmatrix} \tag{3a}$$

$$\begin{bmatrix} j_1 & 0 \\ 0 & j_2 \end{bmatrix} \begin{bmatrix} \ddot{\theta}_1 \\ \ddot{\theta}_2 \end{bmatrix} + \begin{bmatrix} b_1 & 0 \\ 0 & b_2 \end{bmatrix} \begin{bmatrix} \dot{\theta}_1 \\ \dot{\theta}_2 \end{bmatrix} + \begin{bmatrix} k_1 & 0 \\ 0 & k_2 \end{bmatrix} \begin{bmatrix} \theta_1 - q_1 \\ \theta_2 - q_2 \end{bmatrix} = \begin{bmatrix} \tau_{a_1} \\ \tau_{a_2} \end{bmatrix} \tag{3b}$$

3.7 Flexible-Joint Robot Interacting with Environment (FJRE)

The dynamics of an n-link flexible-joint robot interacting with the environment can be described by

$$\mathbf{D}_x(\mathbf{x})\ddot{\mathbf{x}} + \mathbf{C}_x(\mathbf{x}, \dot{\mathbf{x}})\dot{\mathbf{x}} + \mathbf{g}_x(\mathbf{x}) = \mathbf{J}_a^{-T}(\mathbf{q})\mathbf{K}(\boldsymbol{\theta} - \mathbf{q}) - \mathbf{F}_{ext} \tag{1a}$$

$$\mathbf{J}\ddot{\boldsymbol{\theta}} + \mathbf{B}\dot{\boldsymbol{\theta}} + \mathbf{K}(\boldsymbol{\theta} - \mathbf{q}) = \boldsymbol{\tau}_a \tag{1b}$$

Example 3.7: A 2-D rigid-link flexible-joint robot interacting with environment

When the robot in Example 3.6 performs compliant motion control, its dynamic equation in the Cartesian space becomes

$$
\begin{bmatrix} d_{x11} & d_{x12} \\ d_{x21} & d_{x22} \end{bmatrix} \begin{bmatrix} \ddot{x} \\ \ddot{y} \end{bmatrix} + \begin{bmatrix} c_{x11} & c_{x12} \\ c_{x21} & c_{x22} \end{bmatrix} \begin{bmatrix} \dot{x} \\ \dot{y} \end{bmatrix} + \begin{bmatrix} g_{x1} \\ g_{x2} \end{bmatrix} =
$$
$$
\begin{bmatrix} J_{a11} & J_{a12} \\ J_{a21} & J_{a22} \end{bmatrix}^{-T} \begin{bmatrix} k_1 & 0 \\ 0 & k_2 \end{bmatrix} \begin{bmatrix} \theta_1 - q_1 \\ \theta_2 - q_2 \end{bmatrix} + \begin{bmatrix} f_{ext} \\ 0 \end{bmatrix} \tag{2a}
$$

$$
\begin{bmatrix} j_1 & 0 \\ 0 & j_2 \end{bmatrix} \begin{bmatrix} \ddot{\theta}_1 \\ \ddot{\theta}_2 \end{bmatrix} + \begin{bmatrix} b_1 & 0 \\ 0 & b_2 \end{bmatrix} \begin{bmatrix} \dot{\theta}_1 \\ \dot{\theta}_2 \end{bmatrix} + \begin{bmatrix} k_1 & 0 \\ 0 & k_2 \end{bmatrix} \begin{bmatrix} \theta_1 - q_1 \\ \theta_2 - q_2 \end{bmatrix} = \begin{bmatrix} \tau_{a_1} \\ \tau_{a_2} \end{bmatrix} \tag{2b}
$$

3.8 Electrically-Driven Flexible Joint Robot (EDFJR)

The dynamics of an electrically-driven flexible-joint robot can be described by including the motor dynamics to (3.6-1) as

$$
\mathbf{D}(\mathbf{q})\ddot{\mathbf{q}} + \mathbf{C}(\mathbf{q},\dot{\mathbf{q}})\dot{\mathbf{q}} + \mathbf{g}(\mathbf{q}) = \mathbf{K}(\boldsymbol{\theta} - \mathbf{q}) \tag{1a}
$$

$$
\mathbf{J}\ddot{\boldsymbol{\theta}} + \mathbf{B}\dot{\boldsymbol{\theta}} + \mathbf{K}(\boldsymbol{\theta} - \mathbf{q}) = \mathbf{H}\mathbf{i} \tag{1b}
$$

$$
\mathbf{L}\dot{\mathbf{i}} + \mathbf{R}\mathbf{i} + \mathbf{K}_b\dot{\mathbf{q}} = \mathbf{u} \tag{1c}
$$

Example 3.8: A 2-D electrically-driven flexible-joint robot

When considering the actuator dynamics, the robot in Example 3.6 is able to be modified in the form

$$
\begin{bmatrix} d_{11} & d_{12} \\ d_{21} & d_{22} \end{bmatrix} \begin{bmatrix} \ddot{q}_1 \\ \ddot{q}_2 \end{bmatrix} + \begin{bmatrix} c_{11} & c_{12} \\ c_{21} & c_{22} \end{bmatrix} \begin{bmatrix} \dot{q}_1 \\ \dot{q}_2 \end{bmatrix} + \begin{bmatrix} g_1 \\ g_2 \end{bmatrix} = \begin{bmatrix} k_1 & 0 \\ 0 & k_2 \end{bmatrix} \begin{bmatrix} \theta_1 - q_1 \\ \theta_2 - q_2 \end{bmatrix} \tag{2a}
$$

$$
\begin{bmatrix} j_1 & 0 \\ 0 & j_2 \end{bmatrix} \begin{bmatrix} \ddot{\theta}_1 \\ \ddot{\theta}_2 \end{bmatrix} + \begin{bmatrix} b_1 & 0 \\ 0 & b_2 \end{bmatrix} \begin{bmatrix} \dot{\theta}_1 \\ \dot{\theta}_2 \end{bmatrix} + \begin{bmatrix} k_1 & 0 \\ 0 & k_2 \end{bmatrix} \begin{bmatrix} \theta_1 - q_1 \\ \theta_2 - q_2 \end{bmatrix} = \begin{bmatrix} h_1 & 0 \\ 0 & h_2 \end{bmatrix} \begin{bmatrix} i_1 \\ i_2 \end{bmatrix} \tag{2b}
$$

$$
\begin{bmatrix} L_1 & 0 \\ 0 & L_2 \end{bmatrix} \begin{bmatrix} \dot{i}_1 \\ \dot{i}_2 \end{bmatrix} + \begin{bmatrix} r_1 & 0 \\ 0 & r_2 \end{bmatrix} \begin{bmatrix} i_1 \\ i_2 \end{bmatrix} + \begin{bmatrix} k_{b_1} & 0 \\ 0 & k_{b_2} \end{bmatrix} \begin{bmatrix} \dot{q}_1 \\ \dot{q}_2 \end{bmatrix} = \begin{bmatrix} u_1 \\ u_2 \end{bmatrix} \tag{2c}
$$

3.9 Electrically-Driven Flexible-Joint Robot Interacting with Environment (EDFJRE)

The dynamics of an electrically-driven flexible-joint robot interacting with the environment in the Cartesian space can be described by

$$\mathbf{D}_x(\mathbf{x})\ddot{\mathbf{x}} + \mathbf{C}_x(\mathbf{x},\dot{\mathbf{x}})\dot{\mathbf{x}} + \mathbf{g}_x(\mathbf{x}) = \mathbf{J}_a^{-T}(\mathbf{q})\mathbf{K}(\boldsymbol{\theta}-\mathbf{q}) - \mathbf{F}_{ext} \qquad (1a)$$

$$\mathbf{J}\ddot{\boldsymbol{\theta}} + \mathbf{B}\dot{\boldsymbol{\theta}} + \mathbf{K}(\boldsymbol{\theta}-\mathbf{q}) = \mathbf{H}\mathbf{i} \qquad (1b)$$

$$\mathbf{L}\dot{\mathbf{i}} + \mathbf{R}\mathbf{i} + \mathbf{K}_b\dot{\mathbf{q}} = \mathbf{u} \qquad (1c)$$

It is the most complex system considered in this book. For each joint, a 5th order differential equation is needed to model its dynamics.

Example 3.9: A 2-D electrically-driven flexible-joint robot interacting with environment

When considering the actuator dynamics, the robot in Example 3.7 can be rewritten into the form

$$\begin{bmatrix} d_{x11} & d_{x12} \\ d_{x21} & d_{x22} \end{bmatrix}\begin{bmatrix} \ddot{x} \\ \ddot{y} \end{bmatrix} + \begin{bmatrix} c_{x11} & c_{x12} \\ c_{x21} & c_{x22} \end{bmatrix}\begin{bmatrix} \dot{x} \\ \dot{y} \end{bmatrix} + \begin{bmatrix} g_{x1} \\ g_{x2} \end{bmatrix} =$$
$$\begin{bmatrix} J_{a11} & J_{a12} \\ J_{a21} & J_{a22} \end{bmatrix}^{-T}\begin{bmatrix} k_1 & 0 \\ 0 & k_2 \end{bmatrix}\begin{bmatrix} \theta_1 - q_1 \\ \theta_2 - q_2 \end{bmatrix} + \begin{bmatrix} f_{ext} \\ 0 \end{bmatrix} \qquad (2a)$$

$$\begin{bmatrix} j_1 & 0 \\ 0 & j_2 \end{bmatrix}\begin{bmatrix} \ddot{\theta}_1 \\ \ddot{\theta}_2 \end{bmatrix} + \begin{bmatrix} b_1 & 0 \\ 0 & b_2 \end{bmatrix}\begin{bmatrix} \dot{\theta}_1 \\ \dot{\theta}_2 \end{bmatrix} + \begin{bmatrix} k_1 & 0 \\ 0 & k_2 \end{bmatrix}\begin{bmatrix} \theta_1 - q_1 \\ \theta_2 - q_2 \end{bmatrix} = \begin{bmatrix} h_1 & 0 \\ 0 & h_2 \end{bmatrix}\begin{bmatrix} i_1 \\ i_2 \end{bmatrix} \qquad (2b)$$

$$\begin{bmatrix} L_1 & 0 \\ 0 & L_2 \end{bmatrix}\begin{bmatrix} \dot{i}_1 \\ \dot{i}_2 \end{bmatrix} + \begin{bmatrix} r_1 & 0 \\ 0 & r_2 \end{bmatrix}\begin{bmatrix} i_1 \\ i_2 \end{bmatrix} + \begin{bmatrix} k_{b_1} & 0 \\ 0 & k_{b_2} \end{bmatrix}\begin{bmatrix} \dot{q}_1 \\ \dot{q}_2 \end{bmatrix} = \begin{bmatrix} u_1 \\ u_2 \end{bmatrix} \qquad (2c)$$

3.10 Conclusions

In this chapter, eight robot models have been introduced. Some of them consider the actuator dynamics, some take the joint flexibility into account, and some allow the robot to interact with the environment. These robot models are summarized in Table 3-1 for comparison. In the following chapters, controllers will be designed for these robots when the models are known followed by derivations of adaptive controllers for uncertain robots.

Table 3.1 Robot models considered in this book

Systems	Dynamics Models	Equation Numbers
RR	$\mathbf{D}(\mathbf{q})\ddot{\mathbf{q}} + \mathbf{C}(\mathbf{q},\dot{\mathbf{q}})\dot{\mathbf{q}} + \mathbf{g}(\mathbf{q}) = \boldsymbol{\tau}$	(3.2-1)
RRE	$\mathbf{D}_x(\mathbf{x})\ddot{\mathbf{x}} + \mathbf{C}_x(\mathbf{x},\dot{\mathbf{x}})\dot{\mathbf{x}} + \mathbf{g}_x(\mathbf{x}) = \mathbf{J}_a^{-T}(\mathbf{q})\boldsymbol{\tau} - \mathbf{F}_{ext}$	(3.3-2)
EDRR	$\mathbf{D}(\mathbf{q})\ddot{\mathbf{q}} + \mathbf{C}(\mathbf{q},\dot{\mathbf{q}})\dot{\mathbf{q}} + \mathbf{g}(\mathbf{q}) = \mathbf{H}\mathbf{i}$ $\mathbf{L}\dot{\mathbf{i}} + \mathbf{R}\mathbf{i} + \mathbf{K}_b\dot{\mathbf{q}} = \mathbf{u}$	(3.4-1)
EDRRE	$\mathbf{D}_x(\mathbf{x})\ddot{\mathbf{x}} + \mathbf{C}_x(\mathbf{x},\dot{\mathbf{x}})\dot{\mathbf{x}} + \mathbf{g}_x(\mathbf{x}) = \mathbf{J}_a^{-T}(\mathbf{q})\mathbf{H}\mathbf{i} - \mathbf{F}_{ext}$ $\mathbf{L}\dot{\mathbf{i}} + \mathbf{R}\mathbf{i} + \mathbf{K}_b\dot{\mathbf{q}} = \mathbf{u}$	(3.5-2)
FJR	$\mathbf{D}(\mathbf{q})\ddot{\mathbf{q}} + \mathbf{C}(\mathbf{q},\dot{\mathbf{q}})\dot{\mathbf{q}} + \mathbf{g}(\mathbf{q}) = \mathbf{K}(\boldsymbol{\theta} - \mathbf{q})$ $\mathbf{J}\ddot{\boldsymbol{\theta}} + \mathbf{B}\dot{\boldsymbol{\theta}} + \mathbf{K}(\boldsymbol{\theta} - \mathbf{q}) = \boldsymbol{\tau}_a$	(3.6-1)
FJRE	$\mathbf{D}_x(\mathbf{x})\ddot{\mathbf{x}} + \mathbf{C}_x(\mathbf{x},\dot{\mathbf{x}})\dot{\mathbf{x}} + \mathbf{g}_x(\mathbf{x})$ $\quad = \mathbf{J}_a^{-T}(\mathbf{q})\mathbf{K}(\boldsymbol{\theta} - \mathbf{q}) - \mathbf{F}_{ext}$ $\mathbf{J}\ddot{\boldsymbol{\theta}} + \mathbf{B}\dot{\boldsymbol{\theta}} + \mathbf{K}(\boldsymbol{\theta} - \mathbf{q}) = \boldsymbol{\tau}_a$	(3.7-1)
EDFJR	$\mathbf{D}(\mathbf{q})\ddot{\mathbf{q}} + \mathbf{C}(\mathbf{q},\dot{\mathbf{q}})\dot{\mathbf{q}} + \mathbf{g}(\mathbf{q}) = \mathbf{K}(\boldsymbol{\theta} - \mathbf{q})$ $\mathbf{J}\ddot{\boldsymbol{\theta}} + \mathbf{B}\dot{\boldsymbol{\theta}} + \mathbf{K}(\boldsymbol{\theta} - \mathbf{q}) = \mathbf{H}\mathbf{i}$ $\mathbf{L}\dot{\mathbf{i}} + \mathbf{R}\mathbf{i} + \mathbf{K}_b\dot{\mathbf{q}} = \mathbf{u}$	(3.8-1)
EDFJRE	$\mathbf{D}_x(\mathbf{x})\ddot{\mathbf{x}} + \mathbf{C}_x(\mathbf{x},\dot{\mathbf{x}})\dot{\mathbf{x}} + \mathbf{g}_x(\mathbf{x})$ $\quad = \mathbf{J}_a^{-T}(\mathbf{q})\mathbf{K}(\boldsymbol{\theta} - \mathbf{q}) - \mathbf{F}_{ext}$ $\mathbf{J}\ddot{\boldsymbol{\theta}} + \mathbf{B}\dot{\boldsymbol{\theta}} + \mathbf{K}(\boldsymbol{\theta} - \mathbf{q}) = \mathbf{H}\mathbf{i}$ $\mathbf{L}\dot{\mathbf{i}} + \mathbf{R}\mathbf{i} + \mathbf{K}_b\dot{\mathbf{q}} = \mathbf{u}$	(3.9-1)

Chapter 4

Adaptive Control of Rigid Robots

4.1 Introduction

The dynamics of a rigid robot is well-known to be modeled by a set of coupled highly nonlinear differential equations. Its controller design is generally not easy even when the system model is precisely known. In practical operations of an industrial robot, since the mathematical model inevitably contains various uncertainties and disturbances, the widely used computed-torque controller may not give high precision performance. Under this circumstance, several robust control schemes (Abdallah et. al. 1991) and adaptive control strategies (Ortega and Spong 1988, Pagilla and Tomizuka 2001) are suggested.

For the adaptive approaches, although these control laws can give proper tracking performance under various uncertainties, most of them require computation of the regressor matrix. This is because, with the regressor matrix, the robot dynamics is able to be expressed in a linearly parameterized form so that a proper Lyapunov function candidate can be found to give stable update laws for uncertain parameters. Since the regressor matrix depends on the joint position, velocity and acceleration, it should be updated in every control cycle. Due to the complexity in the regressor computation, these approaches may have difficulties in practical implementation. Sadegh and Horowitz (1990) proposed a method to allow off-line computation of the regressor using the desired trajectories instead of actual measurements. Sometimes a large memory space should be allocated to store the look-up table containing the regressor. Lu and Meng (1991a, 1993) proposed some recursive algorithms for general n DOF robots. Kawasaki et al. (1996) presented a model-based adaptive control for a robot manipulator whose regressor was computed explicitly by a recursive algorithm based on the Newton-Euler formulation. Yang (1999) proposed a robust adaptive tracking controller for manipulators whose regressor depends only on the desired trajectory and hence can be calculated off-line.

Some regressor-free approaches for the adaptive control of robot manipulators are available. In Qu and Dorsey (1991), a non-regressor based controller was proposed using linear state feedback. To confirm robust stability of the closed loop system, one of their controller parameters should be determined based on variation bounds of some complex system dynamics. However, it is generally not easy to find such a parameter for robots with more than 3 DOF. Song (1994) suggested an adaptive controller for robot motion control without using the regressor. In his design, some bounds of the system dynamics should be found, and the tracking error can not be driven to arbitrary small in the steady state. Park et al. (1994) designed an adaptive sliding controller which does not require computation of the regressor matrix, but some critical bounded time functions are to be determined to have bounded tracking error performance. Yuan and Stepanenko (1993) suggested an adaptive PD controller for flexible joint robots without using the high-order regressor, but the usual regressor is still needed. Su and Stepanenko (1996) designed a robust adaptive controller for constrained robots without using the regressor matrix, but bounds of some system dynamics should be available. Huang et al. (2006) proposed an adaptive controller for robot manipulators without computation of the regressor matrix. Chien and Huang (2007b) designed a regressor-free adaptive controller for electrically driven robots.

In this chapter, we are going to study the regressor-free adaptive control strategies for rigid robot manipulators. We firstly review the conventional adaptive control laws for rigid robots in Section 4.2 whose regressor matrix depends on the joint position, velocity and acceleration which is inconvenient in real-time implementation. In addition, in the process of updating the inertia matrix, there might be some singularity problem which greatly limits the effectiveness of the approach. The famous Slotine and Li approach reviewed in Section 4.3 eliminates the requirement for the acceleration feedback and avoids the singularity problem. However, it is still based on the regressor matrix. In Section 4.4, we investigate the entries in the regressor matrix and the parameter vector to justify the necessity for the regressor-free approach. The regressor-free adaptive control strategy is then designed in Section 4.5 based on the function approximation technique. In Section 4.6, the regressor-free design is extended to the system considering the actuator dynamics. Significant performance improvement can be seen in the simulation results to verify the efficacy of the regressor-free design.

4.2 Review of Conventional Adaptive Control for Rigid Robots

Let us consider the rigid robot described in (3.2-1)

$$\mathbf{D}(\mathbf{q})\ddot{\mathbf{q}} + \mathbf{C}(\mathbf{q}, \dot{\mathbf{q}})\dot{\mathbf{q}} + \mathbf{g}(\mathbf{q}) = \boldsymbol{\tau} \qquad (1)$$

If all parameters are available, a PD controller can be designed as

$$\boldsymbol{\tau} = \mathbf{D}[\ddot{\mathbf{q}}_d - \mathbf{K}_d(\dot{\mathbf{q}} - \dot{\mathbf{q}}_d) - \mathbf{K}_p(\mathbf{q} - \mathbf{q}_d)] + \mathbf{C}\dot{\mathbf{q}} + \mathbf{g} \qquad (2)$$

where $\mathbf{q}_d \in \mathfrak{R}^n$ is the desired trajectory, and gain matrices $\mathbf{K}_d, \mathbf{K}_p \in \mathfrak{R}^{n \times n}$ are selected such that the closed loop dynamics

$$\ddot{\mathbf{e}} + \mathbf{K}_d \dot{\mathbf{e}} + \mathbf{K}_p \mathbf{e} = \mathbf{0}, \quad \mathbf{e} = \mathbf{q} - \mathbf{q}_d \qquad (3)$$

is asymptotically stable. It is obvious that realization of controller (2) needs the knowledge of the system model. Now, let us consider the case when some of the parameters are not available and an adaptive controller is to be designed.

As indicated in (3.2-2), the left hand side of equation (1) can be linearly parameterized as a known regressor matrix $\mathbf{Y}(\mathbf{q}, \dot{\mathbf{q}}, \ddot{\mathbf{q}}) \in \mathfrak{R}^{n \times r}$ multiplied by an unknown parameter vector $\mathbf{p} \in \mathfrak{R}^r$, i.e.,

$$\mathbf{D}(\mathbf{q})\ddot{\mathbf{q}} + \mathbf{C}(\mathbf{q}, \dot{\mathbf{q}})\dot{\mathbf{q}} + \mathbf{g}(\mathbf{q}) = \mathbf{Y}(\mathbf{q}, \dot{\mathbf{q}}, \ddot{\mathbf{q}})\mathbf{p} \qquad (4)$$

An intuitive controller can be designed based on (2) as

$$\boldsymbol{\tau} = \hat{\mathbf{D}}[\ddot{\mathbf{q}}_d - \mathbf{K}_d(\dot{\mathbf{q}} - \dot{\mathbf{q}}_d) - \mathbf{K}_p(\mathbf{q} - \mathbf{q}_d)] + \hat{\mathbf{C}}\dot{\mathbf{q}} + \hat{\mathbf{g}} \qquad (5)$$

where $\hat{\mathbf{D}}, \hat{\mathbf{C}}$ and $\hat{\mathbf{g}}$ are estimates of $\mathbf{D}$, $\mathbf{C}$ and $\mathbf{g}$, respectively. With the controller defined in (5), the closed loop system becomes

$$\hat{\mathbf{D}}(\ddot{\mathbf{e}} + \mathbf{K}_d \dot{\mathbf{e}} + \mathbf{K}_p \mathbf{e}) = -\tilde{\mathbf{D}}\ddot{\mathbf{q}} - \tilde{\mathbf{C}}\dot{\mathbf{q}} - \tilde{\mathbf{g}} \qquad (6)$$

where $\tilde{\mathbf{D}} = \mathbf{D} - \hat{\mathbf{D}}, \tilde{\mathbf{C}} = \mathbf{C} - \hat{\mathbf{C}}$, and $\tilde{\mathbf{g}} = \mathbf{g} - \hat{\mathbf{g}}$ are estimation errors. By defining $\tilde{\mathbf{p}} = \mathbf{p} - \hat{\mathbf{p}}$ with $\hat{\mathbf{p}} \in \mathfrak{R}^r$ an estimate of $\mathbf{p}$ in (4) and assuming that $\hat{\mathbf{D}}$ is invertible for all time $t \geq 0$, equation (6) can be further written to be

$$\ddot{\mathbf{e}} + \mathbf{K}_d \dot{\mathbf{e}} + \mathbf{K}_p \mathbf{e} = -\hat{\mathbf{D}}^{-1}\mathbf{Y}(\mathbf{q}, \dot{\mathbf{q}}, \ddot{\mathbf{q}})\tilde{\mathbf{p}} \qquad (7)$$

The above error dynamics implies that if we may find a proper update law for $\hat{\mathbf{p}}$ such that $\hat{\mathbf{p}} \to \mathbf{p}$ asymptotically, then (7) converges to (3) as $t \to \infty$, and hence, we may have convergence of the output tracking error. To this end, let us denote $\mathbf{x} = [\mathbf{e}^T \; \dot{\mathbf{e}}^T]^T \in \mathfrak{R}^{2n}$ to represent equation (7) in its state space form

$$\dot{\mathbf{x}} = \mathbf{A}\mathbf{x} - \mathbf{B}\hat{\mathbf{D}}^{-1}\mathbf{Y}\tilde{\mathbf{p}} \tag{8}$$

where $\mathbf{A} = \begin{bmatrix} \mathbf{0} & \mathbf{I}_n \\ -\mathbf{K}_p & -\mathbf{K}_d \end{bmatrix} \in \mathfrak{R}^{2n \times 2n}$ and $\mathbf{B} = \begin{bmatrix} \mathbf{0} \\ \mathbf{I}_n \end{bmatrix} \in \mathfrak{R}^{2n \times n}$. To design an

update law for $\hat{\mathbf{p}}$ to ensure closed loop stability, a Lyapunov-like function candidate can be selected as

$$V(\mathbf{x}, \tilde{\mathbf{p}}) = \frac{1}{2}\mathbf{x}^T\mathbf{P}\mathbf{x} + \frac{1}{2}\tilde{\mathbf{p}}^T\mathbf{\Gamma}\tilde{\mathbf{p}} \tag{9}$$

where $\mathbf{\Gamma} \in \mathfrak{R}^{r \times r}$ is a positive definite matrix and $\mathbf{P} = \mathbf{P}^T \in \mathfrak{R}^{2n \times 2n}$ is a positive definite solution to the Lyapunov equation $\mathbf{A}^T\mathbf{P} + \mathbf{P}\mathbf{A} = -\mathbf{Q}$ for a given positive definite matrix $\mathbf{Q} = \mathbf{Q}^T \in \mathfrak{R}^{2n \times 2n}$. Along the trajectory of (8), the time derivative of V can be computed to be

$$\dot{V} = -\frac{1}{2}\mathbf{x}^T\mathbf{Q}\mathbf{x} - \tilde{\mathbf{p}}^T[(\hat{\mathbf{D}}^{-1}\mathbf{Y})^T\mathbf{B}^T\mathbf{P}\mathbf{x} + \mathbf{\Gamma}\dot{\hat{\mathbf{p}}}] \tag{10}$$

By selecting the update law as

$$\dot{\hat{\mathbf{p}}} = -\mathbf{\Gamma}^{-1}[\hat{\mathbf{D}}^{-1}\mathbf{Y}(\mathbf{q},\dot{\mathbf{q}},\ddot{\mathbf{q}})]^T\mathbf{B}^T\mathbf{P}\mathbf{x} \tag{11}$$

equation (10) becomes

$$\dot{V} = -\frac{1}{2}\mathbf{x}^T\mathbf{Q}\mathbf{x} \leq 0 \tag{12}$$

Hence, we have proved that $\mathbf{x} \in L_\infty^{2n}$ and $\tilde{\mathbf{p}} \in L_\infty^r$. Since it is easy to have

$$\int_0^\infty (\sqrt{\mathbf{Q}}\mathbf{x})^T(\sqrt{\mathbf{Q}}\mathbf{x})dt = \int_0^\infty \mathbf{x}^T\mathbf{Q}\mathbf{x}dt = \int_0^\infty -2\dot{V}dt = 2[V(0) - V(\infty)] < \infty,$$

we may conclude $\mathbf{x} \in L_2^{2n}$. Because $\mathbf{A}$ is Hurwitz and $\mathbf{x}$ is bounded, equation (8) gives boundedness of $\dot{\mathbf{x}}$ if $\hat{\mathbf{D}}$ is nonsingular. Therefore, asymptotic convergence of **x** can be concluded by Barbalat's lemma. This further implies asymptotic convergence of the output tracking error **e**. It can also be proved that convergence of the parameter vector is dependent to the PE condition of the reference input signal. To realize the control law (5) and update law (11), the joint accelerations are required to be available. However, their measurements are generally costly and subject to noise. In addition, although the inertia matrix **D** is nonsingular for all $t>0$, its estimate $\hat{\mathbf{D}}$ is not guaranteed to be invertible. Hence, (11) might suffer the singularity problem when the determinant of $\hat{\mathbf{D}}$ gets very close to 0, and some projection modification should be applied. To solve these problems, a well-known strategy proposed by Slotine and Li (1988, 1991) based on the passivity design for rigid robots will be introduced in next section.

4.3 Slotine and Li's Approach

To get rid of the need for the joint acceleration feedback and to avoid the possible singularity problem stated above, Slotine and Li proposed the following design strategy. Define an error vector $\mathbf{s} = \dot{\mathbf{e}} + \Lambda \mathbf{e}$ where $\Lambda = diag(\lambda_1, \lambda_2, ..., \lambda_n)$ with $\lambda_i > 0$ for all $i=1,...,n$. By this definition, convergence of **s** implies convergence of the output error **e**. Rewrite the robot model (4.2-1) into the form

$$\mathbf{D}\dot{\mathbf{s}} + \mathbf{C}\mathbf{s} + \mathbf{g} + \mathbf{D}\ddot{\mathbf{q}}_d - \mathbf{D}\Lambda\dot{\mathbf{e}} + \mathbf{C}\dot{\mathbf{q}}_d - \mathbf{C}\Lambda\mathbf{e} = \tau \tag{1}$$

Suppose the robot model is precisely known, then we may pick an intuitive controller for (1) as

$$\tau = \mathbf{D}\ddot{\mathbf{q}}_d - \mathbf{D}\Lambda\dot{\mathbf{e}} + \mathbf{C}\dot{\mathbf{q}}_d - \mathbf{C}\Lambda\mathbf{e} + \mathbf{g} - \mathbf{K}_d\mathbf{s} \tag{2}$$

where $\mathbf{K}_d$ is a positive definite matrix. Hence, the closed loop system becomes

$$\mathbf{D}\dot{\mathbf{s}} + \mathbf{C}\mathbf{s} + \mathbf{K}_d\mathbf{s} = 0 \tag{3}$$

To justify the feasibility of the controller (2), let us define a Lyapunov-like function candidate as $V = \frac{1}{2}\mathbf{s}^T\mathbf{D}\mathbf{s}$. Its time derivative along the trajectory of (3) can be computed as

$$\dot{V} = -\mathbf{s}^T\mathbf{K}_d\mathbf{s} + \frac{1}{2}\mathbf{s}^T(\dot{\mathbf{D}} - 2\mathbf{C})\mathbf{s}$$

Since $\dot{\mathbf{D}} - 2\mathbf{C}$ can be proved to be skew-symmetric, the above equation becomes

$$\dot{V} = -\mathbf{s}^T \mathbf{K}_d \mathbf{s} \le 0 \tag{4}$$

It is easy to prove that $\mathbf{s}$ is uniformly bounded and square integrable, and $\dot{\mathbf{s}}$ is also uniformly bounded. Hence, $\mathbf{s} \to \mathbf{0}$ as $t \to \infty$, or we may conclude that the tracking error $\mathbf{e}$ converges asymptotically. It is noted that the above design is valid if all robot parameters are known.

Now let us consider the case when $\mathbf{D}$, $\mathbf{C}$ and $\mathbf{g}$ in (1) are not available, and controller (2) cannot be realized. A controller can be constructed based on (2) as

$$\boldsymbol{\tau} = \hat{\mathbf{D}}\ddot{\mathbf{q}}_d - \hat{\mathbf{D}}\boldsymbol{\Lambda}\dot{\mathbf{e}} + \hat{\mathbf{C}}\dot{\mathbf{q}}_d - \hat{\mathbf{C}}\boldsymbol{\Lambda}\mathbf{e} + \hat{\mathbf{g}} - \mathbf{K}_d \mathbf{s} \tag{5}$$

if some update laws for the estimates $\hat{\mathbf{D}}$, $\hat{\mathbf{C}}$ and $\hat{\mathbf{g}}$ can be properly designed. The above control law can be rewritten into the form

$$\boldsymbol{\tau} = \hat{\mathbf{D}}\dot{\mathbf{v}} + \hat{\mathbf{C}}\mathbf{v} + \hat{\mathbf{g}} - \mathbf{K}_d \mathbf{s} \tag{6}$$

where $\mathbf{v} = \dot{\mathbf{q}}_d - \boldsymbol{\Lambda}\mathbf{e}$ is a known signal vector. With this control law, the closed loop system can be represented in the form

$$\mathbf{D}\dot{\mathbf{s}} + \mathbf{C}\mathbf{s} + \mathbf{K}_d \mathbf{s} = -\tilde{\mathbf{D}}\dot{\mathbf{v}} - \tilde{\mathbf{C}}\mathbf{v} - \tilde{\mathbf{g}} \tag{7}$$

The right hand side of the above equation can be further expressed in the linearly parameterized form

$$\mathbf{D}\dot{\mathbf{s}} + \mathbf{C}\mathbf{s} + \mathbf{K}_d \mathbf{s} = -\mathbf{Y}(\mathbf{q}, \dot{\mathbf{q}}, \mathbf{v}, \dot{\mathbf{v}})\tilde{\mathbf{p}} \tag{8}$$

It is worth to mention that unlike the regressor matrix $\mathbf{Y}(\mathbf{q}, \dot{\mathbf{q}}, \ddot{\mathbf{q}})$ in (4.2-7), the regressor matrix $\mathbf{Y}(\mathbf{q}, \dot{\mathbf{q}}, \mathbf{v}, \dot{\mathbf{v}})$ in (8) is independent to the joint accelerations. On the other hand, if we may find an appropriate update law for $\hat{\mathbf{p}}$ such that $\tilde{\mathbf{p}} \to \mathbf{0}$ as $t \to \infty$, then (8) converges to (3) asymptotically, and the closed loop stability can be ensured. To find the update law, let us define a Lyapunov-like function candidate as

$$V(\mathbf{x}, \tilde{\mathbf{p}}) = \frac{1}{2}\mathbf{s}^T \mathbf{D}\mathbf{s} + \frac{1}{2}\tilde{\mathbf{p}}^T \boldsymbol{\Gamma}\tilde{\mathbf{p}} \tag{9}$$

Its time derivative along the trajectory of (8) can be derived as

$$\dot{V} = -\mathbf{s}^T \mathbf{K}_d \mathbf{s} - \tilde{\mathbf{p}}^T (\boldsymbol{\Gamma}\dot{\tilde{\mathbf{p}}} + \mathbf{Y}^T \mathbf{s}) \tag{10}$$

Hence, the update law can be picked as

$$\dot{\hat{\mathbf{p}}} = -\mathbf{\Gamma}^{-1}\mathbf{Y}^T(\mathbf{q}, \dot{\mathbf{q}}, \mathbf{v}, \dot{\mathbf{v}})\mathbf{s} \tag{11}$$

and (10) becomes

$$\dot{V} = -\mathbf{s}^T\mathbf{K}_d\mathbf{s} \leq 0. \tag{12}$$

This is the same result we have in (4) and same convergence performance can thus be concluded for the tracking error. To implement the control law (6) and update law (11), we do not need the information of joint accelerations and it is free from the singularity problem in the estimation of the inertia matrix.

4.4 The Regressor Matrix

In the above development, the robot model has to be represented as a linear parametric form so that an adaptive controller can be designed. However, derivation of the regressor matrix for a high-DOF robot is tedious. In the real-time realization, the regressor matrix has to be computed in every control cycle, and its complexity results in a considerable burden to the control computer. Besides, to satisfy the limitation of the traditional adaptive design that the uncertainties should be time-invariant, all time varying terms in the robot dynamics are collected inside the regressor matrix. It can be seen that all entries in the parameter vector are unknown constants and most of them are relatively easy to obtain. For example, the 2-D robot in (3.2-3) can be represented into a linear parameterization form in (4.2-4) by defining the parameter vector as (Spong and Vidyasagar 1989)

$$\mathbf{p} = \begin{bmatrix} m_1 l_{c1}^2 \\ m_2 l_1^2 \\ m_2 l_{c2}^2 \\ m_2 l_1 l_{c2} \\ I_1 \\ I_2 \\ m_1 l_{c1} g \\ m_2 l_1 g \\ m_2 l_{c2} g \end{bmatrix} \tag{1}$$

and the regressor matrix is in the form

$$
\mathbf{Y}(\mathbf{q},\dot{\mathbf{q}},\ddot{\mathbf{q}}) = \begin{bmatrix} \ddot{q}_1 & \ddot{q}_2 & \ddot{q}_1 + \ddot{q}_2 & y_{14} & \ddot{q}_1 & \ddot{q}_1 + \ddot{q}_2 & \cos q_1 & \cos q_1 & \cos(q_1 + q_2) \\ 0 & 0 & \ddot{q}_1 + \ddot{q}_2 & y_{24} & \ddot{q}_2 & \ddot{q}_2 & 0 & 0 & \cos(q_1 + q_2) \end{bmatrix}
$$

$$(2)$$

where

$$
y_{14} = 2\cos q_2 \ddot{q}_1 + \cos q_2 \ddot{q}_2 - 2\sin q_2 \dot{q}_1 \dot{q}_2 - \sin q_2 \dot{q}_2^2
$$
$$
y_{24} = \cos q_2 \ddot{q}_1 + \sin q_2.
$$

For the acceleration-free regressor used in (4.3-8), one realization can be given as

$$
\mathbf{Y}(\mathbf{q},\dot{\mathbf{q}},\mathbf{v},\dot{\mathbf{v}}) = \begin{bmatrix} \dot{v}_1 & \dot{v}_2 & \dot{v}_2 \cos q_2 & y'_{14} & \cos q_1 & \cos(q_1 + q_2) \\ 0 & \dot{v}_1 + \dot{v}_2 & 0 & y'_{24} & 0 & \cos(q_1 + q_2) \end{bmatrix} \quad (3)
$$

where

$$
y'_{14} = \dot{v}_2 \cos q_2 + v_1 \dot{q}_2 \sin q_2 + (\dot{q}_1 + \dot{q}_2)v_2 \sin q_2
$$
$$
y'_{24} = v_1 \cos q_2 + v_1 \dot{q}_1 \sin q_2
$$

and the corresponding parameter vector is

$$
\mathbf{p} = \begin{bmatrix} m_1 l_{c1}^2 + m_2 l_1^2 + m_2 l_{c2}^2 + I_1 + I_2 \\ m_2 l_{c2}^2 + I_2 \\ 2m_2 l_1 l_{c2} \\ m_2 l_1 l_{c2} \\ m_1 l_{c1} g + m_2 l_1 g \\ m_2 l_{c2} g \end{bmatrix} \quad (4)
$$

In (1) and (4), the entries are some combinations of system parameters such as link lengths, masses, and moments of inertia,…, etc. Obviously, these quantities are relatively easy to measure in practical applications compared to the derivation of the regressor matrix. However, in traditional robot adaptive control designs, we are required to know the complex regressor matrix, but update the easy-to-obtain parameter vector. To ease the controller design and implementation, it is suggested to consider the regressor-free adaptive control strategies. In next section, a FAT-based adaptive controller is designed for a rigid robot without the need for the regressor matrix.

4.5 FAT-Based Adaptive Controller Design

The same controller (4.3-6) is employed in this approach

$$\boldsymbol{\tau} = \hat{\mathbf{D}}\dot{\mathbf{v}} + \hat{\mathbf{C}}\mathbf{v} + \hat{\mathbf{g}} - \mathbf{K}_d\mathbf{s} \tag{1a}$$

and hence the closed-loop dynamics can still be represented as

$$\mathbf{D}\dot{\mathbf{s}} + \mathbf{C}\mathbf{s} + \mathbf{K}_d\mathbf{s} = -\tilde{\mathbf{D}}\dot{\mathbf{v}} - \tilde{\mathbf{C}}\mathbf{v} - \tilde{\mathbf{g}} \tag{1b}$$

This implies that $\mathbf{s}$ is an output of a stable first order filter driven by the approximation errors and tracking errors. If some proper update laws can be found so that $\hat{\mathbf{D}} \to \mathbf{D}$, $\hat{\mathbf{C}} \to \mathbf{C}$ and $\hat{\mathbf{g}} \to \mathbf{g}$, then $\mathbf{e} \to 0$ can be concluded from (1b). Since $\mathbf{D}$, $\mathbf{C}$ and $\mathbf{g}$ are functions of states and hence functions of time, traditional adaptive controllers are not applicable to give proper update laws except that the linearly parameterization assumption as shown in (4.3-8) is feasible. On the other hand, since their variation bounds are not given, conventional robust designs do not work either. Here, we would like to use FAT to representation $\mathbf{D}$, $\mathbf{C}$ and $\mathbf{g}$ with the assumption that proper numbers of basis functions are employed

$$\mathbf{D} = \mathbf{W}_{\mathbf{D}}^T \mathbf{Z}_{\mathbf{D}} + \boldsymbol{\varepsilon}_{\mathbf{D}}$$

$$\mathbf{C} = \mathbf{W}_{\mathbf{C}}^T \mathbf{Z}_{\mathbf{C}} + \boldsymbol{\varepsilon}_{\mathbf{C}} \tag{2a}$$

$$\mathbf{g} = \mathbf{W}_{\mathbf{g}}^T \mathbf{z}_{\mathbf{g}} + \boldsymbol{\varepsilon}_{\mathbf{g}}$$

where $\mathbf{W}_{\mathbf{D}} \in \Re^{n^2\beta_D \times n}$, $\mathbf{W}_{\mathbf{C}} \in \Re^{n^2\beta_C \times n}$ and $\mathbf{W}_{\mathbf{g}} \in \Re^{n\beta_g \times n}$ are weighting matrices and $\mathbf{Z}_{\mathbf{D}} \in \Re^{n^2\beta_D \times n}$, $\mathbf{Z}_{\mathbf{C}} \in \Re^{n^2\beta_C \times n}$ and $\mathbf{z}_{\mathbf{g}} \in \Re^{n\beta_g \times 1}$ are matrices of basis functions. The number $\beta_{(\cdot)}$ represents the number of basis functions used. Using the same set of basis functions, the corresponding estimates can be represented as

$$\hat{\mathbf{D}} = \hat{\mathbf{W}}_{\mathbf{D}}^T \mathbf{Z}_{\mathbf{D}}$$

$$\hat{\mathbf{C}} = \hat{\mathbf{W}}_{\mathbf{C}}^T \mathbf{Z}_{\mathbf{C}} \tag{2b}$$

$$\hat{\mathbf{g}} = \hat{\mathbf{W}}_{\mathbf{g}}^T \mathbf{z}_{\mathbf{g}}$$

Therefore, the controller (1) becomes

$$\boldsymbol{\tau} = \hat{\mathbf{W}}_{\mathbf{D}}^T \mathbf{Z}_{\mathbf{D}} \dot{\mathbf{v}} + \hat{\mathbf{W}}_{\mathbf{C}}^T \mathbf{Z}_{\mathbf{C}} \mathbf{v} + \hat{\mathbf{W}}_{\mathbf{g}}^T \mathbf{z}_{\mathbf{g}} - \mathbf{K}_d \mathbf{s} \tag{3}$$

and the closed loop system dynamics can be represented as

$$\mathbf{D}\dot{\mathbf{s}} + \mathbf{C}\mathbf{s} + \mathbf{K}_d \mathbf{s} = -\tilde{\mathbf{W}}_{\mathbf{D}}^T \mathbf{Z}_{\mathbf{D}} \dot{\mathbf{v}} - \tilde{\mathbf{W}}_{\mathbf{C}}^T \mathbf{Z}_{\mathbf{C}} \mathbf{v} - \tilde{\mathbf{W}}_{\mathbf{g}}^T \mathbf{z}_{\mathbf{g}} + \boldsymbol{\varepsilon}_1 \tag{4}$$

where $\tilde{\mathbf{W}}_{(\cdot)} = \mathbf{W}_{(\cdot)} - \hat{\mathbf{W}}_{(\cdot)}$ and $\boldsymbol{\varepsilon}_1 = \boldsymbol{\varepsilon}_1(\boldsymbol{\varepsilon}_{\mathbf{D}}, \boldsymbol{\varepsilon}_{\mathbf{C}}, \boldsymbol{\varepsilon}_{\mathbf{g}}, \mathbf{s}, \ddot{\mathbf{q}}_d) \in \mathfrak{R}^n$ is a lumped approximation error vector. Since $\mathbf{W}_{(\cdot)}$ are constant vectors, their update laws can be easily found by proper selection of a Lyapunov-like function. Let us consider a candidate

$$V(\mathbf{s}, \tilde{\mathbf{W}}_{\mathbf{D}}, \tilde{\mathbf{W}}_{\mathbf{C}}, \tilde{\mathbf{W}}_{\mathbf{g}}) = \frac{1}{2} \mathbf{s}^T \mathbf{D} \mathbf{s}$$
$$+ \frac{1}{2} Tr(\tilde{\mathbf{W}}_{\mathbf{D}}^T \mathbf{Q}_{\mathbf{D}} \tilde{\mathbf{W}}_{\mathbf{D}} + \tilde{\mathbf{W}}_{\mathbf{C}}^T \mathbf{Q}_{\mathbf{C}} \tilde{\mathbf{W}}_{\mathbf{C}} + \tilde{\mathbf{W}}_{\mathbf{g}}^T \mathbf{Q}_{\mathbf{g}} \tilde{\mathbf{W}}_{\mathbf{g}}) \tag{5}$$

where $\mathbf{Q}_{\mathbf{D}} \in \mathfrak{R}^{n^2 \beta_{\mathbf{D}} \times n^2 \beta_{\mathbf{D}}}$, $\mathbf{Q}_{\mathbf{C}} \in \mathfrak{R}^{n^2 \beta_C \times n^2 \beta_C}$ and $\mathbf{Q}_{\mathbf{g}} \in \mathfrak{R}^{n\beta_g \times n\beta_g}$ are positive definite weighting matrices. The time derivative of V along the trajectory of (4) can be computed as

$$\dot{V} = \mathbf{s}^T[-\mathbf{C}\mathbf{s} - \mathbf{K}_d \mathbf{s} - \tilde{\mathbf{W}}_{\mathbf{D}}^T \mathbf{Z}_{\mathbf{D}} \dot{\mathbf{v}} - \tilde{\mathbf{W}}_{\mathbf{C}}^T \mathbf{Z}_{\mathbf{C}} \mathbf{v} - \tilde{\mathbf{W}}_{\mathbf{g}}^T \mathbf{z}_{\mathbf{g}}]$$
$$+ \frac{1}{2} \mathbf{s}^T \dot{\mathbf{D}} \mathbf{s} - Tr(\tilde{\mathbf{W}}_{\mathbf{D}}^T \mathbf{Q}_{\mathbf{D}} \dot{\hat{\mathbf{W}}}_{\mathbf{D}} + \tilde{\mathbf{W}}_{\mathbf{C}}^T \mathbf{Q}_{\mathbf{C}} \dot{\hat{\mathbf{W}}}_{\mathbf{C}} + \tilde{\mathbf{W}}_{\mathbf{g}}^T \mathbf{Q}_{\mathbf{g}} \dot{\hat{\mathbf{W}}}_{\mathbf{g}}) + \mathbf{s}^T \boldsymbol{\varepsilon}_1$$

Using the fact that the matrix $\dot{\mathbf{D}} - 2\mathbf{C}$ is skew-symmetric, we further have

$$\dot{V} = -\mathbf{s}^T \mathbf{K}_d \mathbf{s} - Tr[\tilde{\mathbf{W}}_{\mathbf{D}}^T (\mathbf{Z}_{\mathbf{D}} \dot{\mathbf{v}} \mathbf{s}^T + \mathbf{Q}_{\mathbf{D}} \dot{\hat{\mathbf{W}}}_{\mathbf{D}})]$$
$$- Tr[\tilde{\mathbf{W}}_{\mathbf{C}}^T (\mathbf{Z}_{\mathbf{C}} \mathbf{v} \mathbf{s}^T + \mathbf{Q}_{\mathbf{C}} \dot{\hat{\mathbf{W}}}_{\mathbf{C}})] - Tr[\tilde{\mathbf{W}}_{\mathbf{g}}^T (\mathbf{z}_{\mathbf{g}} \mathbf{s}^T + \mathbf{Q}_{\mathbf{g}} \dot{\hat{\mathbf{W}}}_{\mathbf{g}})] + \mathbf{s}^T \boldsymbol{\varepsilon}_1 \tag{6}$$

Let us select the update laws with σ-modifications to be

$$\dot{\hat{\mathbf{W}}}_{\mathbf{D}} = -\mathbf{Q}_{\mathbf{D}}^{-1}(\mathbf{Z}_{\mathbf{D}} \dot{\mathbf{v}} \mathbf{s}^T + \sigma_{\mathbf{D}} \hat{\mathbf{W}}_{\mathbf{D}})$$
$$\dot{\hat{\mathbf{W}}}_{\mathbf{C}} = -\mathbf{Q}_{\mathbf{C}}^{-1}(\mathbf{Z}_{\mathbf{C}} \mathbf{v} \mathbf{s}^T + \sigma_{\mathbf{C}} \hat{\mathbf{W}}_{\mathbf{C}}) \tag{7}$$
$$\dot{\hat{\mathbf{W}}}_{\mathbf{g}} = -\mathbf{Q}_{\mathbf{g}}^{-1}(\mathbf{z}_{\mathbf{g}} \mathbf{s}^T + \sigma_{\mathbf{g}} \hat{\mathbf{W}}_{\mathbf{g}})$$

where $\sigma_{(\cdot)}$ are positive numbers. Hence, equation (6) becomes

$$\dot{V} = -\mathbf{s}^T \mathbf{K}_d \mathbf{s} + \mathbf{s}^T \boldsymbol{\varepsilon}_1 + \sigma_\mathbf{D} Tr(\tilde{\mathbf{W}}_\mathbf{D}^T \hat{\mathbf{W}}_\mathbf{D})$$
$$+ \sigma_\mathbf{C} Tr(\tilde{\mathbf{W}}_\mathbf{C}^T \hat{\mathbf{W}}_\mathbf{C}) + \sigma_\mathbf{g} Tr(\tilde{\mathbf{W}}_\mathbf{g}^T \hat{\mathbf{W}}_\mathbf{g}) \tag{8}$$

Remark 1: Suppose a sufficient number of basis functions are used and the approximation error can be ignored, then it is not necessary to include the σ-modification terms in (7). Hence, (8) can be reduced to (4.3-12), and convergence of $\mathbf{s}$ can be further proved by Barbalat's lemma.

Remark 2: If the approximation error cannot be ignored, but we can find a positive number δ such that $\|\boldsymbol{\varepsilon}_1\| \leq \delta$, then a robust term $\boldsymbol{\tau}_{robust}$ can be added into (1a) to have a new control law

$$\boldsymbol{\tau} = \hat{\mathbf{D}}\dot{\mathbf{v}} + \hat{\mathbf{C}}\mathbf{v} + \hat{\mathbf{g}} - \mathbf{K}_d \mathbf{s} + \boldsymbol{\tau}_{robust} \tag{9}$$

Consider the Lyapunov-like function candidate (5) again, and the update law (7) without σ-modification; then the time derivative of V becomes

$$\dot{V} \leq -\mathbf{s}^T \mathbf{K}_d \mathbf{s} + \delta \|\mathbf{s}\| + \mathbf{s}^T \boldsymbol{\tau}_{robust} \tag{10}$$

If we select $\boldsymbol{\tau}_{robust} = -\delta[\mathrm{sgn}(s_1) \quad \mathrm{sgn}(s_2) \quad \cdots \quad \mathrm{sgn}(s_n)]^T$, where s_i, $i = 1, \ldots, n$ is the i-th entry in $\mathbf{s}$, then we may have $\dot{V} \leq -\mathbf{s}^T \mathbf{K}_d \mathbf{s} \leq 0$ which is similar to the result of (4.3-12). This will further give convergence of the output error by Barbalat's lemma.

With the existence of the approximation error $\boldsymbol{\varepsilon}_1$ and the σ-modification terms, equation (8) may not conclude its definiteness as the one we have in (4.3-12). The following two inequalities are very useful in further derivation

$$-\mathbf{s}^T \mathbf{K}_d \mathbf{s} + \mathbf{s}^T \boldsymbol{\varepsilon}_1 \leq -\frac{1}{2}\left[\lambda_{\min}(\mathbf{K}_d)\|\mathbf{s}\|^2 - \frac{\|\boldsymbol{\varepsilon}_1\|^2}{\lambda_{\min}(\mathbf{K}_d)} \right] \tag{11a}$$

$$Tr(\tilde{\mathbf{W}}_{(\cdot)}^T \hat{\mathbf{W}}_{(\cdot)}) \leq \frac{1}{2} Tr(\mathbf{W}_{(\cdot)}^T \mathbf{W}_{(\cdot)}) - \frac{1}{2} Tr(\tilde{\mathbf{W}}_{(\cdot)}^T \tilde{\mathbf{W}}_{(\cdot)}) \tag{11b}$$

The proof for the first inequality is straightforward, and the proof for the second one can be found in the Appendix. Together with the relationship

$$
V = \frac{1}{2} [\mathbf{s}^T \mathbf{D} \mathbf{s} + Tr(\tilde{\mathbf{W}}_{\mathbf{D}}^T \mathbf{Q}_{\mathbf{D}} \tilde{\mathbf{W}}_{\mathbf{D}} + \tilde{\mathbf{W}}_{\mathbf{C}}^T \mathbf{Q}_{\mathbf{C}} \tilde{\mathbf{W}}_{\mathbf{C}} + \tilde{\mathbf{W}}_{\mathbf{g}}^T \mathbf{Q}_{\mathbf{g}} \tilde{\mathbf{W}}_{\mathbf{g}})]
$$

$$
\leq \frac{1}{2} [\lambda_{\max}(\mathbf{D}) \|\mathbf{s}\|^2 + \lambda_{\max}(\mathbf{Q}_{\mathbf{D}}) Tr(\tilde{\mathbf{W}}_{\mathbf{D}}^T \tilde{\mathbf{W}}_{\mathbf{D}})
$$

$$
+ \lambda_{\max}(\mathbf{Q}_{\mathbf{C}}) Tr(\tilde{\mathbf{W}}_{\mathbf{C}}^T \tilde{\mathbf{W}}_{\mathbf{C}}) + \lambda_{\max}(\mathbf{Q}_{\mathbf{g}}) Tr(\tilde{\mathbf{W}}_{\mathbf{g}}^T \tilde{\mathbf{W}}_{\mathbf{g}})] \qquad (12)
$$

we may rewrite (8) into the form

$$
\dot{V} \leq -\alpha V + \frac{\|\boldsymbol{\varepsilon}_1\|^2}{2\lambda_{\min}(\mathbf{K}_d)}
$$

$$
+ \frac{1}{2} \{ [\alpha \lambda_{\max}(\mathbf{D}) - \lambda_{\min}(\mathbf{K}_d)] \|\mathbf{s}\|^2 + [\alpha \lambda_{\max}(\mathbf{Q}_{\mathbf{D}})
$$

$$
- \sigma_{\mathbf{D}}] Tr(\tilde{\mathbf{W}}_{\mathbf{D}}^T \tilde{\mathbf{W}}_{\mathbf{D}}) \} + \frac{1}{2} \{ [\alpha \lambda_{\max}(\mathbf{Q}_{\mathbf{C}}) - \sigma_{\mathbf{C}}] Tr(\tilde{\mathbf{W}}_{\mathbf{C}}^T \tilde{\mathbf{W}}_{\mathbf{C}})
$$

$$
+ [\alpha \lambda_{\max}(\mathbf{Q}_{\mathbf{g}}) - \sigma_{\mathbf{g}}] Tr(\tilde{\mathbf{W}}_{\mathbf{g}}^T \tilde{\mathbf{W}}_{\mathbf{g}}) \} + \frac{1}{2} [\sigma_{\mathbf{D}} Tr(\mathbf{W}_{\mathbf{D}}^T \mathbf{W}_{\mathbf{D}})
$$

$$
+ \sigma_{\mathbf{C}} Tr(\mathbf{W}_{\mathbf{C}}^T \mathbf{W}_{\mathbf{C}}) + \sigma_{\mathbf{g}} Tr(\mathbf{W}_{\mathbf{g}}^T \mathbf{W}_{\mathbf{g}})] \qquad (13)
$$

where α is a constant to be selected as

$$
\alpha \leq \min \left\{ \frac{\lambda_{\min}(\mathbf{K}_d)}{\lambda_{\max}(\mathbf{D})}, \frac{\sigma_{\mathbf{D}}}{\lambda_{\max}(\mathbf{Q}_{\mathbf{D}})}, \frac{\sigma_{\mathbf{C}}}{\lambda_{\max}(\mathbf{Q}_{\mathbf{C}})}, \frac{\sigma_{\mathbf{g}}}{\lambda_{\max}(\mathbf{Q}_{\mathbf{g}})} \right\} \qquad (14)
$$

then (13) becomes

$$
\dot{V} \leq -\alpha V + \frac{\|\boldsymbol{\varepsilon}_1\|^2}{2\lambda_{\min}(\mathbf{K}_d)} + \frac{1}{2} [\sigma_{\mathbf{D}} Tr(\mathbf{W}_{\mathbf{D}}^T \mathbf{W}_{\mathbf{D}}) +
$$

$$
\sigma_{\mathbf{C}} Tr(\mathbf{W}_{\mathbf{C}}^T \mathbf{W}_{\mathbf{C}}) + \sigma_{\mathbf{g}} Tr(\mathbf{W}_{\mathbf{g}}^T \mathbf{W}_{\mathbf{g}})] \qquad (15)
$$

This implies $\dot{V} < 0$ whenever

$$
V > \frac{1}{2\alpha \lambda_{\min}(\mathbf{K}_d)} \sup_{\tau \geq t_0} \|\boldsymbol{\varepsilon}_1(\tau)\|^2 + \frac{1}{2\alpha} [\sigma_{\mathbf{D}} Tr(\mathbf{W}_{\mathbf{D}}^T \mathbf{W}_{\mathbf{D}})
$$

$$
+ \sigma_{\mathbf{C}} Tr(\mathbf{W}_{\mathbf{C}}^T \mathbf{W}_{\mathbf{C}}) + \sigma_{\mathbf{g}} Tr(\mathbf{W}_{\mathbf{g}}^T \mathbf{W}_{\mathbf{g}})] \qquad (16)
$$

It should be noted that selection of α in (14) depends on the maximum eigenvalue of the inertia matrix $\mathbf{D}$ which is not available. However, according to Property 1 in Section 3.2, it is easy to prove that $\exists\ \eta_D, \underline{\eta}_D > 0$ such that $\lambda_{max}(\mathbf{D}) \leq \eta_D$ and $\lambda_{min}(\mathbf{D}) \geq \underline{\eta}_D$ and (14) can be rewritten as

$$\alpha \leq \min\left\{\frac{\lambda_{min}(\mathbf{K}_d)}{\eta_D}, \frac{\sigma_D}{\lambda_{max}(\mathbf{Q_D})}, \frac{\sigma_C}{\lambda_{max}(\mathbf{Q_C})}, \frac{\sigma_g}{\lambda_{max}(\mathbf{Q_g})}\right\}$$

Since α will not be used in the realization of the control law or the update law, we are going to use similar treatment in (14) in later chapters to simplify the derivation.

It can be seen that all terms in the right side of (16) are constants. By proper selection of the parameters there, we may adjust the set where $\dot{V} \geq 0$ to be sufficiently small. Hence, we have proved that $\mathbf{s}$, $\tilde{\mathbf{W}}_D$, $\tilde{\mathbf{W}}_C$ and $\tilde{\mathbf{W}}_g$ are uniformly ultimately bounded. In addition, we may also compute the upper bound for V by solving the differential inequality of V in (15) as

$$V(t) \leq e^{-\alpha(t-t_0)}V(t_0) + \frac{1}{2\alpha\lambda_{min}(\mathbf{K}_d)} \sup_{t_0 < \tau < t} \|\varepsilon_1(\tau)\|^2$$
$$+ \frac{1}{2\alpha}[\sigma_D Tr(\mathbf{W}_D^T \mathbf{W}_D) + \sigma_C Tr(\mathbf{W}_C^T \mathbf{W}_C) + \sigma_g Tr(\mathbf{W}_g^T \mathbf{W}_g)] \quad (17)$$

Using the inequality

$$V \geq \frac{1}{2}[\lambda_{min}(\mathbf{D})\|\mathbf{s}\|^2 + \lambda_{min}(\mathbf{Q_D})Tr(\tilde{\mathbf{W}}_D^T \tilde{\mathbf{W}}_D)$$
$$+ \lambda_{min}(\mathbf{Q_C})Tr(\tilde{\mathbf{W}}_C^T \tilde{\mathbf{W}}_C) + \lambda_{min}(\mathbf{Q_g})Tr(\tilde{\mathbf{W}}_g^T \tilde{\mathbf{W}}_g)] \quad (18)$$

we may find the upper bound for $\|\mathbf{s}\|^2$ as

$$\|\mathbf{s}\|^2 \leq \frac{1}{\lambda_{min}(\mathbf{D})}[2V - \lambda_{min}(\mathbf{Q_D})Tr(\tilde{\mathbf{W}}_D^T \tilde{\mathbf{W}}_D)$$
$$- \lambda_{min}(\mathbf{Q_C})Tr(\tilde{\mathbf{W}}_C^T \tilde{\mathbf{W}}_C) - \lambda_{min}(\mathbf{Q_g})Tr(\tilde{\mathbf{W}}_g^T \tilde{\mathbf{W}}_g)]$$
$$\leq \frac{2}{\lambda_{min}(\mathbf{D})}V$$

With (17), this can be further written as

$$\|\mathbf{s}\|^2 \le \frac{2}{\lambda_{\min}(\mathbf{D})} e^{-\alpha(t-t_0)} V(t_0) + \frac{1}{\alpha\lambda_{\min}(\mathbf{D})\lambda_{\min}(\mathbf{K}_d)} \sup_{t_0 < \tau < t} \|\boldsymbol{\varepsilon}_1(\tau)\|^2$$

$$+ \frac{1}{\alpha\lambda_{\min}(\mathbf{D})} [\sigma_\mathbf{D} Tr(\mathbf{W}_\mathbf{D}^T \mathbf{W}_\mathbf{D}) + \sigma_\mathbf{C} Tr(\mathbf{W}_\mathbf{C}^T \mathbf{W}_\mathbf{C}) + \sigma_\mathbf{g} Tr(\mathbf{W}_\mathbf{g}^T \mathbf{W}_\mathbf{g})] \quad (19)$$

We may also write the bound for the error signal **s** as

$$\|\mathbf{s}\| \le \sqrt{\frac{2V(t_0)}{\lambda_{\min}(\mathbf{D})}} e^{-\frac{\alpha}{2}(t-t_0)} + \frac{1}{\sqrt{\alpha\lambda_{\min}(\mathbf{D})\lambda_{\min}(\mathbf{K}_d)}} \sup_{t_0 < \tau < t} \|\boldsymbol{\varepsilon}_1(\tau)\|$$

$$+ \frac{1}{\sqrt{\alpha\lambda_{\min}(\mathbf{D})}} [\sigma_\mathbf{D} Tr(\mathbf{W}_\mathbf{D}^T \mathbf{W}_\mathbf{D}) + \sigma_\mathbf{C} Tr(\mathbf{W}_\mathbf{C}^T \mathbf{W}_\mathbf{C}) + \sigma_\mathbf{g} Tr(\mathbf{W}_\mathbf{g}^T \mathbf{W}_\mathbf{g})]^{\frac{1}{2}} \quad (20)$$

Inequality (20) implies that the time history of the error signal **s** is bounded by an exponential function plus some constants. This completes the transient performance analysis.

Table 4.1 summarizes the adaptive control laws derived in this section. Two columns are arranged to present the regressor-based and regressor-free designs respectively according to their controller forms, update laws, and implementation issues.

Table 4.1 Summary of the adaptive control for RR

	Rigid-Link Rigid-Joint Robot $\mathbf{D}(\mathbf{q})\ddot{\mathbf{q}} + \mathbf{C}(\mathbf{q},\dot{\mathbf{q}})\dot{\mathbf{q}} + \mathbf{g}(\mathbf{q}) = \boldsymbol{\tau}$ (4.2-1)	
	Regressor-based	Regressor-free
Controller	$\boldsymbol{\tau} = \hat{\mathbf{D}}\dot{\mathbf{v}} + \hat{\mathbf{C}}\mathbf{v} + \hat{\mathbf{g}} - \mathbf{K}_d\mathbf{s}$ $= \mathbf{Y}(\mathbf{q},\dot{\mathbf{q}},\mathbf{v},\dot{\mathbf{v}})\hat{\mathbf{p}} - \mathbf{K}_d\mathbf{s}$ (4.3-6)	$\boldsymbol{\tau} = \hat{\mathbf{D}}\dot{\mathbf{v}} + \hat{\mathbf{C}}\mathbf{v} + \hat{\mathbf{g}} - \mathbf{K}_d\mathbf{s}$ $= \hat{\mathbf{W}}_\mathbf{g}^T \mathbf{z}_\mathbf{g} + \hat{\mathbf{W}}_\mathbf{D}^T \mathbf{Z}_\mathbf{D}\dot{\mathbf{v}}$ $+ \hat{\mathbf{W}}_\mathbf{C}^T \mathbf{Z}_\mathbf{C}\mathbf{v} - \mathbf{K}_d\mathbf{s}$ (4.5-1)
Adaptive Law	$\dot{\hat{\mathbf{p}}} = -\boldsymbol{\Gamma}^{-1}\mathbf{Y}^T\mathbf{s}$ (4.3-11)	$\dot{\hat{\mathbf{W}}}_\mathbf{D} = -\mathbf{Q}_\mathbf{D}^{-1}(\mathbf{Z}_\mathbf{D}\dot{\mathbf{v}}\mathbf{s}^T + \sigma_\mathbf{D}\hat{\mathbf{W}}_\mathbf{D})$ $\dot{\hat{\mathbf{W}}}_\mathbf{C} = -\mathbf{Q}_\mathbf{C}^{-1}(\mathbf{Z}_\mathbf{C}\mathbf{v}\mathbf{s}^T + \sigma_\mathbf{C}\hat{\mathbf{W}}_\mathbf{C})$ $\dot{\hat{\mathbf{W}}}_\mathbf{g} = -\mathbf{Q}_\mathbf{g}^{-1}(\mathbf{z}_\mathbf{g}\mathbf{s}^T + \sigma_\mathbf{g}\hat{\mathbf{W}}_\mathbf{g})$ (4.5-7)
Realization Issue	Need regressor matrix	Does not need regressor matrix

Example 4.1:

Consider a 2-DOF planar robot represented in example 3.1, and we are going to verify the control strategy developed in this section by using computer simulations. Actual values of link parameters are selected as $m_1 = m_2 = 0.5(kg)$, $l_1 = l_2 = 0.75(m)$, $l_{c1} = l_{c2} = 0.375(m)$, and $I_1 = I_2 = 0.0234(kg\text{-}m^2)$. We would like the endpoint to track a $0.2m$ radius circle centered at $(0.8m, 1.0m)$ in 10 seconds without knowing its precise model. The initial condition of the generalized coordinate vector is set to be $\mathbf{q}(0) = [0.0022 \quad 1.5019 \quad 0 \quad 0]^T$, i.e., the endpoint is initially at $(0.8m, 0,75m)$. Since it is away from the desired initial endpoint position $(0.8m, 0,8m)$, some significant transient response can be observed. The controller in (4.5-1a) is applied with the gain matrices

$$\mathbf{K}_d = \begin{bmatrix} 20 & 0 \\ 0 & 20 \end{bmatrix}, \text{ and } \mathbf{\Lambda} = \begin{bmatrix} 10 & 0 \\ 0 & 10 \end{bmatrix}.$$

Since we have assumed that the entries of $\mathbf{D}$, $\mathbf{C}$ and $\mathbf{g}$ are all unavailable, and their variation bounds are not known, we employ the FAT to have the representations in (2). The 11-term Fourier series is selected as the basis function for the approximation. Therefore, $\hat{\mathbf{W}}_\mathbf{D}$ and $\hat{\mathbf{W}}_\mathbf{C}$ are in $\mathfrak{R}^{44 \times 2}$, and $\hat{\mathbf{W}}_\mathbf{g}$ is in $\mathfrak{R}^{22 \times 2}$. The initial weighting vectors for the entries are assigned to be

$$\hat{\mathbf{w}}_{D_{11}}(0) = [0.05 \quad 0 \quad \cdots \quad 0]^T \in \mathfrak{R}^{11 \times 1}$$

$$\hat{\mathbf{w}}_{D_{12}}(0) = \hat{\mathbf{w}}_{D_{21}}(0) = [-0.05 \quad 0 \quad \cdots \quad 0]^T \in \mathfrak{R}^{11 \times 1}$$

$$\hat{\mathbf{w}}_{D_{22}}(0) = [0.1 \quad 0 \quad \cdots \quad 0]^T \in \mathfrak{R}^{11 \times 1}$$

$$\hat{\mathbf{w}}_{C_{11}}(0) = [0.05 \quad 0 \quad \cdots \quad 0]^T \in \mathfrak{R}^{11 \times 1}$$

$$\hat{\mathbf{w}}_{C_{12}}(0) = \hat{\mathbf{w}}_{C_{21}}(0) = [-0.05 \quad 0 \quad \cdots \quad 0]^T \in \mathfrak{R}^{11 \times 1}$$

$$\hat{\mathbf{w}}_{C_{22}}(0) = [0.1 \quad 0 \quad \cdots \quad 0]^T \in \mathfrak{R}^{11 \times 1}$$

$$\hat{\mathbf{w}}_{g_1}(0) = \hat{\mathbf{w}}_{g_2}(0) = [0 \quad 0 \quad \cdots \quad 0]^T \in \mathfrak{R}^{11 \times 1}$$

The gain matrices in the update laws (4.5-7) are selected as

$$\mathbf{Q}_\mathbf{D}^{-1} = \mathbf{I}_{44}, \ \mathbf{Q}_\mathbf{C}^{-1} = \mathbf{I}_{44} \text{ and } \mathbf{Q}_\mathbf{g}^{-1} = 100\mathbf{I}_{22}.$$

In this simulation, we assume that the approximation error can be neglected, and hence the σ-modification parameters are chosen as $\sigma_{(\cdot)} = 0$. The simulation results are shown in Figure 4.1 to 4.6. Figure 4.1 shows the tracking performance of the robot endpoint and its desired trajectory in the Cartesian space. It is observed that the endpoint trajectory converges nicely to the desired trajectory, although the initial position error is quite large. After the transient state, the tracking error is small regardless of the time-varying uncertainties in **D**, **C** and **g**. Computation of the complex regressor is avoided in this strategy which greatly simplifies the design and implementation of the control law. Figure 4.2 presents the time history of the joint space tracking performance. The transient states converge very fast without unwanted oscillations. The control efforts to the two joints are reasonable that can be verified in Figure 4.3. Figure 4.4 to 4.6 are the performance of function approximation. Although most parameters do not converge to their actual values, they still remain bounded as desired.

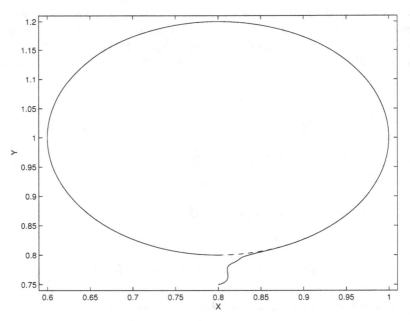

Figure 4.1 Tracking performance in the Cartesian space
(— actual trajectory; --- desired trajectory)

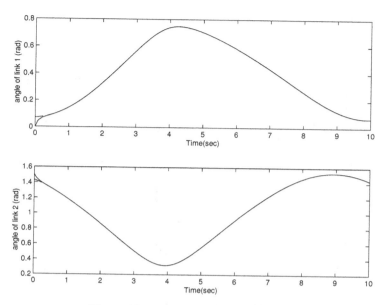

Figure 4.2 Joint space tracking performance
(— actual trajectory; --- desired trajectory)

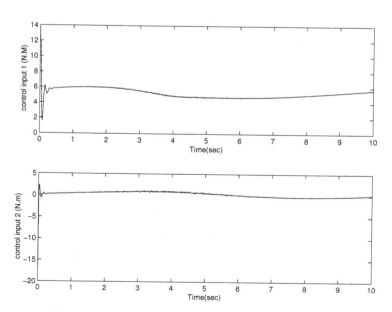

Figure 4.3 The control efforts for both joints are all reasonable

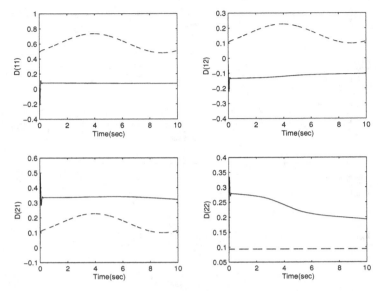

Figure 4.4 Approximation of **D**
(—estimate; --- actual value)

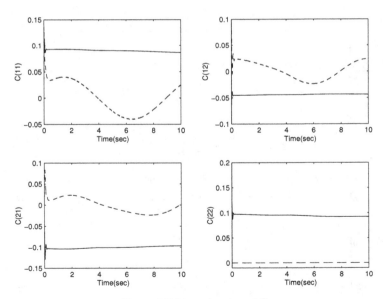

Figure 4.5 Approximation of **C**
(—estimate; --- actual value)

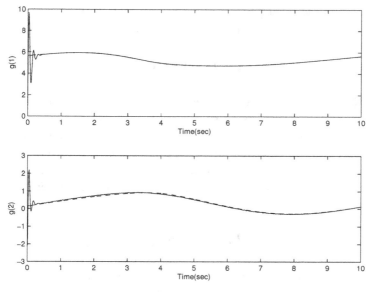

Figure 4.6 Approximation of **g**
(—estimate; --- actual value)

4.6 Consideration of Actuator Dynamics

In this section, we are going to derive adaptive controllers for the rigid-link electrically-driven robot described in (3.4-1) as

$$\mathbf{D(q)\ddot{q}} + \mathbf{C(q,\dot{q})\dot{q}} + \mathbf{g(q)} = \mathbf{Hi} \tag{1a}$$

$$\mathbf{L\dot{i}} + \mathbf{Ri} + \mathbf{K}_b\mathbf{\dot{q}} = \mathbf{u} \tag{1b}$$

We firstly consider the case when all robot parameters are known, and then the regressor-based adaptive controller is derived if the robot parameters are not available. Finally, a regressor-free adaptive controller is introduced. Several simulation results will be presented to justify the necessity for the consideration of the actuator dynamics, and to evaluate the performance of the controllers designed here.

With the definitions of **s** and **v** in Section 4.2, we may rewrite (1a) into the form

$$\mathbf{D\dot{s}} + \mathbf{Cs} + \mathbf{g} + \mathbf{D\dot{v}} + \mathbf{Cv} = \mathbf{Hi} \tag{2}$$

Suppose all of the robot parameters are known, then we may design a proper control law $\mathbf{u}$ such that the current $\mathbf{i}$ in (1b) follows the trajectory

$$\mathbf{i} = \mathbf{H}^{-1}(\mathbf{g} + \mathbf{D}\dot{\mathbf{v}} + \mathbf{C}\mathbf{v} - \mathbf{K}_d\mathbf{s}) \tag{3}$$

where $\mathbf{K}_d$ is a positive definite matrix. Substituting (3) into (2), the closed loop dynamics becomes $\mathbf{D}\dot{\mathbf{s}} + \mathbf{C}\mathbf{s} + \mathbf{K}_d\mathbf{s} = \mathbf{0}$. It is exactly the same as the one in (4.3-3), and hence convergence of $\mathbf{s}$ follows. To realize the perfect current vector in (3), we have to design a control input $\mathbf{u}$ in (1b) to ensure that the actual current can converge to the perfect one. Since all parameters are assumed to be known at the present stage, we may construct the control input in the form

$$\mathbf{u} = \mathbf{L}\dot{\mathbf{i}}_d + \mathbf{R}\mathbf{i} + \mathbf{K}_b\dot{\mathbf{q}} - \mathbf{K}_c\mathbf{e}_i \tag{4}$$

where $\mathbf{e}_i = \mathbf{i} - \mathbf{i}_d$ is the current error, and $\mathbf{i}_d$ is the desired current trajectory which is equivalent to the perfect current in (3). The gain matrix $\mathbf{K}_c$ is selected to be positive definite. Substituting (4) into (1b), the dynamics for the *current tracking loop* becomes

$$\mathbf{L}\dot{\mathbf{e}}_i + \mathbf{K}_c\mathbf{e}_i = \mathbf{0} \tag{5}$$

On the other hand, since the desired current is defined according to (3) as

$$\mathbf{i}_d = \mathbf{H}^{-1}(\mathbf{g} + \mathbf{D}\dot{\mathbf{v}} + \mathbf{C}\mathbf{v} - \mathbf{K}_d\mathbf{s}) \tag{6}$$

we may rewrite (2) into the form below to represent the dynamics of the *output tracking loop*

$$\mathbf{D}\dot{\mathbf{s}} + \mathbf{C}\mathbf{s} + \mathbf{K}_d\mathbf{s} = \mathbf{H}(\mathbf{i} - \mathbf{i}_d) \tag{7}$$

To prove the close loop stability, let us consider the Lyapunov-like function candidate

$$V(\mathbf{s}, \mathbf{e}_i) = \frac{1}{2}\mathbf{s}^T\mathbf{D}\mathbf{s} + \frac{1}{2}\mathbf{e}_i^T\mathbf{L}\mathbf{e}_i \tag{8}$$

Along the trajectory of (5) and (7), we may compute the time derivative of V as

$$\dot{V} = -\mathbf{s}^T \mathbf{K}_d \mathbf{s} + \frac{1}{2}\mathbf{s}^T (\dot{\mathbf{D}} - 2\mathbf{C})\mathbf{s} + \mathbf{s}^T \mathbf{H}\mathbf{e}_i - \mathbf{e}_i^T \mathbf{K}_c \mathbf{e}_i$$

$$= -[\mathbf{s}^T \quad \mathbf{e}_i^T]\mathbf{Q}\begin{bmatrix}\mathbf{s} \\ \mathbf{e}_i\end{bmatrix} \leq 0 \tag{9}$$

where $\mathbf{Q} = \begin{bmatrix} \mathbf{K}_d & -\dfrac{1}{2}\mathbf{H} \\ -\dfrac{1}{2}\mathbf{H} & \mathbf{K}_c \end{bmatrix}$ is positive definite by proper selection of $\mathbf{K}_c$ and

$\mathbf{K}_d$. It is easy to further prove that $\mathbf{s}$ and $\mathbf{e}_i$ are also square integrable, and their time derivatives are uniformly bounded. Hence, by Barbalat's lemma, we may conclude asymptotic convergence of $\mathbf{s}$ and $\mathbf{e}_i$.

In summary, if all parameters in the EDRR (1) are available, the controller (4) with the perfect current trajectory (6) can give asymptotic convergence of the output error.

Remark 1: It has to be noted that in controller (4), we have to find the time derivative of the desired current trajectory which implies that we need to feedback the joint accelerations to complete that computation. This necessity will be eliminated in the following design of the FAT-based regressor-free adaptive controller in section 4.6.2.

In next step, we would like to derive a regressor-based adaptive controller for EDRR. The regressor-free design will be developed in section 4.6.2.

4.6.1 Regressor-based adaptive control

Let us consider the EDRR in (1) and (2) again

$$\mathbf{D}\dot{\mathbf{s}} + \mathbf{C}\mathbf{s} + \mathbf{g} + \mathbf{D}\dot{\mathbf{v}} + \mathbf{C}\mathbf{v} = \mathbf{H}\mathbf{i} \tag{10a}$$

$$\mathbf{L}\dot{\mathbf{i}} + \mathbf{R}\mathbf{i} + \mathbf{K}_b\dot{\mathbf{q}} = \mathbf{u} \tag{10b}$$

Suppose $\mathbf{D}$, $\mathbf{C}$, $\mathbf{g}$, $\mathbf{L}$, $\mathbf{R}$, and $\mathbf{K}_b$ are not available, and we may not realize the control laws in (4) and (6). Instead, let us consider the desired current trajectory

$$\mathbf{i}_d = \mathbf{H}^{-1}(\hat{\mathbf{g}} + \hat{\mathbf{D}}\dot{\mathbf{v}} + \hat{\mathbf{C}}\mathbf{v} - \mathbf{K}_d\mathbf{s})$$

$$= \mathbf{H}^{-1}[\mathbf{Y}(\mathbf{q}, \dot{\mathbf{q}}, \mathbf{v}, \dot{\mathbf{v}})\hat{\mathbf{p}} - \mathbf{K}_d\mathbf{s}] \qquad (11)$$

which is a modification of (6) with $\hat{\mathbf{D}}$, $\hat{\mathbf{C}}$, $\hat{\mathbf{g}}$ and $\hat{\mathbf{p}}$ the estimates of $\mathbf{D}$, $\mathbf{C}$, $\mathbf{g}$ and $\mathbf{p}$, respectively. With this desired current trajectory, equation (10a) can be written as

$$\mathbf{D}\dot{\mathbf{s}} + \mathbf{C}\mathbf{s} + \mathbf{K}_d\mathbf{s} = -\tilde{\mathbf{D}}\dot{\mathbf{v}} - \tilde{\mathbf{C}}\mathbf{v} - \tilde{\mathbf{g}} + \mathbf{H}(\mathbf{i} - \mathbf{i}_d)$$

$$= -\mathbf{Y}(\mathbf{q}, \dot{\mathbf{q}}, \mathbf{v}, \dot{\mathbf{v}})\tilde{\mathbf{p}} + \mathbf{H}(\mathbf{i} - \mathbf{i}_d) \qquad (12)$$

where $\tilde{\mathbf{D}} = \mathbf{D} - \hat{\mathbf{D}}$, $\tilde{\mathbf{C}} = \mathbf{C} - \hat{\mathbf{C}}$, $\tilde{\mathbf{g}} = \mathbf{g} - \hat{\mathbf{g}}$ and $\tilde{\mathbf{p}} = \mathbf{p} - \hat{\mathbf{p}}$. If we may design a controller $\mathbf{u}$ and an update law such that $\mathbf{i} \to \mathbf{i}_d$ and $\hat{\mathbf{p}} \to \mathbf{p}$, then (12) implies convergence of the output error vector. Let us consider the controller which is a modification of (4) as

$$\mathbf{u} = \hat{\mathbf{L}}\mathbf{i}_d + \hat{\mathbf{R}}\mathbf{i} + \hat{\mathbf{K}}_b\dot{\mathbf{q}} - \mathbf{K}_c\mathbf{e}_i \qquad (13)$$

where $\hat{\mathbf{L}}$, $\hat{\mathbf{R}}$ and $\hat{\mathbf{K}}_b$ are estimates of $\mathbf{L}$, $\mathbf{R}$ and $\mathbf{K}_b$, respectively. For convenience, let use define

$$\boldsymbol{\varphi} = [\mathbf{i}_d^T \quad \mathbf{i}^T \quad \dot{\mathbf{q}}^T]^T \in \mathfrak{R}^{3n}$$

$$\mathbf{p}_i = [\mathbf{L}^T \quad \mathbf{R}^T \quad \mathbf{K}_b^T]^T \in \mathfrak{R}^{3n \times n}$$

$$\hat{\mathbf{p}}_i = [\hat{\mathbf{L}}^T \quad \hat{\mathbf{R}}^T \quad \hat{\mathbf{K}}_b^T]^T \in \mathfrak{R}^{3n \times n}$$

Therefore, the controller (13) becomes

$$\mathbf{u} = \hat{\mathbf{p}}_i^T \boldsymbol{\varphi} - \mathbf{K}_c\mathbf{e}_i \qquad (14)$$

Substituting (14) into (10b), we may have the dynamics in the current tracking loop

$$\mathbf{L}\dot{\mathbf{e}}_i + \mathbf{K}_c\mathbf{e}_i = -\tilde{\mathbf{p}}_i^T \boldsymbol{\varphi} \qquad (15)$$

where $\tilde{\mathbf{p}}_i = \mathbf{p}_i - \hat{\mathbf{p}}_i$. To proof the closed loop stability and to find appropriate update laws, let us consider the Lyapunov-like function candidate

$$V(\mathbf{s}, \mathbf{e}_i, \tilde{\mathbf{p}}, \tilde{\mathbf{p}}_i) = \frac{1}{2}\mathbf{s}^T\mathbf{D}\mathbf{s} + \frac{1}{2}\mathbf{e}_i^T\mathbf{L}\mathbf{e}_i + \frac{1}{2}\tilde{\mathbf{p}}^T\boldsymbol{\Gamma}\tilde{\mathbf{p}} + \frac{1}{2}Tr(\tilde{\mathbf{p}}_i^T\boldsymbol{\Gamma}_i\tilde{\mathbf{p}}_i) \qquad (16)$$

where $\mathbf{\Gamma} \in \Re^{r \times r}$ and $\mathbf{\Gamma}_i \in \Re^{3n \times 3n}$ are positive definite matrices. Along the trajectories of (12) and (15), the time derivative of V is computed as

$$\dot{V} = -\mathbf{s}^T \mathbf{K}_d \mathbf{s} + \mathbf{s}^T \mathbf{H} \mathbf{e}_i - \mathbf{e}_i^T \mathbf{K}_c \mathbf{e}_i$$
$$-\tilde{\mathbf{p}}^T (\mathbf{\Gamma} \dot{\hat{\mathbf{p}}} + \mathbf{Y}^T \mathbf{s}) - Tr[\tilde{\mathbf{p}}_i^T (\mathbf{\Gamma}_i \dot{\hat{\mathbf{p}}}_i + \boldsymbol{\varphi} \mathbf{e}_i^T)] \tag{17}$$

The update laws can thus be picked as

$$\dot{\hat{\mathbf{p}}} = -\mathbf{\Gamma}^{-1} \mathbf{Y}^T \mathbf{s}$$
$$\dot{\hat{\mathbf{p}}}_i = -\mathbf{\Gamma}_i^{-1} \boldsymbol{\varphi} \mathbf{e}_i^T \tag{18}$$

Therefore, (17) can be further written as

$$\dot{V} = -[\mathbf{s}^T \quad \mathbf{e}_i^T] \mathbf{Q} \begin{bmatrix} \mathbf{s} \\ \mathbf{e}_i \end{bmatrix} \le 0 \tag{19}$$

where $\mathbf{Q} = \begin{bmatrix} \mathbf{K}_d & -\dfrac{1}{2} \mathbf{H} \\ -\dfrac{1}{2} \mathbf{H} & \mathbf{K}_c \end{bmatrix}$ is positive definite by proper selection of $\mathbf{K}_d$

and $\mathbf{K}_c$. Equation (19) implies that $\mathbf{s}$ and $\mathbf{e}_i$ are uniformly bounded and square integrable. Their time derivatives can also be proved to be bounded. Hence, asymptotic convergence of $\mathbf{s}$ and $\mathbf{e}_i$ can be obtained by the Barbalat's lemma. This further implies $\mathbf{q} \to \mathbf{q}_d$ and $\mathbf{i} \to \mathbf{i}_d$ as $t \to \infty$.

Remark 2: Realization of controller (14) and update law (18) needs to know the time derivative of the desired current which implies the need for the joint acceleration feedback and the time derivative of the regressor matrix. All of these requirements will be eliminated in the following design of the FAT-based regressor-free adaptive controller.

4.6.2 Regressor-free adaptive control

Now, let us consider the case when $\mathbf{D}$, $\mathbf{C}$, $\mathbf{g}$, $\mathbf{L}$, $\mathbf{R}$ and $\mathbf{K}_b$ are not available, and $\ddot{\mathbf{q}}$ is not easy to measure, we would like to design a desired current $\mathbf{i}_d$ so that a FAT-based adaptive controller $\mathbf{u}$ can be constructed without using the regressor matrix to have $\mathbf{i} \to \mathbf{i}_d$ which further implies convergence of the output error as desired. Instead of (6), let us consider the desired current $\mathbf{i}_d$ in (11) as

$$\mathbf{i}_d = \mathbf{H}^{-1}(\hat{\mathbf{g}} + \hat{\mathbf{D}}\dot{\mathbf{v}} + \hat{\mathbf{C}}\mathbf{v} - \mathbf{K}_d\mathbf{s}) \tag{20}$$

where $\hat{\mathbf{D}}$, $\hat{\mathbf{C}}$ and $\hat{\mathbf{g}}$ are respectively estimates of $\mathbf{D}$, $\mathbf{C}$ and $\mathbf{g}$. The dynamics of the output tracking loop can thus be found as

$$\begin{aligned}
\mathbf{D}\dot{\mathbf{s}} + \mathbf{C}\mathbf{s} + \mathbf{K}_d\mathbf{s} = \mathbf{H}(\mathbf{i} - \mathbf{i}_d) + (\hat{\mathbf{D}} - \mathbf{D})\dot{\mathbf{v}} \\
+ (\hat{\mathbf{C}} - \mathbf{C})\mathbf{v} + (\hat{\mathbf{g}} - \mathbf{g})
\end{aligned} \tag{21}$$

If a proper controller $\mathbf{u}$ and update laws for $\hat{\mathbf{D}}$, $\hat{\mathbf{C}}$ and $\hat{\mathbf{g}}$ can be designed, we may have $\mathbf{i} \to \mathbf{i}_d$, $\hat{\mathbf{D}} \to \mathbf{D}$, $\hat{\mathbf{C}} \to \mathbf{C}$ and $\hat{\mathbf{g}} \to \mathbf{g}$ so that (21) can give desired performance. Here, according to (4), let us select the control input to be

$$\mathbf{u} = \hat{\mathbf{f}} - \mathbf{K}_c\mathbf{e}_i \tag{22}$$

where $\hat{\mathbf{f}}$ is an estimate of the function $\mathbf{f}(\mathbf{i}_d, \mathbf{i}, \dot{\mathbf{q}}) = \mathbf{L}\dot{\mathbf{i}}_d + \mathbf{R}\mathbf{i} + \mathbf{K}_b\dot{\mathbf{q}}$. Substituting (22) into (1b), we may have the dynamics of the current tracking loop

$$\mathbf{L}\dot{\mathbf{e}}_i + \mathbf{K}_c\mathbf{e}_i = \hat{\mathbf{f}} - \mathbf{f} \tag{23}$$

If an appropriate update law for $\hat{\mathbf{f}}$ can be designed, we may ensure $\mathbf{i} \to \mathbf{i}_d$ as $t \to \infty$. Let us apply the function approximation representation for $\mathbf{D}$, $\mathbf{C}$, $\mathbf{g}$ and $\mathbf{f}$ as

$$\begin{aligned}
\mathbf{D} &= \mathbf{W}_{\mathbf{D}}^T \mathbf{Z}_{\mathbf{D}} + \varepsilon_{\mathbf{D}} \\
\mathbf{C} &= \mathbf{W}_{\mathbf{C}}^T \mathbf{Z}_{\mathbf{C}} + \varepsilon_{\mathbf{C}} \\
\mathbf{g} &= \mathbf{W}_{\mathbf{g}}^T \mathbf{z}_{\mathbf{g}} + \varepsilon_{\mathbf{g}} \\
\mathbf{f} &= \mathbf{W}_{\mathbf{f}}^T \mathbf{z}_{\mathbf{f}} + \varepsilon_{\mathbf{f}}
\end{aligned} \tag{24}$$

where $\mathbf{W}_{\mathbf{D}} \in \mathfrak{R}^{n^2 \beta_D \times n}$, $\mathbf{W}_{\mathbf{C}} \in \mathfrak{R}^{n^2 \beta_C \times n}$, $\mathbf{W}_{\mathbf{g}} \in \mathfrak{R}^{n \beta_g \times n}$, and $\mathbf{W}_{\mathbf{f}} \in \mathfrak{R}^{n \beta_f \times n}$ are weighting matrices, $\mathbf{Z}_{\mathbf{D}} \in \mathfrak{R}^{n^2 \beta_D \times n}$, $\mathbf{Z}_{\mathbf{C}} \in \mathfrak{R}^{n^2 \beta_C \times n}$, $\mathbf{z}_{\mathbf{g}} \in \mathfrak{R}^{n \beta_g \times 1}$, and

$\mathbf{z_f} \in \Re^{n\beta_f \times 1}$ are matrices of basis functions, and $\boldsymbol{\varepsilon}_{(.)}$ are approximation error matrices. The number $\beta_{(.)}$ represents the number of basis functions used. Using respectively the same set of basis functions, the corresponding estimates can also be represented as

$$\hat{\mathbf{D}} = \hat{\mathbf{W}}_{\mathbf{D}}^T \mathbf{Z}_{\mathbf{D}}$$

$$\hat{\mathbf{C}} = \hat{\mathbf{W}}_{\mathbf{C}}^T \mathbf{Z}_{\mathbf{C}}$$

$$\hat{\mathbf{g}} = \hat{\mathbf{W}}_{\mathbf{g}}^T \mathbf{z}_{\mathbf{g}}$$

$$\hat{\mathbf{f}} = \hat{\mathbf{W}}_{\mathbf{f}}^T \mathbf{z}_{\mathbf{f}}$$

(25)

then equation (21) and (23) becomes

$$\mathbf{D}\dot{\mathbf{s}} + \mathbf{C}\mathbf{s} + \mathbf{K}_d\mathbf{s} = \mathbf{H}(\mathbf{i} - \mathbf{i}_d) - \tilde{\mathbf{W}}_{\mathbf{D}}^T \mathbf{Z}_{\mathbf{D}}\dot{\mathbf{v}} - \tilde{\mathbf{W}}_{\mathbf{C}}^T \mathbf{Z}_{\mathbf{C}}\mathbf{v} - \tilde{\mathbf{W}}_{\mathbf{g}}^T \mathbf{z}_{\mathbf{g}} + \boldsymbol{\varepsilon}_1 \qquad (26a)$$

$$\mathbf{L}\dot{\mathbf{e}}_i + \mathbf{K}_c\mathbf{e}_i = -\tilde{\mathbf{W}}_{\mathbf{f}}^T \mathbf{z}_{\mathbf{f}} + \boldsymbol{\varepsilon}_2 \qquad (26b)$$

where $\boldsymbol{\varepsilon}_1 = \boldsymbol{\varepsilon}_1(\boldsymbol{\varepsilon}_{\mathbf{D}}, \boldsymbol{\varepsilon}_{\mathbf{C}}, \boldsymbol{\varepsilon}_{\mathbf{g}}, \mathbf{s}, \ddot{\mathbf{q}}_d)$ and $\boldsymbol{\varepsilon}_2 = \boldsymbol{\varepsilon}_2(\boldsymbol{\varepsilon}_{\mathbf{f}}, \mathbf{e}_i)$ are lumped approximation errors. Since $\mathbf{W}_{(.)}$ are constant matrices, their update laws can be easily found by proper selection of the Lyapunov-like function. Let us consider a candidate

$$V(\mathbf{s}, \mathbf{e}_i, \tilde{\mathbf{W}}_{\mathbf{D}}, \tilde{\mathbf{W}}_{\mathbf{C}}, \tilde{\mathbf{W}}_{\mathbf{g}}, \tilde{\mathbf{W}}_{\mathbf{f}}) = \frac{1}{2}\mathbf{s}^T\mathbf{D}\mathbf{s} + \frac{1}{2}\mathbf{e}_i^T\mathbf{L}\mathbf{e}_i$$

$$+ \frac{1}{2}Tr(\tilde{\mathbf{W}}_{\mathbf{D}}^T\mathbf{Q}_{\mathbf{D}}\tilde{\mathbf{W}}_{\mathbf{D}} + \tilde{\mathbf{W}}_{\mathbf{C}}^T\mathbf{Q}_{\mathbf{C}}\tilde{\mathbf{W}}_{\mathbf{C}} + \tilde{\mathbf{W}}_{\mathbf{g}}^T\mathbf{Q}_{\mathbf{g}}\tilde{\mathbf{W}}_{\mathbf{g}} + \tilde{\mathbf{W}}_{\mathbf{f}}^T\mathbf{Q}_{\mathbf{f}}\tilde{\mathbf{W}}_{\mathbf{f}}) \qquad (27)$$

The matrices $\mathbf{Q}_{\mathbf{D}} \in \Re^{n^2\beta_D \times n^2\beta_D}$, $\mathbf{Q}_{\mathbf{C}} \in \Re^{n^2\beta_C \times n^2\beta_C}$, $\mathbf{Q}_{\mathbf{g}} \in \Re^{n\beta_g \times n\beta_g}$, and $\mathbf{Q}_{\mathbf{f}} \in \Re^{n\beta_f \times n\beta_f}$ are all positive definite. The time derivative of V along the trajectory of (26) can be computed as

$$\dot{V} = -\mathbf{s}^T\mathbf{K}_d\mathbf{s} + \mathbf{s}^T\mathbf{H}\mathbf{e}_i - \mathbf{e}_i^T\mathbf{K}_c\mathbf{e}_i + \mathbf{s}^T\boldsymbol{\varepsilon}_1 + \mathbf{e}_i^T\boldsymbol{\varepsilon}_2$$

$$- Tr[\tilde{\mathbf{W}}_{\mathbf{D}}^T(\mathbf{Z}_{\mathbf{D}}\dot{\mathbf{v}}\mathbf{s}^T + \mathbf{Q}_{\mathbf{D}}\dot{\hat{\mathbf{W}}}_{\mathbf{D}}) + \tilde{\mathbf{W}}_{\mathbf{D}}^T(\mathbf{Z}_{\mathbf{C}}\mathbf{v}\mathbf{s}^T + \mathbf{Q}_{\mathbf{C}}\dot{\hat{\mathbf{W}}}_{\mathbf{C}})$$

$$+ \tilde{\mathbf{W}}_{\mathbf{g}}^T(\mathbf{z}_{\mathbf{g}}\mathbf{s}^T + \mathbf{Q}_{\mathbf{g}}\dot{\hat{\mathbf{W}}}_{\mathbf{g}}) + \tilde{\mathbf{W}}_{\mathbf{f}}^T(\mathbf{z}_{\mathbf{f}}\mathbf{e}_i^T + \mathbf{Q}_{\mathbf{f}}\dot{\hat{\mathbf{W}}}_{\mathbf{f}})] \qquad (28)$$

By selecting the update laws as

$$\dot{\hat{\mathbf{W}}}_{\mathbf{D}} = -\mathbf{Q}_{\mathbf{D}}^{-1}(\mathbf{Z}_{\mathbf{D}}\dot{\mathbf{v}}\mathbf{s}^{T} + \sigma_{\mathbf{D}}\hat{\mathbf{W}}_{\mathbf{D}})$$

$$\dot{\hat{\mathbf{W}}}_{\mathbf{C}} = -\mathbf{Q}_{\mathbf{C}}^{-1}(\mathbf{Z}_{\mathbf{C}}\mathbf{v}\mathbf{s}^{T} + \sigma_{\mathbf{C}}\hat{\mathbf{W}}_{\mathbf{C}})$$

$$\dot{\hat{\mathbf{W}}}_{\mathbf{g}} = -\mathbf{Q}_{\mathbf{g}}^{-1}(\mathbf{z}_{\mathbf{g}}\mathbf{s}^{T} + \sigma_{\mathbf{g}}\hat{\mathbf{W}}_{\mathbf{g}})$$

$$\dot{\hat{\mathbf{W}}}_{\mathbf{f}} = -\mathbf{Q}_{\mathbf{f}}^{-1}(\mathbf{z}_{\mathbf{f}}\mathbf{e}_{i}^{T} + \sigma_{\mathbf{f}}\hat{\mathbf{W}}_{\mathbf{f}})$$

$$(29)$$

where $\sigma_{(.)}$ are positive numbers. With these selections, equation (28) becomes

$$\dot{V} = -[\mathbf{s}^{T} \quad \mathbf{e}_{i}^{T}]\mathbf{Q}\begin{bmatrix} \mathbf{s} \\ \mathbf{e}_{i} \end{bmatrix} + [\mathbf{s}^{T} \quad \mathbf{e}_{i}^{T}]\begin{bmatrix} \boldsymbol{\varepsilon}_{1} \\ \boldsymbol{\varepsilon}_{2} \end{bmatrix} + \sigma_{\mathbf{D}}Tr(\tilde{\mathbf{W}}_{\mathbf{D}}^{T}\hat{\mathbf{W}}_{\mathbf{D}})$$

$$+ \sigma_{\mathbf{C}}Tr(\tilde{\mathbf{W}}_{\mathbf{C}}^{T}\hat{\mathbf{W}}_{\mathbf{C}}) + \sigma_{\mathbf{g}}Tr(\tilde{\mathbf{W}}_{\mathbf{g}}^{T}\hat{\mathbf{W}}_{\mathbf{g}}) + \sigma_{\mathbf{f}}Tr(\tilde{\mathbf{W}}_{\mathbf{f}}^{T}\hat{\mathbf{W}}_{\mathbf{f}}) \qquad (30)$$

where $\mathbf{Q} = \begin{bmatrix} \mathbf{K}_{d} & -\dfrac{1}{2}\mathbf{H} \\ -\dfrac{1}{2}\mathbf{H} & \mathbf{K}_{c} \end{bmatrix}$ is positive definite which can be achieved by

proper selections of $\mathbf{K}_{d}$ and $\mathbf{K}_{c}$.

Remark 3: Realization of the control law (22) and update laws (29) does not need the information of joint accelerations which largely simplifies its implementation.

Remark 4: Suppose a sufficient number of basis functions are used and the approximation error can be ignored, then it is not necessary to include the σ-modification terms in (29). Hence, (30) can be reduced to (9), and convergence of $\mathbf{s}$ and $\mathbf{e}_{i}$ can be further proved by Barbalat's lemma.

Remark 5: If the approximation error cannot be ignored, but we can find positive numbers δ_{1} and δ_{2} such that $\|\boldsymbol{\varepsilon}_{1}\| \leq \delta_{1}$ and $\|\boldsymbol{\varepsilon}_{2}\| \leq \delta_{2}$, then robust terms $\tau_{robust1}$ and $\tau_{robust2}$ can be included into (20) and (22) to have

$$\mathbf{i}_{d} = \mathbf{H}^{-1}(\hat{\mathbf{g}} + \hat{\mathbf{D}}\dot{\mathbf{v}} + \hat{\mathbf{C}}\mathbf{v} - \mathbf{K}_{d}\mathbf{s} + \tau_{robust1})$$

$$\mathbf{u} = \hat{\mathbf{f}} - \mathbf{K}_{c}\mathbf{e}_{i} + \tau_{robust2}$$

Consider the Lyapunov-like function candidate (27) again, and the update law (29) without σ-modification; then the time derivative of V becomes

$$\dot{V} \le -[\mathbf{s}^T \quad \mathbf{e}_i^T]\mathbf{Q}\begin{bmatrix} \mathbf{s} \\ \mathbf{e}_i \end{bmatrix} + \delta_1\|\mathbf{s}\| + \delta_2\|\mathbf{e}_i\| + \mathbf{s}^T\boldsymbol{\tau}_{robust\,1} + \mathbf{e}_i^T\boldsymbol{\tau}_{robust\,2}$$

If we select the vector $\boldsymbol{\tau}_{robust\,1} = -\delta_1[\mathrm{sgn}(s_1) \quad \mathrm{sgn}(s_2) \quad \cdots \quad \mathrm{sgn}(s_n)]^T$, where s_i, $i=1,\dots,n$ is the i-th entry in $\mathbf{s}$, and $\boldsymbol{\tau}_{robust\,2} = -\delta_2[\mathrm{sgn}(e_{i_1}) \quad \cdots \quad \mathrm{sgn}(e_{i_n})]^T$, where e_{i_k}, $k=1,\dots,n$ is the k-th entry in $\mathbf{e}_i$, then we may have (9) again. This will further give convergence of the output error by Barbalat's lemma.

Owing to the existence of $\boldsymbol{\varepsilon}_1$ and $\boldsymbol{\varepsilon}_2$ in (30), the definiteness of $\dot{V}$ cannot be determined. In the following, we would like to investigate closed loop stability in the presence of these approximation errors. It is very easy to prove the inequalities hold

$$-[\mathbf{s}^T \quad \mathbf{e}_i^T]\mathbf{Q}\begin{bmatrix} \mathbf{s} \\ \mathbf{e}_i \end{bmatrix} + [\mathbf{s}^T \quad \mathbf{e}_i^T]\begin{bmatrix} \boldsymbol{\varepsilon}_1 \\ \boldsymbol{\varepsilon}_2 \end{bmatrix}$$

$$\le -\frac{1}{2}\left(\lambda_{\min}(\mathbf{Q})\left\|\begin{bmatrix} \mathbf{s} \\ \mathbf{e}_i \end{bmatrix}\right\|^2 - \frac{1}{\lambda_{\min}(\mathbf{Q})}\left\|\begin{bmatrix} \boldsymbol{\varepsilon}_1 \\ \boldsymbol{\varepsilon}_2 \end{bmatrix}\right\|^2\right) \tag{31}$$

$$Tr(\tilde{\mathbf{W}}_{(\cdot)}^T\hat{\mathbf{W}}_{(\cdot)}) \le \frac{1}{2}Tr(\mathbf{W}_{(\cdot)}^T\mathbf{W}_{(\cdot)}) - \frac{1}{2}Tr(\tilde{\mathbf{W}}_{(\cdot)}^T\tilde{\mathbf{W}}_{(\cdot)})$$

Together with the relationship

$$V = \frac{1}{2}[\mathbf{s}^T\mathbf{D}\mathbf{s} + \mathbf{e}_i^T\mathbf{L}\mathbf{e}_i + Tr(\tilde{\mathbf{W}}_\mathbf{D}^T\mathbf{Q}_\mathbf{D}\tilde{\mathbf{W}}_\mathbf{D} + \tilde{\mathbf{W}}_\mathbf{C}^T\mathbf{Q}_\mathbf{C}\tilde{\mathbf{W}}_\mathbf{C}$$
$$+ \tilde{\mathbf{W}}_\mathbf{g}^T\mathbf{Q}_\mathbf{g}\tilde{\mathbf{W}}_\mathbf{g} + \tilde{\mathbf{W}}_\mathbf{f}^T\mathbf{Q}_\mathbf{f}\tilde{\mathbf{W}}_\mathbf{f})]$$

$$\le \frac{1}{2}[\lambda_{\max}(\mathbf{A})\left\|\begin{bmatrix} \mathbf{s} \\ \mathbf{e}_i \end{bmatrix}\right\|^2 + \lambda_{\max}(\mathbf{Q}_\mathbf{D})Tr(\tilde{\mathbf{W}}_\mathbf{D}^T\tilde{\mathbf{W}}_\mathbf{D})$$

$$+ \lambda_{\max}(\mathbf{Q}_\mathbf{C})Tr(\tilde{\mathbf{W}}_\mathbf{C}^T\tilde{\mathbf{W}}_\mathbf{C}) + \lambda_{\max}(\mathbf{Q}_\mathbf{g})Tr(\tilde{\mathbf{W}}_\mathbf{g}^T\tilde{\mathbf{W}}_\mathbf{g})$$

$$+ \lambda_{\max}(\mathbf{Q}_\mathbf{f})Tr(\tilde{\mathbf{W}}_\mathbf{f}^T\tilde{\mathbf{W}}_\mathbf{f})] \tag{32}$$

where $\mathbf{A} = \begin{bmatrix} \mathbf{D} & \mathbf{0} \\ \mathbf{0} & \mathbf{L} \end{bmatrix}$, we may rewrite (30) into the form

$$\dot{V} \leq -\alpha V + \frac{1}{2}[\alpha\lambda_{\max}(\mathbf{A}) - \lambda_{\min}(\mathbf{Q})]\left\|\begin{bmatrix}\mathbf{s}\\\mathbf{e}_i\end{bmatrix}\right\|^2 + \frac{1}{2\lambda_{\min}(\mathbf{Q})}\left\|\begin{bmatrix}\varepsilon_1\\\varepsilon_2\end{bmatrix}\right\|^2$$

$$+\frac{1}{2}[\alpha\lambda_{\max}(\mathbf{Q_D}) - \sigma_D]Tr(\tilde{\mathbf{W}}_\mathbf{D}^T\tilde{\mathbf{W}}_\mathbf{D}) + \frac{1}{2}[\alpha\lambda_{\max}(\mathbf{Q_C}) - \sigma_C]Tr(\tilde{\mathbf{W}}_\mathbf{C}^T\tilde{\mathbf{W}}_\mathbf{C})$$

$$+\frac{1}{2}[\alpha\lambda_{\max}(\mathbf{Q_g}) - \sigma_g]Tr(\tilde{\mathbf{W}}_\mathbf{g}^T\tilde{\mathbf{W}}_\mathbf{g}) + \frac{1}{2}[\alpha\lambda_{\max}(\mathbf{Q_f}) - \sigma_f]Tr(\tilde{\mathbf{W}}_\mathbf{f}^T\tilde{\mathbf{W}}_\mathbf{f})$$

$$+\frac{1}{2}[\sigma_D Tr(\mathbf{W_D}^T\mathbf{W_D}) + \sigma_C Tr(\mathbf{W_C}^T\mathbf{W_C}) + \sigma_g Tr(\mathbf{W_g}^T\mathbf{W_g}) + \sigma_f Tr(\mathbf{W_f}^T\mathbf{W_f})] \quad (33)$$

where α is a constant to be selected as

$$\alpha \leq \min\left\{\frac{\lambda_{\min}(\mathbf{Q})}{\lambda_{\max}(\mathbf{A})}, \frac{\sigma_D}{\lambda_{\max}(\mathbf{Q_D})}, \frac{\sigma_C}{\lambda_{\max}(\mathbf{Q_C})}, \frac{\sigma_g}{\lambda_{\max}(\mathbf{Q_g})}, \frac{\sigma_f}{\lambda_{\max}(\mathbf{Q_f})}\right\} \quad (34)$$

Then (33) becomes

$$\dot{V} \leq -\alpha V + \frac{1}{2\lambda_{\min}(\mathbf{Q})}\left\|\begin{bmatrix}\varepsilon_1\\\varepsilon_2\end{bmatrix}\right\|^2 + \frac{1}{2}[\sigma_D Tr(\mathbf{W_D}^T\mathbf{W_D})$$

$$+\sigma_C Tr(\mathbf{W_C}^T\mathbf{W_C}) + \sigma_g Tr(\mathbf{W_g}^T\mathbf{W_g}) + \sigma_f Tr(\mathbf{W_f}^T\mathbf{W_f})] \quad (35)$$

This implies $\dot{V} < 0$ whenever

$$V > \frac{1}{2\alpha\lambda_{\min}(\mathbf{Q})}\sup_{\tau \geq t_0}\left\|\begin{bmatrix}\varepsilon_1(\tau)\\\varepsilon_2(\tau)\end{bmatrix}\right\|^2 + \frac{1}{2\alpha}[\sigma_D Tr(\mathbf{W_D}^T\mathbf{W_D})$$

$$+\sigma_C Tr(\mathbf{W_C}^T\mathbf{W_C}) + \sigma_g Tr(\mathbf{W_g}^T\mathbf{W_g}) + \sigma_f Tr(\mathbf{W_f}^T\mathbf{W_f})] \quad (36)$$

Hence, we have proved that $\mathbf{s}$, $\mathbf{e}_i$, $\tilde{\mathbf{W}}_\mathbf{D}$, $\tilde{\mathbf{W}}_\mathbf{C}$, $\tilde{\mathbf{W}}_\mathbf{g}$ and $\tilde{\mathbf{W}}_\mathbf{f}$ are uniformly ultimately bounded. From (35), we may also compute the upper bound for V as

$$V(t) \leq e^{-\alpha(t-t_0)}V(t_0) + \frac{1}{2\alpha\lambda_{\min}(\mathbf{Q})}\sup_{t_0 < \tau < t}\left\|\begin{bmatrix}\varepsilon_1(\tau)\\\varepsilon_2(\tau)\end{bmatrix}\right\|^2$$

$$+\frac{1}{2\alpha}[\sigma_D Tr(\mathbf{W_D}^T\mathbf{W_D}) + \sigma_C Tr(\mathbf{W_C}^T\mathbf{W_C})$$

$$+\sigma_g Tr(\mathbf{W_g}^T\mathbf{W_g}) + \sigma_f Tr(\mathbf{W_f}^T\mathbf{W_f})] \quad (37)$$

Using the inequality

$$V \geq \frac{1}{2}\lambda_{\min}(\mathbf{A}) \left\| \begin{bmatrix} \mathbf{s} \\ \mathbf{e}_i \end{bmatrix} \right\|^2 + \frac{1}{2}[\lambda_{\min}(\mathbf{Q_D})Tr(\tilde{\mathbf{W}}_\mathbf{D}^T\tilde{\mathbf{W}}_\mathbf{D})$$

$$+ \lambda_{\min}(\mathbf{Q_C})Tr(\tilde{\mathbf{W}}_\mathbf{C}^T\tilde{\mathbf{W}}_\mathbf{C}) + \lambda_{\min}(\mathbf{Q_g})Tr(\tilde{\mathbf{W}}_\mathbf{g}^T\tilde{\mathbf{W}}_\mathbf{g})$$

$$+ \lambda_{\min}(\mathbf{Q_f})Tr(\tilde{\mathbf{W}}_\mathbf{f}^T\tilde{\mathbf{W}}_\mathbf{f})] \tag{38}$$

we may find the upper bound for $\left\| [\mathbf{s}^T \quad \mathbf{e}_i^T]^T \right\|^2$ as

$$\left\| \begin{bmatrix} \mathbf{s} \\ \mathbf{e}_i \end{bmatrix} \right\|^2 \leq \frac{2}{\lambda_{\min}(\mathbf{A})}e^{-\alpha(t-t_0)}V(t_0) + \frac{1}{\alpha\lambda_{\min}(\mathbf{A})\lambda_{\min}(\mathbf{Q})}\sup_{t_0 < \tau < t}\left\| \begin{bmatrix} \boldsymbol{\varepsilon}_1(\tau) \\ \boldsymbol{\varepsilon}_2(\tau) \end{bmatrix} \right\|^2$$

$$+ \frac{1}{\alpha\lambda_{\min}(\mathbf{A})}[\sigma_\mathbf{D}Tr(\mathbf{W}_\mathbf{D}^T\mathbf{W}_\mathbf{D}) + \sigma_\mathbf{C}Tr(\mathbf{W}_\mathbf{C}^T\mathbf{W}_\mathbf{C})$$

$$+ \sigma_\mathbf{g}Tr(\mathbf{W}_\mathbf{g}^T\mathbf{W}_\mathbf{g}) + \sigma_\mathbf{f}Tr(\mathbf{W}_\mathbf{f}^T\mathbf{W}_\mathbf{f})]$$

Therefore, we may compute the bound as

$$\left\| \begin{bmatrix} \mathbf{s} \\ \mathbf{e}_i \end{bmatrix} \right\| \leq \sqrt{\frac{2V(t_0)}{\lambda_{\min}(\mathbf{A})}}e^{-\frac{\alpha}{2}(t-t_0)} + \frac{1}{\sqrt{\alpha\lambda_{\min}(\mathbf{A})\lambda_{\min}(\mathbf{Q})}}\sup_{t_0 < \tau < t}\left\| \begin{bmatrix} \boldsymbol{\varepsilon}_1(\tau) \\ \boldsymbol{\varepsilon}_2(\tau) \end{bmatrix} \right\|$$

$$+ \frac{1}{\sqrt{\alpha\lambda_{\min}(\mathbf{A})}}[\sigma_\mathbf{D}Tr(\mathbf{W}_\mathbf{D}^T\mathbf{W}_\mathbf{D}) + \sigma_\mathbf{C}Tr(\mathbf{W}_\mathbf{C}^T\mathbf{W}_\mathbf{C})$$

$$+ \sigma_\mathbf{g}Tr(\mathbf{W}_\mathbf{g}^T\mathbf{W}_\mathbf{g}) + \sigma_\mathbf{f}Tr(\mathbf{W}_\mathbf{f}^T\mathbf{W}_\mathbf{f})]^{\frac{1}{2}}$$

This proves that the time history of the error signal is bounded by an exponential function plus some constants. The transient performance analysis is thus completed.

Table 4.2 summarizes the adaptive control of EDRR derived in this section in terms of their controller forms, update laws and implementation issues.

Table 4.2 Summary of the adaptive control of EDRR

	Electrically Driven Rigid Robot $\mathbf{D\dot{s} + Cs + g + D\dot{v} + Cv = Hi}$ $\mathbf{L\dot{i} + Ri + K_b\dot{q} = u}$ (4.6-1), (4.6-2)	
	Regressor-based	Regressor-free
Controller	$\mathbf{i}_d = \mathbf{H}^{-1}[\mathbf{Y(q,\dot{q},v,\dot{v})\hat{p}} - \mathbf{K}_d\mathbf{s}]$ $\mathbf{u} = \hat{\mathbf{p}}_i^T\boldsymbol{\varphi} - \mathbf{K}_c\mathbf{e}_i$ (4.6-11) (4.6-14)	$\mathbf{i}_d = \mathbf{H}^{-1}(\hat{\mathbf{g}} + \hat{\mathbf{D}}\dot{\mathbf{v}} + \hat{\mathbf{C}}\mathbf{v} - \mathbf{K}_d\mathbf{s})$ $\mathbf{u} = \hat{\mathbf{f}} - \mathbf{K}_c\mathbf{e}_i$ (4.6-20) (4.6-22)
Adaptive Law	$\dot{\hat{\mathbf{p}}} = -\boldsymbol{\Gamma}^{-1}\mathbf{Y}^T\mathbf{s}$ $\dot{\hat{\mathbf{p}}}_i = -\boldsymbol{\Gamma}_i^{-1}\boldsymbol{\varphi}\mathbf{e}_i^T$ (4.6-18)	$\dot{\hat{\mathbf{W}}}_\mathbf{D} = -\mathbf{Q}_\mathbf{D}^{-1}(\mathbf{Z}_\mathbf{D}\dot{\mathbf{v}}\mathbf{s}^T + \sigma_\mathbf{D}\hat{\mathbf{W}}_\mathbf{D})$ $\dot{\hat{\mathbf{W}}}_\mathbf{C} = -\mathbf{Q}_\mathbf{C}^{-1}(\mathbf{Z}_\mathbf{C}\mathbf{v}\mathbf{s}^T + \sigma_\mathbf{C}\hat{\mathbf{W}}_\mathbf{C})$ $\dot{\hat{\mathbf{W}}}_\mathbf{g} = -\mathbf{Q}_\mathbf{g}^{-1}(\mathbf{Z}_\mathbf{g}\mathbf{s}^T + \sigma_\mathbf{g}\hat{\mathbf{W}}_\mathbf{g})$ $\dot{\hat{\mathbf{W}}}_\mathbf{f} = -\mathbf{Q}_\mathbf{f}^{-1}(\mathbf{Z}_\mathbf{f}\mathbf{e}_i^T + \sigma_\mathbf{f}\hat{\mathbf{W}}_\mathbf{f})$ (4.6-29)
Realization Issue	1. need computation of regressor matrix 2. need $\mathbf{i}_d$ to compute $\mathbf{u}$ which implies the need for the joint accelerations and time derivative of $\mathbf{Y}$.	1. no need for regressor matrix or its time derivatives 2. no need for joint accelerations

Example 4.2:

Consider the same 2-DOF planar robot in example 4.1 with the inclusion of the actuator dynamics, and we are going to verify the control strategy developed in this section by using computer simulations. Actual values of link parameters are selected as $m_1=m_2=0.5(kg)$, $l_1=l_2=0.75(m)$, $l_{c1}=l_{c2}=0.375(m)$, and $I_1=I_2=0.0234(kg\text{-}m^2)$. Parameters related to the actuator dynamics are given with $h_1=h_2=10(\text{N-}m/\text{A})$, $L_1=L_2=0.025(\text{H})$, $r_1=r_2=1(\Omega)$, and $k_{b1}=k_{b2}=1(\text{Vol/rad/sec})$. In order to observe the effect of the actuator dynamics, the endpoint is required to track a $0.2m$ radius circle centered at $(0.8m, 1.0m)$ in 2 seconds which is much faster then the case in example 4.1. The initial conditions of the generalized coordinate vector is $\mathbf{q}(0) = [0.0022 \quad 1.5019 \quad 0 \quad 0]^T$, i.e., the endpoint is still at $(0.8m, 0,75m)$ initially. Three cases will be investigated in this example to clarify the significance of the actuator dynamics in the closed loop stability. In case 1, an adaptive controller designed for EDRR robots is applied to a EDRR robot. Since the actuator dynamics is considered in the controller, good performance is to be expected. However, if an adaptive controller for RR is applied to the same

EDRR with the same set of controller parameters, the performance would, of course, be unsatisfactory which will be presented in case 2. In the last case, we consider the same configuration in case 2, but with improvements in the tracking performance via controller gain adjustments. It is seen that although the tracking error can be limited to some range, the control effort would become impractically huge. Hence, we may arrive at a conclusion later that consideration of the actuator dynamics is very important if good performance is required. Table 4.3 summarizes the configuration of the simulation cases.

Table 4.3 Simulation cases

	Plant	Controller	Remark
Case 1	EDRR	Designed for EDRR	-
Case 2	EDRR	Designed for RR	Same controller parameters as in Case 1
Case 3	EDRR	Designed for RR	Same configuration as in Case 2 but with gain adjustments

It has to be emphasized that, in case 2 and 3, the robot models are in the voltage level, i.e., the input to the joint is voltage (Figure 4.7). However, the controllers designed for RR are in the torque level, i.e., their outputs are torque. Hence, some modification is needed so that the robot and the controller are compatible.

Figure 4.7 The input signal for a RR is in the torque level and its controller should also be in the torque level. The input for the EDRR is in the voltage level implying that its controller must be with output in voltage

The adaptive controller in (4.5-1a) is designed for the rigid robot in the torque level. Now, let us introduce a conversion matrix $\mathbf{K}_\tau$ which satisfies $\mathbf{K}_\tau \mathbf{u}(\infty) = \mathbf{Hi}(\infty) = \boldsymbol{\tau}(\infty)$ so that the controller becomes in the voltage level

$$\mathbf{u} = \mathbf{K}_\tau^{-1}(\hat{\mathbf{D}}\dot{\mathbf{v}} + \hat{\mathbf{C}}\mathbf{v} + \hat{\mathbf{g}} - \mathbf{K}_d\mathbf{s})$$
$$= \mathbf{K}_\tau^{-1}(\hat{\mathbf{W}}_\mathbf{D}^T\mathbf{Z}_\mathbf{D}\dot{\mathbf{v}} + \hat{\mathbf{W}}_\mathbf{C}^T\mathbf{Z}_\mathbf{C}\mathbf{v} + \hat{\mathbf{W}}_\mathbf{g}^T\mathbf{z}_\mathbf{g} - \mathbf{K}_d\mathbf{s}) \qquad (39)$$

The update laws can still be derived to be

$$\dot{\hat{\mathbf{W}}}_\mathbf{D} = -\mathbf{Q}_\mathbf{D}^{-1}(\mathbf{Z}_\mathbf{D}\dot{\mathbf{v}}\mathbf{s}^T + \sigma_\mathbf{D}\hat{\mathbf{W}}_\mathbf{D})$$

$$\dot{\hat{\mathbf{W}}}_\mathbf{C} = -\mathbf{Q}_\mathbf{C}^{-1}(\mathbf{Z}_\mathbf{C}\mathbf{v}\mathbf{s}^T + \sigma_\mathbf{C}\hat{\mathbf{W}}_\mathbf{C}) \qquad (40)$$

$$\dot{\hat{\mathbf{W}}}_\mathbf{g} = -\mathbf{Q}_\mathbf{g}^{-1}(\mathbf{z}_\mathbf{g}\mathbf{s}^T + \sigma_\mathbf{g}\hat{\mathbf{W}}_\mathbf{g})$$

Therefore, some more detail in the simulation cases can be summarized as shown in Table 4.4.

Case 1: Controller for EDRR applied to EDRR

The controller in (22) is applied with the gain matrices

$$\mathbf{K}_d = \begin{bmatrix} 20 & 0 \\ 0 & 20 \end{bmatrix},\ \mathbf{\Lambda} = \begin{bmatrix} 10 & 0 \\ 0 & 10 \end{bmatrix}\ \text{and}\ \mathbf{K}_c = \begin{bmatrix} 50 & 0 \\ 0 & 50 \end{bmatrix}.$$

Table 4.4 Realization details in simulation cases

	Plant	Controller	Update Laws
Case 1	EDRR (4.6-1)	$\mathbf{i}_d = \mathbf{H}^{-1}(\hat{\mathbf{g}} + \hat{\mathbf{D}}\dot{\mathbf{v}} + \hat{\mathbf{C}}\mathbf{v} - \mathbf{K}_d\mathbf{s})$ $= \mathbf{H}^{-1}(\hat{\mathbf{W}}_\mathbf{g}^T\mathbf{z}_\mathbf{g} + \hat{\mathbf{W}}_\mathbf{D}^T\mathbf{Z}_\mathbf{D}\dot{\mathbf{v}}$ $+ \hat{\mathbf{W}}_\mathbf{C}^T\mathbf{Z}_\mathbf{C}\mathbf{v} - \mathbf{K}_d\mathbf{s})$ $\mathbf{u} = \hat{\mathbf{f}} - \mathbf{K}_c\mathbf{e}_i$ $= \hat{\mathbf{W}}_\mathbf{f}^T\mathbf{z}_\mathbf{f} - \mathbf{K}_c\mathbf{e}_i$ (4.6-20), (4.6-22)	$\dot{\hat{\mathbf{W}}}_\mathbf{D} = -\mathbf{Q}_\mathbf{D}^{-1}(\mathbf{Z}_\mathbf{D}\dot{\mathbf{v}}\mathbf{s}^T + \sigma_\mathbf{D}\hat{\mathbf{W}}_\mathbf{D})$ $\dot{\hat{\mathbf{W}}}_\mathbf{C} = -\mathbf{Q}_\mathbf{C}^{-1}(\mathbf{Z}_\mathbf{C}\mathbf{v}\mathbf{s}^T + \sigma_\mathbf{C}\hat{\mathbf{W}}_\mathbf{C})$ $\dot{\hat{\mathbf{W}}}_\mathbf{g} = -\mathbf{Q}_\mathbf{g}^{-1}(\mathbf{z}_\mathbf{g}\mathbf{s}^T + \sigma_\mathbf{g}\hat{\mathbf{W}}_\mathbf{g})$ $\dot{\hat{\mathbf{W}}}_\mathbf{f} = -\mathbf{Q}_\mathbf{f}^{-1}(\mathbf{z}_\mathbf{f}\mathbf{e}_i^T + \sigma_\mathbf{f}\hat{\mathbf{W}}_\mathbf{f})$ (4.6-29)
Case 2	EDRR (4.6-1)	$\mathbf{u} = \mathbf{K}_\tau^{-1}(\hat{\mathbf{D}}\dot{\mathbf{v}} + \hat{\mathbf{C}}\mathbf{v} + \hat{\mathbf{g}} - \mathbf{K}_d\mathbf{s})$ $= \mathbf{K}_\tau^{-1}(\hat{\mathbf{W}}_\mathbf{D}^T\mathbf{Z}_\mathbf{D}\dot{\mathbf{v}} + \hat{\mathbf{W}}_\mathbf{C}^T\mathbf{Z}_\mathbf{C}\mathbf{v}$ $+ \hat{\mathbf{W}}_\mathbf{g}^T\mathbf{z}_\mathbf{g} - \mathbf{K}_d\mathbf{s})$ (4.6-39)	$\dot{\hat{\mathbf{W}}}_\mathbf{D} = -\mathbf{Q}_\mathbf{D}^{-1}(\mathbf{Z}_\mathbf{D}\dot{\mathbf{v}}\mathbf{s}^T + \sigma_\mathbf{D}\hat{\mathbf{W}}_\mathbf{D})$ $\dot{\hat{\mathbf{W}}}_\mathbf{C} = -\mathbf{Q}_\mathbf{C}^{-1}(\mathbf{Z}_\mathbf{C}\mathbf{v}\mathbf{s}^T + \sigma_\mathbf{C}\hat{\mathbf{W}}_\mathbf{C})$ $\dot{\hat{\mathbf{W}}}_\mathbf{g} = -\mathbf{Q}_\mathbf{g}^{-1}(\mathbf{z}_\mathbf{g}\mathbf{s}^T + \sigma_\mathbf{g}\hat{\mathbf{W}}_\mathbf{g})$ (4.6-40)
Case 3	EDRR (4.6-1)	$\mathbf{u} = \mathbf{K}_\tau^{-1}(\hat{\mathbf{D}}\dot{\mathbf{v}} + \hat{\mathbf{C}}\mathbf{v} + \hat{\mathbf{g}} - \mathbf{K}_d\mathbf{s})$ $= \mathbf{K}_\tau^{-1}(\hat{\mathbf{W}}_\mathbf{D}^T\mathbf{Z}_\mathbf{D}\dot{\mathbf{v}} + \hat{\mathbf{W}}_\mathbf{C}^T\mathbf{Z}_\mathbf{C}\mathbf{v}$ $+ \hat{\mathbf{W}}_\mathbf{g}^T\mathbf{z}_\mathbf{g} - \mathbf{K}_d\mathbf{s})$ (4.6-39)	$\dot{\hat{\mathbf{W}}}_\mathbf{D} = -\mathbf{Q}_\mathbf{D}^{-1}(\mathbf{Z}_\mathbf{D}\dot{\mathbf{v}}\mathbf{s}^T + \sigma_\mathbf{D}\hat{\mathbf{W}}_\mathbf{D})$ $\dot{\hat{\mathbf{W}}}_\mathbf{C} = -\mathbf{Q}_\mathbf{C}^{-1}(\mathbf{Z}_\mathbf{C}\mathbf{v}\mathbf{s}^T + \sigma_\mathbf{C}\hat{\mathbf{W}}_\mathbf{C})$ $\dot{\hat{\mathbf{W}}}_\mathbf{g} = -\mathbf{Q}_\mathbf{g}^{-1}(\mathbf{z}_\mathbf{g}\mathbf{s}^T + \sigma_\mathbf{g}\hat{\mathbf{W}}_\mathbf{g})$ (4.6-40)

The initial value for the desired current can be found by calculation from (20) as $\mathbf{i}_d(0) = \mathbf{i}(0) = [1.5498 \quad -3.3570]^T$. The 11-term Fourier series is selected as the basis function for the approximation so that $\hat{\mathbf{W}}_\mathbf{D}$ and $\hat{\mathbf{W}}_\mathbf{C}$ are in $\mathfrak{R}^{44 \times 2}$, while $\hat{\mathbf{W}}_\mathbf{g}$ and $\hat{\mathbf{W}}_\mathbf{f}$ are in $\mathfrak{R}^{22 \times 2}$. The initial weighting vectors for the entries are assigned to be

$$\hat{\mathbf{w}}_{D_{11}}(0) = [0.05 \quad 0 \quad \cdots \quad 0]^T \in \mathfrak{R}^{11 \times 1}$$

$$\hat{\mathbf{w}}_{D_{12}}(0) = \hat{\mathbf{w}}_{D_{21}}(0) = [-0.05 \quad 0 \quad \cdots \quad 0]^T \in \mathfrak{R}^{11 \times 1}$$

$$\hat{\mathbf{w}}_{D_{22}}(0) = [0.1 \quad 0 \quad \cdots \quad 0]^T \in \mathfrak{R}^{11 \times 1}$$

$$\hat{\mathbf{w}}_{C_{11}}(0) = [0.05 \quad 0 \quad \cdots \quad 0]^T \in \mathfrak{R}^{11 \times 1}$$

$$\hat{\mathbf{w}}_{C_{12}}(0) = \hat{\mathbf{w}}_{C_{21}}(0) = [-0.05 \quad 0 \quad \cdots \quad 0]^T \in \mathfrak{R}^{11 \times 1}$$

$$\hat{\mathbf{w}}_{C_{22}}(0) = [0.1 \quad 0 \quad \cdots \quad 0]^T \in \mathfrak{R}^{11 \times 1}$$

$$\hat{\mathbf{w}}_{g_1}(0) = \hat{\mathbf{w}}_{g_2}(0) = [0 \quad 0 \quad \cdots \quad 0]^T \in \mathfrak{R}^{11 \times 1}$$

$$\hat{\mathbf{w}}_{f_1}(0) = \hat{\mathbf{w}}_{f_2}(0) = [0 \quad 0 \quad \cdots \quad 0]^T \in \mathfrak{R}^{11 \times 1}$$

The gain matrices in the update laws (29) are selected as

$$\mathbf{Q}_\mathbf{D}^{-1} = \mathbf{I}_{44}, \ \mathbf{Q}_\mathbf{C}^{-1} = \mathbf{I}_{44}, \ \mathbf{Q}_\mathbf{g}^{-1} = 100\mathbf{I}_{22}, \text{ and } \mathbf{Q}_\mathbf{f}^{-1} = 100\mathbf{I}_{22}.$$

The approximation error is assumed to be neglected, and the σ-modification parameters are all zero. The simulation results are shown in Figure 4.8 to 4.15. Figure 4.8 shows the tracking performance of the robot endpoint and its desired trajectory in the Cartesian space. It is observed that the endpoint trajectory converges smoothly to the desired trajectory, although the initial position error is quite large and most plant parameters are uncertain. The transient state takes only about 0.3 seconds which can be justified from the joint space tracking history in Figure 4.9. This justifies the effectiveness of the consideration of the actuator dynamics when high performance control is required. The performance in the current tracking loop is quite good as shown in Figure 4.10. The control efforts to the two joints are reasonable that are presented in Figure 4.11. Figure 4.12 to 4.15 are the performance of function approximation. Although most parameters do not converge to their actual values, they still remain bounded as desired.

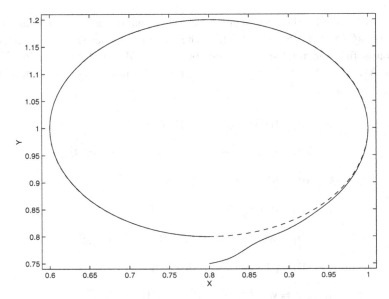

Figure 4.8 Robot endpoint tracking performance in the Cartesian space
(— actual trajectory; --- desired trajectory)

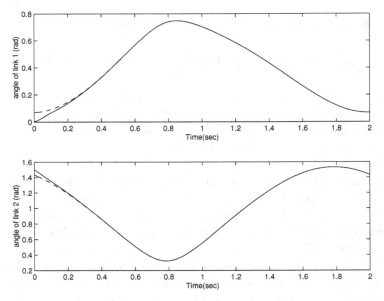

Figure 4.9 Joint space tracking performance
(— actual trajectory; --- desired trajectory)

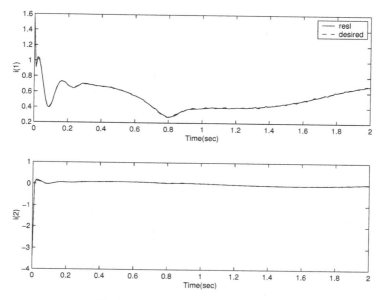

Figure 4.10 Tracking in the current loop
(— actual trajectory; --- desired trajectory)

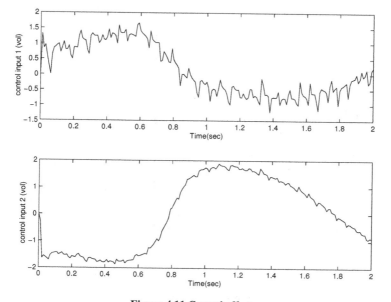

Figure 4.11 Control efforts

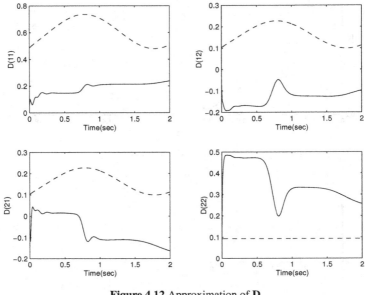

Figure 4.12 Approximation of **D**
(—estimate; --- actual value)

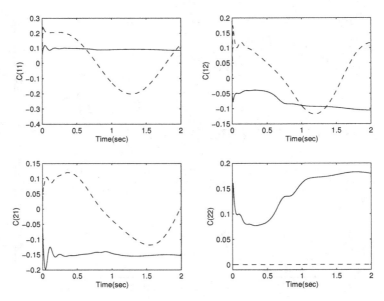

Figure 4.13 Approximation of **C**
(—estimate; --- actual value)

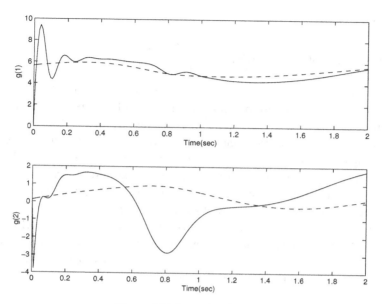

Figure 4.14 Approximation of **g**
(—estimate; --- actual value)

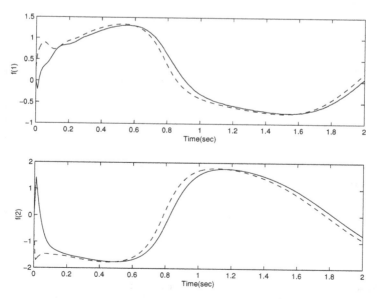

Figure 4.15 Approximation of **f**
(—estimate; --- actual value)

Case 2: Controller for RR applied to EDRR

Controller (39) is used with $\mathbf{K}_\tau = \begin{bmatrix} 10 & 0 \\ 0 & 10 \end{bmatrix}$ (N-m/Vol). All other required

parameters for the controller and update law (40) are the same as those in the previous case. The purpose of using the same set of parameters is to have an effective comparison. In the following figures we may observe that the controller designed for RR is not able to give acceptable performance to a EDRR under the conditions when the actuator dynamics is important such as the fast motion trajectory tested here. The simulation results are presented in Figure 4.16 to 4.22. Figure 4.16 shows that the endpoint motion does not converge to the desired trajectory. Figure 4.17 indicates that the joint space motion deviates from the desired trajectory after 0.6 seconds. Figure 4.18 and 19 presents the motor current and the control effort respectively, and impractically large values can be seen in both curves. Function approximation results shown in Figure 20 to 22 are not satisfactory either.

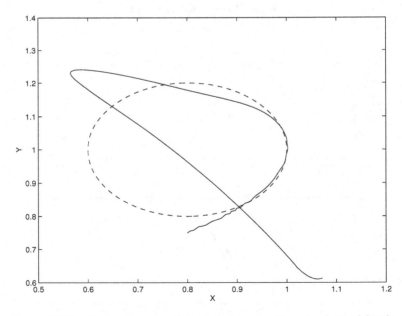

Figure 4.16 Tracking performance in the Cartesian space. It can be observed that the controller designed for RR is not able to give satisfactory performance when applied to EDRR under fast motion condition

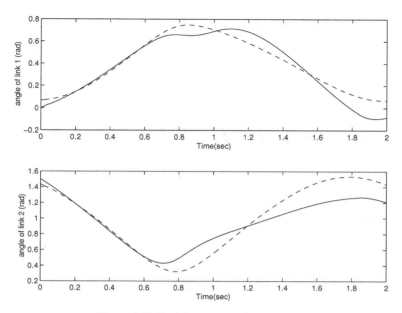

Figure 4.17 The joint space motion trajectory

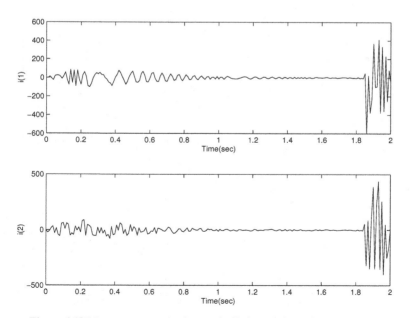

Figure 4.18 Motor currents go to impractically large values after 1.8 seconds

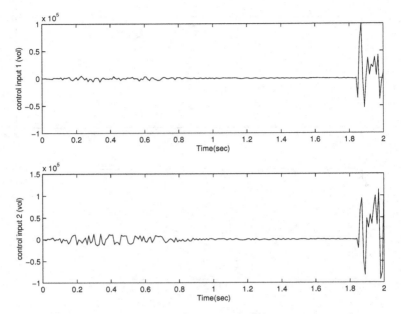

Figure 4.19 The control efforts become very large after 1.8 seconds

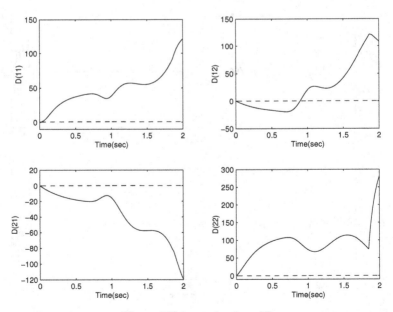

Figure 4.20 Approximation of **D**

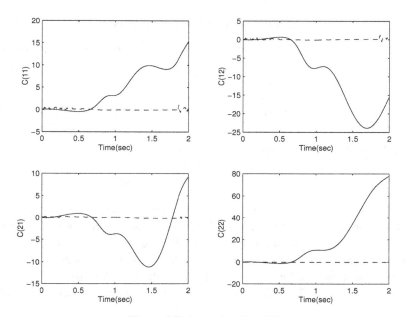

Figure 4.21 Approximation of **C**

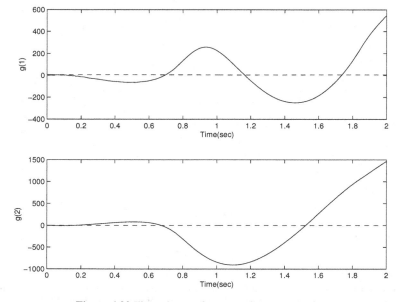

Figure 4.22 The estimate of vector **g** diverges very fast

Case 3: Same as case 2 but with adjusted control parameters

Same configuration as in case 2 is considered here but with adjusted gain matrices

$$\mathbf{K}_d = \begin{bmatrix} 200 & 0 \\ 0 & 200 \end{bmatrix} \text{ and } \mathbf{\Lambda} = \begin{bmatrix} 100 & 0 \\ 0 & 100 \end{bmatrix}.$$

The gain matrices in the update laws are selected as

$$\mathbf{Q}_\mathbf{D}^{-1} = 0.01\mathbf{I}_{44}, \ \mathbf{Q}_\mathbf{C}^{-1} = 0.01\mathbf{I}_{44}, \text{ and } \mathbf{Q}_\mathbf{g}^{-1} = \mathbf{I}_{22}.$$

The simulation results are shown in Figure 4.23 to 29. The Cartesian space tracking performance in Figure 4.23 shows significant improvement (compared with Figure 4.16), but the tracking error is still large (compared with Figure 4.8). The joint space tracking performance is shown in Figure 4.24. The motor currents and control efforts shown in Figure 4.25 and 26 respectively are still unacceptably large. The estimated parameters in Figure 4.27 to 29 are bounded as desired.

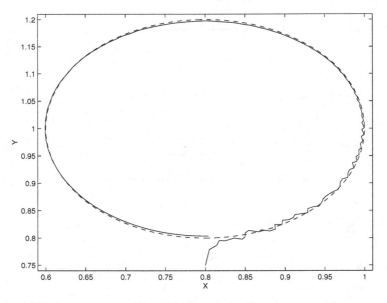

Figure 4.23 Robot endpoint tracking performance in the Cartesian space. After proper gain adjustment, significant improvement in the tracking performance can be observed. However, the tracking error is still unacceptable

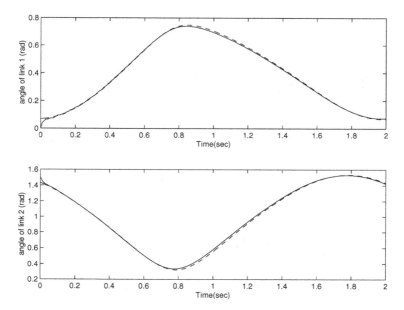

Figure 4.24 The joint space tracking performance

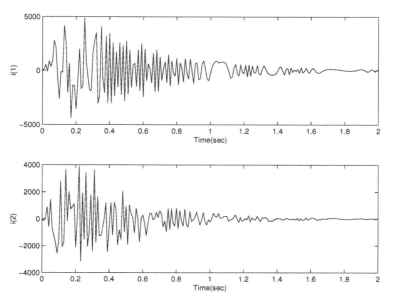

Figure 4.25 Motor currents go to impractically large values

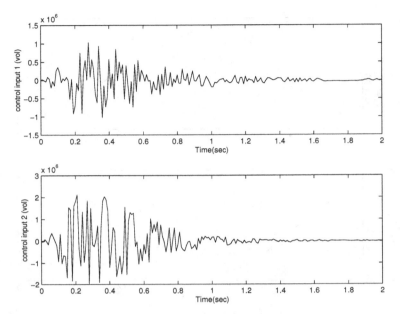

Figure 4.26 The control efforts become very large

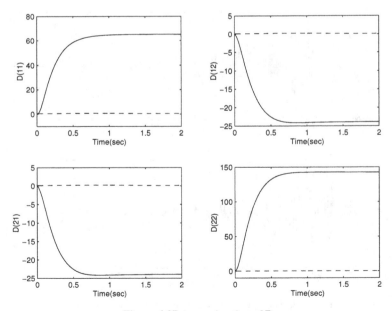

Figure 4.27 Approximation of **D**

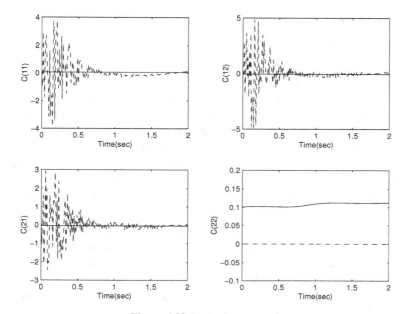

Figure 4.28 Approximation of **C**

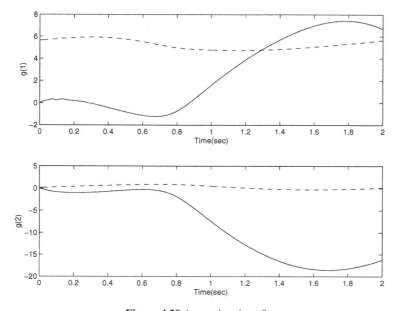

Figure 4.29 Approximation of **g**

4.7 Conclusions

In this chapter, we consider the adaptive control of rigid robots in the free space. In Section 4.2, a regressor based adaptive controller is derived. To implement the update law, the estimate of the inertia matrix is required to be nonsingular and the joint accelerations should be known. Slotine and Li's approach derived in Section 4.3 is well-known to be free from the singularity problem in the estimate of the inertia matrix, and its realization does not need the information of the joint accelerations. All of these approaches need to calculate of the regressor matrix. We have seen in Section 4.4 that computation of the regressor matrix is tedious in general; a regressor-free adaptive controller is derived in Section 4.5 based on the FAT. In some operation conditions, the actuator dynamics should be carefully considered to give better control performance. A regressor-based adaptive controller is derived in Section 4.6.1 for a EDRR. However, it not only needs the joint acceleration feedback but also the calculation of the time derivative of the regressor matrix. A regressor-free adaptive controller for EDRR is then introduced in Section 4.6.2, whose implementation is similar to those in Section 4.4. Finally, example 4.2 investigates the necessity for the consideration of the actuator dynamics in three cases. The simulation results show that, under the fast motion condition, only the controller designed with consideration of actuator dynamics can give good performance with reasonable control efforts for a EDRR.

Chapter 5

Adaptive Impedance Control of Rigid Robots

5.1 Introduction

The impedance control of robot manipulators is to maintain a desired dynamic relationship between the end-effector and the environment where a second order mass-spring-damper system is used to specify the target behavior. It gives a unified approach for controlling the robot in both free space and constrained motion phases. Following the work of Hogan (1985), several studies of the impedance control have been proposed. Anderson and Spong (1988) combined the impedance control with the hybrid control. Goldenberg (1988) used feedback and feedforward compensation for both force and impedance control. Mills and Liu (1991) proposed an impedance control method to control the generalized contact force and position. Gonzalez and Widmann (1995) presented a hybrid impedance control scheme which uses force commands to replace desired trajectory. Yoshikawa (2000) surveyed the force control for robot manipulators.

In Hogan's design, the entire robot dynamics is required to be known, and the impedance controller is derived so that the closed loop system behaves like the target impedance which can interact with the environment compliantly. In practical applications, however, the dynamics of the robot manipulator and the environment inevitably contains various uncertainties and disturbances. Under this circumstance, one of the effective ways to deal with this difficulty is to apply the adaptive strategy to the impedance control. Slotine and Li (1987) extended the adaptive free motion control to the constrained manipulators. Kelly et al. (1987) suggested two adaptive impedance controllers to reduce model uncertainties. Lu and Meng (1991b) presented a concept of target-impedance reference trajectories to deal with the problems of imperfect sensor feedback and uncertain robot parameters. Based on singular motion robot representation, Carelli and Kelly (1991) designed an adaptive position/force controller for constrained robots to achieve global stability results. Colbaugh et al. (1991)

proposed a direct adaptive impedance control scheme without the knowledge of the structure or the parameters of the robot dynamics. Since joint acceleration is difficult to measure precisely, Zhen and Goldenberg (1995) designed an adaptive impedance controller without requiring measurements or estimates of acceleration. Using the camera-in-hand, Mut et. al. (2000) proposed an adaptive impedance tracking controller with visual feedback.

Most of the existing adaptive impedance designs require computation of the regressor matrix. In this chapter, we would like to introduce an adaptive impedance controller based on FAT without using the regressor matrix (Chien and Huang 2004). This chapter is organized as following: Section 5.2 reviews the traditional impedance control and adaptive impedance control strategies. Section 5.3 presents regressor-based adaptive impedance control designs. Section 5.4 gives the regressor–free adaptive impedance controller based on FAT with rigorous proof of closed loop stability. Section 5.5 considers the case of inclusion of actuator dynamics. Simulation cases are also presented for verifying the effectiveness of the scheme introduced.

5.2 Impedance Control and Adaptive Impedance Control

The dynamics of an n-link rigid robot interacting with the environment can be described by (3.3-1) as

$$\mathbf{D}(\mathbf{q})\ddot{\mathbf{q}} + \mathbf{C}(\mathbf{q},\dot{\mathbf{q}})\dot{\mathbf{q}} + \mathbf{g}(\mathbf{q}) = \boldsymbol{\tau} - \mathbf{J}_a^T \mathbf{F}_{ext} \tag{1}$$

where $\mathbf{J}_a(\mathbf{q}) \in \Re^{n \times n}$ is the Jacobian matrix, which is assumed to be nonsingular, and $\mathbf{F}_{ext} \in \Re^n$ is the external force exerted by the end-effector on the environment which is assumed to be measured precisely by a wrist force sensor. It is often more convenient to describe the dynamics of the robot in the Cartesian space when interacting with the environment. Let $\mathbf{x} \in \Re^n$ be the position vector of the end-effector in the Cartesian space and we may rewrite equation (1) as

$$\mathbf{D}_x(\mathbf{x})\ddot{\mathbf{x}} + \mathbf{C}_x(\mathbf{x},\dot{\mathbf{x}})\dot{\mathbf{x}} + \mathbf{g}_x(\mathbf{x}) = \mathbf{J}_a^{-T}\boldsymbol{\tau} - \mathbf{F}_{ext} \tag{2a}$$

where

$$\mathbf{D}_x(\mathbf{x}) = \mathbf{J}_a^{-T}\mathbf{D}(\mathbf{q})\mathbf{J}_a^{-1}$$
$$\mathbf{C}_x(\mathbf{x},\dot{\mathbf{x}}) = \mathbf{J}_a^{-T}[\mathbf{C}(\mathbf{q},\dot{\mathbf{q}}) - \mathbf{D}(\mathbf{q})\mathbf{J}_a^{-1}\dot{\mathbf{J}}_a]\mathbf{J}_a^{-1} \tag{2b}$$
$$\mathbf{g}_x(\mathbf{x}) = \mathbf{J}_a^{-T}\mathbf{g}(\mathbf{q})$$

In the traditional impedance control, all system parameters are given and a controller is designed so that the closed-loop system behaves like the target impedance

$$\mathbf{M}_i(\ddot{\mathbf{x}} - \ddot{\mathbf{x}}_d) + \mathbf{B}_i(\dot{\mathbf{x}} - \dot{\mathbf{x}}_d) + \mathbf{K}_i(\mathbf{x} - \mathbf{x}_d) = -\mathbf{F}_{ext} \tag{3}$$

where $\mathbf{x}_d \in \Re^n$ is the desired trajectory, and $\mathbf{M}_i \in \Re^{n \times n}$, $\mathbf{B}_i \in \Re^{n \times n}$, and $\mathbf{K}_i \in \Re^{n \times n}$ are diagonal matrices representing the desired apparent inertia, damping, and stiffness, respectively. Equation (3) implies that, in the free space tracking phase of the operation, i.e. $\mathbf{F}_{ext} = 0$, the system trajectory converges to the desired trajectory asymptotically. On the other hand, in the constrained motion phase, equation (3) represents a stable 2nd order LTI system driven by the external force. Conceptually, we may regard the impedance controller as a model reference controller and the target impedance plays the role of the reference model. The impedance controller drives the robot to follow the dynamics of the reference model in both the free space tracking and constrained motion phases without any switching activity.

Since all quantities in equation (2) are known, we may design the impedance controller as below such that the closed loop system behaves like the target dynamics shown in (3)

$$\begin{aligned} \boldsymbol{\tau} &= \mathbf{J}_a^T(\mathbf{F}_{ext} + \mathbf{C}_x\dot{\mathbf{x}} + \mathbf{g}_x) \\ &+ \mathbf{J}_a^T\mathbf{D}_x\{\ddot{\mathbf{x}}_d - \mathbf{M}_i^{-1}[\mathbf{B}_i(\dot{\mathbf{x}} - \dot{\mathbf{x}}_d) + \mathbf{K}_i(\mathbf{x} - \mathbf{x}_d) + \mathbf{F}_{ext}]\} \end{aligned} \tag{4}$$

where the terms in the first parenthesis is to cancel the corresponding dynamics in the robot model, while the rest of the terms are for completing the target impedance in (3). It is obvious that by plugging (4) into (2), the closed loop system becomes exactly the target impedance (3).

Suppose some of the system parameters are not available and the above impedance controller cannot be realized. An adaptive controller can thus be designed by referring (4) as

$$\begin{aligned} \boldsymbol{\tau} &= \mathbf{J}_a^T(\mathbf{F}_{ext} + \hat{\mathbf{C}}_x\dot{\mathbf{x}} + \hat{\mathbf{g}}_x) \\ &+ \mathbf{J}_a^T\hat{\mathbf{D}}_x\{\ddot{\mathbf{x}}_d - \mathbf{M}_i^{-1}[\mathbf{B}_i(\dot{\mathbf{x}} - \dot{\mathbf{x}}_d) + \mathbf{K}_i(\mathbf{x} - \mathbf{x}_d) + \mathbf{F}_{ext}]\} \end{aligned} \tag{5}$$

where quantities with hats are respective estimates. Substituting (5) into (2) and after some straightforward manipulations, we may have

$$\mathbf{M}_i(\ddot{\mathbf{x}} - \ddot{\mathbf{x}}_d) + \mathbf{B}_i(\dot{\mathbf{x}} - \dot{\mathbf{x}}_d) + \mathbf{K}_i(\mathbf{x} - \mathbf{x}_d)$$
$$= \mathbf{M}_i\hat{\mathbf{D}}_x^{-1}[(\hat{\mathbf{D}}_x - \mathbf{D}_x)\ddot{\mathbf{x}} + (\hat{\mathbf{C}}_x - \mathbf{C}_x)\dot{\mathbf{x}} + (\hat{\mathbf{g}}_x - \mathbf{g}_x)] - \mathbf{F}_{ext} \tag{6}$$

Define $\tilde{\mathbf{D}}_x = \mathbf{D}_x - \hat{\mathbf{D}}_x$, $\tilde{\mathbf{C}}_x = \mathbf{C}_x - \hat{\mathbf{C}}_x$ and $\tilde{\mathbf{g}}_x = \mathbf{g}_x - \hat{\mathbf{g}}_x$, then (6) can be further written as

$$\ddot{\mathbf{e}} + \mathbf{M}_i^{-1}\mathbf{B}_i\dot{\mathbf{e}} + \mathbf{M}_i^{-1}\mathbf{K}_i\mathbf{e} = -\hat{\mathbf{D}}_x^{-1}(\tilde{\mathbf{D}}_x\ddot{\mathbf{x}} + \tilde{\mathbf{C}}_x\dot{\mathbf{x}} + \tilde{\mathbf{g}}_x) - \mathbf{M}_i^{-1}\mathbf{F}_{ext} \tag{7}$$

Represent the right side of (7) into a linearly parameterized regressor form to have

$$\ddot{\mathbf{e}} + \mathbf{M}_i^{-1}\mathbf{B}_i\dot{\mathbf{e}} + \mathbf{M}_i^{-1}\mathbf{K}_i\mathbf{e} = -\hat{\mathbf{D}}_x^{-1}\mathbf{Y}(\mathbf{x}, \dot{\mathbf{x}}, \ddot{\mathbf{x}})\tilde{\mathbf{p}}_x - \mathbf{M}_i^{-1}\mathbf{F}_{ext} \tag{8}$$

Denote $\mathbf{x} = [\mathbf{e}^T \ \dot{\mathbf{e}}^T]^T \in \Re^{2n}$ and then equation (8) is rewritten as

$$\dot{\mathbf{x}} = \mathbf{A}_x\mathbf{x} - \mathbf{B}_x(\hat{\mathbf{D}}_x^{-1}\mathbf{Y}\tilde{\mathbf{p}}_x - \mathbf{M}_i^{-1}\mathbf{F}_{ext}) \tag{9}$$

where $\mathbf{A}_x = \begin{bmatrix} \mathbf{0}_{n \times n} & \mathbf{I}_n \\ -\mathbf{M}_i^{-1}\mathbf{K}_i & -\mathbf{M}_i^{-1}\mathbf{B}_i \end{bmatrix} \in \Re^{2n \times 2n}$ and $\mathbf{B}_x = \begin{bmatrix} \mathbf{0}_{n \times n} \\ \mathbf{I}_n \end{bmatrix} \in \Re^{2n \times n}$. To

design an update law for $\hat{\mathbf{p}}_x$ to ensure closed loop stability, a Lyapunov-like function candidate can be selected as

$$V(\mathbf{x}, \tilde{\mathbf{p}}) = \frac{1}{2}\mathbf{x}^T\mathbf{P}\mathbf{x} + \frac{1}{2}\tilde{\mathbf{p}}_x^T\mathbf{\Gamma}\tilde{\mathbf{p}}_x \tag{10}$$

where $\mathbf{\Gamma} \in \Re^{r \times r}$ is a positive definite matrix and $\mathbf{P} = \mathbf{P}^T \in \Re^{2n \times 2n}$ is a positive definite solution to the Lyapunov equation $\mathbf{A}_x^T\mathbf{P} + \mathbf{P}\mathbf{A}_x = -\mathbf{Q}$ for a given positive definite matrix $\mathbf{Q} = \mathbf{Q}^T \in \Re^{2n \times 2n}$. Along the trajectory of (9), the time derivative of V can be computed to be

$$\dot{V} = -\frac{1}{2}\mathbf{x}^T\mathbf{Q}\mathbf{x} - \tilde{\mathbf{p}}_x^T[(\hat{\mathbf{D}}_x^{-1}\mathbf{Y})^T\mathbf{B}_x^T\mathbf{P}\mathbf{x} + \mathbf{\Gamma}\dot{\hat{\mathbf{p}}}_x] + \mathbf{x}^T\mathbf{P}\mathbf{B}_x\mathbf{M}_i^{-1}\mathbf{F}_{ext} \tag{11}$$

In the free space tracking phase, the external force is zero, i.e. $\mathbf{F}_{ext} = \mathbf{0}$, and the selection of the update law

$$\dot{\hat{\mathbf{p}}}_x = -\mathbf{\Gamma}^{-1}(\hat{\mathbf{D}}_x^{-1}\mathbf{Y})^T\mathbf{B}_x^T\mathbf{P}\mathbf{x} \tag{12}$$

gives the result

$$\dot{V} = -\frac{1}{2}\mathbf{x}^T\mathbf{Q}\mathbf{x} \le 0. \tag{13}$$

Hence, we have $\mathbf{x} \in L_\infty^{2n}$ and $\tilde{\mathbf{p}}_x \in L_\infty^r$. It is also very easy to have $\mathbf{x} \in L_2^{2n}$, and $\dot{\mathbf{x}} \in L_\infty^{2n}$; therefore, by Barbalat's lemma we may conclude asymptotic convergence of $\mathbf{x}$. This further implies asymptotic convergence of the tracking error $\mathbf{e}$ in the free space tracking phase. However, in the constrained motion phase, i.e. $\mathbf{F}_{ext} \ne \mathbf{0}$, the selection of the update law in (12) will give the result

$$\dot{V} = -\frac{1}{2}\mathbf{x}^T\mathbf{Q}\mathbf{x} + \mathbf{x}^T\mathbf{P}\mathbf{B}_x\mathbf{M}_i^{-1}\mathbf{F}_{ext} \tag{14}$$

Therefore, we may not conclude any stability property for the system states. A possible modification to the controller (5) is to include an additional term as

$$\boldsymbol{\tau} = \mathbf{J}_a^T(\mathbf{F}_{ext} + \hat{\mathbf{C}}_x\dot{\mathbf{x}} + \hat{\mathbf{g}}_x) + \mathbf{J}_a^T\hat{\mathbf{D}}_x\{\ddot{\mathbf{x}}_d - \mathbf{M}_i^{-1}[\mathbf{B}_i(\dot{\mathbf{x}} - \dot{\mathbf{x}}_d) \\ + \mathbf{K}_i(\mathbf{x} - \mathbf{x}_d) - \mathbf{F}_{ext}]\} + \mathbf{J}_a^T\boldsymbol{\tau}_1 \tag{15}$$

where $\boldsymbol{\tau}_1$ is to be designed. Substituting (15) into (2) and we may have

$$\mathbf{M}_i(\ddot{\mathbf{x}} - \ddot{\mathbf{x}}_d) + \mathbf{B}_i(\dot{\mathbf{x}} - \dot{\mathbf{x}}_d) + \mathbf{K}_i(\mathbf{x} - \mathbf{x}_d) \\ = \mathbf{M}_i\hat{\mathbf{D}}_x^{-1}[(\hat{\mathbf{D}}_x - \mathbf{D}_x)\ddot{\mathbf{x}} + (\hat{\mathbf{C}}_x - \mathbf{C}_x)\dot{\mathbf{x}} + (\hat{\mathbf{g}}_x - \mathbf{g}_x)] - \mathbf{F}_{ext} + \mathbf{M}_i\hat{\mathbf{D}}_x^{-1}\boldsymbol{\tau}_1 \tag{16}$$

Let $\boldsymbol{\tau}_1 = \hat{\mathbf{D}}_x\mathbf{M}_i^{-1}\mathbf{F}_{ext}$, then (16) becomes

$$\ddot{\mathbf{e}} + \mathbf{M}_i^{-1}\mathbf{B}_i\dot{\mathbf{e}} + \mathbf{M}_i^{-1}\mathbf{K}_i\mathbf{e} = -\hat{\mathbf{D}}_x^{-1}(\tilde{\mathbf{D}}_x\ddot{\mathbf{x}} + \tilde{\mathbf{C}}_x\dot{\mathbf{x}} + \tilde{\mathbf{g}}_x) \\ = -\hat{\mathbf{D}}_x^{-1}\mathbf{Y}(\mathbf{x}, \dot{\mathbf{x}}, \ddot{\mathbf{x}})\tilde{\mathbf{p}}_x \tag{17}$$

Similar to (9), we may represent (17) into

$$\dot{\mathbf{x}} = \mathbf{A}_x\mathbf{x} - \mathbf{B}_x\hat{\mathbf{D}}_x^{-1}\mathbf{Y}\tilde{\mathbf{p}}_x \tag{18}$$

Select the same Lyapunov-like function candidate in (10), and then we may have

$$\dot{V} = -\frac{1}{2}\mathbf{x}^T\mathbf{Q}\mathbf{x} - \tilde{\mathbf{p}}_x^T[(\hat{\mathbf{D}}_x^{-1}\mathbf{Y})^T\mathbf{B}_x^T\mathbf{P}\mathbf{x} + \boldsymbol{\Gamma}\dot{\hat{\mathbf{p}}}_x] \tag{19}$$

With the selection of the update law in (12), equation (19) becomes (13). Therefore, the system is stable for both the free space tracking and constrained motion phases. However, it should be noted that, according to (16), the modified controller (15) can only ensure convergence of the closed loop system to the dynamics

$$\mathbf{M}_i(\ddot{\mathbf{x}} - \ddot{\mathbf{x}}_d) + \mathbf{B}_i(\dot{\mathbf{x}} - \dot{\mathbf{x}}_d) + \mathbf{K}_i(\mathbf{x} - \mathbf{x}_d) = \mathbf{0} \tag{20}$$

if all parameters converge to their exact values. It is obvious that (20) is different from the target impedance in (3).

Remark 1: To realize the control law (15) and update law (12), we need to feedback the acceleration in the Cartesian space for the calculation of the regressor. In addition, the update law also suffers the singularity problem of $\hat{\mathbf{D}}_x$, and some projection technique should be applied.

5.3 Regressor-Based Adaptive Impedance Controller Design

In the previous section, direct extension of the approach in section 4.2 to the adaptive impedance control does not ensure convergence of the closed loop system to the target impedance. In this section, we would like to apply Slotine and Li's approach introduced in section 4.3 to facilitate the design of the adaptive impedance controller. Similar to the result in section 4.3, the design is free from the feedback of acceleration information and free from the singularity problem in parameter estimations. In addition, the closed loop system will converge to the target impedance once the parameters go to their actual values. Instead of (5.2-3), we consider a new target impedance

$$\mathbf{M}_i(\ddot{\mathbf{x}}_i - \ddot{\mathbf{x}}_d) + \mathbf{B}_i(\dot{\mathbf{x}}_i - \dot{\mathbf{x}}_d) + \mathbf{K}_i(\mathbf{x}_i - \mathbf{x}_d) = -\mathbf{F}_{ext} \tag{1a}$$

where $\mathbf{x}_i \in \mathfrak{R}^n$ is the state vector of the reference model

$$\mathbf{M}_i\ddot{\mathbf{x}}_i + \mathbf{B}_i\dot{\mathbf{x}}_i + \mathbf{K}_i\mathbf{x}_i = \mathbf{M}_i\ddot{\mathbf{x}}_d + \mathbf{B}_i\dot{\mathbf{x}}_d + \mathbf{K}_i\mathbf{x}_d - \mathbf{F}_{ext} \tag{1b}$$

If an adaptive controller can be designed such that $\mathbf{x} \to \mathbf{x}_i$ asymptotically, then the new target impedance (1a) converges to (5.2-3) as desired. Meanwhile, because $\mathbf{x}$ converges to $\mathbf{x}_i$, the dynamics of $\mathbf{x}$ will also converge to the dynamics of $\mathbf{x}_i$ in (1b), i.e., the target impedance.

Define an error vector $\mathbf{s} = \dot{\mathbf{e}} + \Lambda \mathbf{e}$ where $\mathbf{e} = \mathbf{x} - \mathbf{x}_i$ is the state error in the Cartesian space and $\Lambda = diag(\lambda_1, \lambda_2, ..., \lambda_n)$ with $\lambda_i > 0$ for all $i = 1, ..., n$. Rewrite the robot model (5.2-2) into the form

$$\mathbf{D}_x \dot{\mathbf{s}} + \mathbf{C}_x \mathbf{s} + \mathbf{g}_x + \mathbf{D}_x \dot{\mathbf{v}} + \mathbf{C}_x \mathbf{v} = \mathbf{J}_a^{-T} \boldsymbol{\tau} - \mathbf{F}_{ext} \tag{2}$$

where $\mathbf{v} = \dot{\mathbf{x}}_i - \Lambda \mathbf{e}$. The adaptive control law is designed as

$$\boldsymbol{\tau} = \mathbf{J}_a^T (\hat{\mathbf{g}}_x + \hat{\mathbf{D}}_x \dot{\mathbf{v}} + \hat{\mathbf{C}}_x \mathbf{v} + \mathbf{F}_{ext} - \mathbf{K}_d \mathbf{s}) \tag{3}$$

Then the closed loop system can be represented in the form

$$\mathbf{D}_x \dot{\mathbf{s}} + \mathbf{C}_x \mathbf{s} + \mathbf{K}_d \mathbf{s} = -\tilde{\mathbf{D}}_x \dot{\mathbf{v}} - \tilde{\mathbf{C}}_x \mathbf{v} - \tilde{\mathbf{g}}_x \tag{4}$$

Represent the right side of (4) into a linearly parameterized regressor form to have

$$\mathbf{D}_x \dot{\mathbf{s}} + \mathbf{C}_x \mathbf{s} + \mathbf{K}_d \mathbf{s} = -\mathbf{Y}(\mathbf{x}, \dot{\mathbf{x}}, \mathbf{v}, \dot{\mathbf{v}}) \tilde{\mathbf{p}}_x \tag{5}$$

It is noted that the regressor matrix here is not a function of Cartesian space accelerations. To find the update law, define the Lyapunov-like function candidate as

$$V(\mathbf{x}, \tilde{\mathbf{p}}_x) = \frac{1}{2} \mathbf{s}^T \mathbf{D}_x \mathbf{s} + \frac{1}{2} \tilde{\mathbf{p}}_x^T \Gamma \tilde{\mathbf{p}}_x \tag{6}$$

Its time derivative along the trajectory of (5) can be derived as

$$\dot{V} = -\mathbf{s}^T \mathbf{K}_d \mathbf{s} - \tilde{\mathbf{p}}_x^T (\Gamma \dot{\hat{\mathbf{p}}}_x + \mathbf{Y}^T \mathbf{s}) \tag{7}$$

Hence, the update law is selected as

$$\dot{\hat{\mathbf{p}}}_x = -\Gamma^{-1} \mathbf{Y}^T (\mathbf{x}, \dot{\mathbf{x}}, \mathbf{v}, \dot{\mathbf{v}}) \mathbf{s} \tag{8}$$

and (7) becomes

$$\dot{V} = -\mathbf{s}^T \mathbf{K}_d \mathbf{s} \leq 0. \tag{9}$$

It is very easy to prove by Barbalat's lemma that $\mathbf{s} \to \mathbf{0}$ as $t \to \infty$, and hence $\mathbf{x} \to \mathbf{x}_i$ as $t \to \infty$. This further implies the dynamics of $\mathbf{x}$ will converge to the reference model in (1b), and hence the closed loop system will converge to the target impedance in (5.2-3). This solves one of the difficulties encountered in previous section.

Remark 2: To implement the control (3) and update law (8), we do not need the information of Cartesian space accelerations and there is no singularity problem in the inertia matrix estimation. However, the regressor matrix is still needed in the update law.

5.4 FAT-Based Adaptive Impedance Controller Design

In this section, a FAT-based adaptive impedance controller is designed without requiring the knowledge of the regressor. Let us consider the controller (5.3-3),

$$\boldsymbol{\tau} = \mathbf{J}_a^T (\hat{\mathbf{g}}_x + \hat{\mathbf{D}}_x \dot{\mathbf{v}} + \hat{\mathbf{C}}_x \mathbf{v} + \mathbf{F}_{ext} - \mathbf{K}_d \mathbf{s}) \tag{1}$$

and the closed loop dynamics (5.3-4) again

$$\mathbf{D}_x \dot{\mathbf{s}} + \mathbf{C}_x \mathbf{s} + \mathbf{K}_d \mathbf{s} = -\tilde{\mathbf{D}}_x \dot{\mathbf{v}} - \tilde{\mathbf{C}}_x \mathbf{v} - \tilde{\mathbf{g}}_x \tag{2}$$

If we may design appropriate update laws such that $\hat{\mathbf{D}}_x \to \mathbf{D}_x$, $\hat{\mathbf{C}}_x \to \mathbf{C}_x$, and $\hat{\mathbf{g}}_x \to \mathbf{g}_x$, then (2) becomes

$$\mathbf{D}_x \dot{\mathbf{s}} + \mathbf{C}_x \mathbf{s} + \mathbf{K}_d \mathbf{s} = 0 \tag{3}$$

and with proper selection of $\mathbf{K}_d$ we may have asymptotic convergence of $\mathbf{s}$ which further implies convergence of the closed loop system to the target impedance. To design the update laws without utilizing the regressor matrix, let us apply the function approximation representation

$$\mathbf{D}_x = \mathbf{W}_{\mathbf{D}_x}^T \mathbf{Z}_{\mathbf{D}_x} + \boldsymbol{\varepsilon}_{\mathbf{D}_x}$$

$$\mathbf{C}_x = \mathbf{W}_{\mathbf{C}_x}^T \mathbf{Z}_{\mathbf{C}_x} + \boldsymbol{\varepsilon}_{\mathbf{C}_x} \tag{4}$$

$$\mathbf{g}_x = \mathbf{W}_{\mathbf{g}_x}^T \mathbf{z}_{\mathbf{g}_x} + \boldsymbol{\varepsilon}_{\mathbf{g}_x}$$

where $\mathbf{W}_{\mathbf{D}_x} \in \mathfrak{R}^{n^2 \beta_D \times n}$, $\mathbf{W}_{\mathbf{C}_x} \in \mathfrak{R}^{n^2 \beta_C \times n}$ and $\mathbf{W}_{\mathbf{g}_x} \in \mathfrak{R}^{n \beta_g \times n}$ are weighting matrices, $\mathbf{Z}_{\mathbf{D}_x} \in \mathfrak{R}^{n^2 \beta_D \times n}$, $\mathbf{Z}_{\mathbf{C}_x} \in \mathfrak{R}^{n^2 \beta_C \times n}$ and $\mathbf{z}_{\mathbf{g}_x} \in \mathfrak{R}^{n \beta_g \times 1}$ are matrices of basis functions, and $\varepsilon_{(\cdot)}$ are approximation error matrices. The number $\beta_{(\cdot)}$ represents the number of basis functions used. Using the same set of basis functions, the corresponding estimates can also be represented as

$$\hat{\mathbf{D}}_x = \hat{\mathbf{W}}_{\mathbf{D}_x}^T \mathbf{Z}_{\mathbf{D}_x}$$

$$\hat{\mathbf{C}}_x = \hat{\mathbf{W}}_{\mathbf{C}_x}^T \mathbf{Z}_{\mathbf{C}_x} \qquad (5)$$

$$\hat{\mathbf{g}}_x = \hat{\mathbf{W}}_{\mathbf{g}_x}^T \mathbf{z}_{\mathbf{g}_x}$$

where $\hat{\mathbf{W}}_{\mathbf{D}_x}$, $\hat{\mathbf{W}}_{\mathbf{C}_x}$ and $\hat{\mathbf{W}}_{\mathbf{g}_x}$ are respectively the estimates of $\mathbf{W}_{\mathbf{D}_x}$, $\mathbf{W}_{\mathbf{C}_x}$ and $\mathbf{W}_{\mathbf{g}_x}$. With these representations, equation (2) becomes

$$\mathbf{D}_x\dot{\mathbf{s}} + \mathbf{C}_x\mathbf{s} + \mathbf{K}_d\mathbf{s} = -\tilde{\mathbf{W}}_{\mathbf{D}_x}^T \mathbf{Z}_{\mathbf{D}_x}\dot{\mathbf{v}} - \tilde{\mathbf{W}}_{\mathbf{C}_x}^T \mathbf{Z}_{\mathbf{C}_x}\mathbf{v} - \tilde{\mathbf{W}}_{\mathbf{g}_x}^T \mathbf{z}_{\mathbf{g}_x} + \varepsilon_1 \qquad (6)$$

where $\tilde{\mathbf{W}}_{(\cdot)} = \mathbf{W}_{(\cdot)} - \hat{\mathbf{W}}_{(\cdot)}$ and $\varepsilon_1 = \varepsilon_1(\varepsilon_{\mathbf{D}_x}, \varepsilon_{\mathbf{C}_x}, \varepsilon_{\mathbf{g}_x}, \mathbf{s}, \ddot{\mathbf{x}}_i) \in \Re^n$ is a lumped vector of approximation errors. Since $\mathbf{W}_{(\cdot)}$ are constant vectors, their update laws can be easily found by proper selection of the Lyapunov-like function. Let us consider a candidate

$$V(\mathbf{s}, \tilde{\mathbf{W}}_{\mathbf{D}_x}, \tilde{\mathbf{W}}_{\mathbf{C}_x}, \tilde{\mathbf{W}}_{\mathbf{g}_x}) = \frac{1}{2}\mathbf{s}^T \mathbf{D}_x\mathbf{s}$$

$$+\frac{1}{2}Tr(\tilde{\mathbf{W}}_{\mathbf{D}_x}^T \mathbf{Q}_{\mathbf{D}_x} \tilde{\mathbf{W}}_{\mathbf{D}_x} + \tilde{\mathbf{W}}_{\mathbf{C}_x}^T \mathbf{Q}_{\mathbf{C}_x} \tilde{\mathbf{W}}_{\mathbf{C}_x} + \tilde{\mathbf{W}}_{\mathbf{g}_x}^T \mathbf{Q}_{\mathbf{g}_x} \tilde{\mathbf{W}}_{\mathbf{g}_x}) \qquad (7)$$

where $\mathbf{Q}_{\mathbf{D}_x} \in \Re^{n^2\beta_D \times n^2\beta_D}$, $\mathbf{Q}_{\mathbf{C}_x} \in \Re^{n^2\beta_C \times n^2\beta_C}$ and $\mathbf{Q}_{\mathbf{g}_x} \in \Re^{n\beta_g \times n\beta_g}$ are positive definite weighting matrices. The time derivative of V along the trajectory of (6) can be computed as

$$\dot{V} = -\mathbf{s}^T\mathbf{K}_d\mathbf{s} + \mathbf{s}^T\varepsilon_1 - Tr[\tilde{\mathbf{W}}_{\mathbf{D}_x}^T (\mathbf{Z}_{\mathbf{D}_x}\dot{\mathbf{v}}\mathbf{s}^T + \mathbf{Q}_{\mathbf{D}_x}\dot{\hat{\mathbf{W}}}_{\mathbf{D}_x})$$

$$+ \tilde{\mathbf{W}}_{\mathbf{C}_x}^T (\mathbf{Z}_{\mathbf{C}_x}\mathbf{v}\mathbf{s}^T + \mathbf{Q}_{\mathbf{C}_x}\dot{\hat{\mathbf{W}}}_{\mathbf{C}_x}) + \tilde{\mathbf{W}}_{\mathbf{g}_x}^T (\mathbf{z}_{\mathbf{g}_x}\mathbf{s}^T + \mathbf{Q}_{\mathbf{g}_x}\dot{\hat{\mathbf{W}}}_{\mathbf{g}_x})] \qquad (8)$$

Choosing the update laws to be

$$\dot{\hat{\mathbf{W}}}_{\mathbf{D}_x} = -\mathbf{Q}_{\mathbf{D}_x}^{-1}(\mathbf{Z}_{\mathbf{D}_x}\dot{\mathbf{v}}\mathbf{s}^T + \sigma_{\mathbf{D}_x}\hat{\mathbf{W}}_{\mathbf{D}_x})$$

$$\dot{\hat{\mathbf{W}}}_{\mathbf{C}_x} = -\mathbf{Q}_{\mathbf{C}_x}^{-1}(\mathbf{Z}_{\mathbf{C}_x}\mathbf{v}\mathbf{s}^T + \sigma_{\mathbf{C}_x}\hat{\mathbf{W}}_{\mathbf{C}_x}) \qquad (9)$$

$$\dot{\hat{\mathbf{W}}}_{\mathbf{g}_x} = -\mathbf{Q}_{\mathbf{g}_x}^{-1}(\mathbf{z}_{\mathbf{g}_x}\mathbf{s}^T + \sigma_{\mathbf{g}_x}\hat{\mathbf{W}}_{\mathbf{g}_x})$$

where $\sigma_{(\cdot)}$ are positive constants, then equation (8) becomes

$$\dot{V} = -\mathbf{s}^T \mathbf{K}_d \mathbf{s} + \mathbf{s}^T \boldsymbol{\varepsilon}_1 + \sigma_{\mathbf{D}_x} Tr(\tilde{\mathbf{W}}_{\mathbf{D}_x}^T \hat{\mathbf{W}}_{\mathbf{D}_x})$$
$$+ \sigma_{\mathbf{C}_x} Tr(\tilde{\mathbf{W}}_{\mathbf{C}_x}^T \hat{\mathbf{W}}_{\mathbf{C}_x}) + \sigma_{\mathbf{g}_x} Tr(\tilde{\mathbf{W}}_{\mathbf{g}_x}^T \hat{\mathbf{W}}_{\mathbf{g}_x}) \qquad (10)$$

It can further be derived to

$$\dot{V} \leq -\alpha V + \frac{1}{2}\{[\alpha\lambda_{\max}(\mathbf{D}_x) - \lambda_{\min}(\mathbf{K}_d)]\|\mathbf{s}\|^2 + [\alpha\lambda_{\max}(\mathbf{Q}_{\mathbf{D}_x}) - \sigma_{\mathbf{D}_x}]Tr(\tilde{\mathbf{W}}_{\mathbf{D}_x}^T \tilde{\mathbf{W}}_{\mathbf{D}_x})$$
$$+ [\alpha\lambda_{\max}(\mathbf{Q}_{\mathbf{C}_x}) - \sigma_{\mathbf{C}_x}]Tr(\tilde{\mathbf{W}}_{\mathbf{C}_x}^T \tilde{\mathbf{W}}_{\mathbf{C}_x}) + [\alpha\lambda_{\max}(\mathbf{Q}_{\mathbf{g}_x}) - \sigma_{\mathbf{g}_x}]Tr(\tilde{\mathbf{W}}_{\mathbf{g}_x}^T \tilde{\mathbf{W}}_{\mathbf{g}_x})$$
$$+ \frac{\|\boldsymbol{\varepsilon}_1\|^2}{\lambda_{\min}(\mathbf{K}_d)} + [\sigma_{\mathbf{D}_x} Tr(\mathbf{W}_{\mathbf{D}_x}^T \mathbf{W}_{\mathbf{D}_x}) + \sigma_{\mathbf{C}_x} Tr(\mathbf{W}_{\mathbf{C}_x}^T \mathbf{W}_{\mathbf{C}_x}) + \sigma_{\mathbf{g}_x} Tr(\mathbf{W}_{\mathbf{g}_x}^T \mathbf{W}_{\mathbf{g}_x})]\}$$

By selecting

$$\alpha \leq \min\left\{ \frac{\lambda_{\min}(\mathbf{K}_d)}{\lambda_{\max}(\mathbf{D}_x)}, \frac{\sigma_{\mathbf{D}_x}}{\lambda_{\max}(\mathbf{Q}_{\mathbf{D}_x})}, \frac{\sigma_{\mathbf{C}_x}}{\lambda_{\max}(\mathbf{Q}_{\mathbf{C}_x})}, \frac{\sigma_{\mathbf{g}_x}}{\lambda_{\max}(\mathbf{Q}_{\mathbf{g}_x})} \right\},$$

we may have

$$\dot{V} \leq -\alpha V + \frac{\|\boldsymbol{\varepsilon}_1\|^2}{2\lambda_{\min}(\mathbf{K}_d)} + \frac{1}{2}[\sigma_{\mathbf{D}_x} Tr(\mathbf{W}_{\mathbf{D}_x}^T \mathbf{W}_{\mathbf{D}_x})$$
$$+ \sigma_{\mathbf{C}_x} Tr(\mathbf{W}_{\mathbf{C}_x}^T \mathbf{W}_{\mathbf{C}_x}) + \sigma_{\mathbf{g}_x} Tr(\mathbf{W}_{\mathbf{g}_x}^T \mathbf{W}_{\mathbf{g}_x})] \qquad (11)$$

Therefore, $\dot{V} < 0$ whenever

$$V > \frac{\sup_{\tau \geq t_0}\|\boldsymbol{\varepsilon}_1(\tau)\|^2}{2\alpha\lambda_{\min}(\mathbf{K}_d)} + \frac{1}{2\alpha}[\sigma_{\mathbf{D}_x} Tr(\mathbf{W}_{\mathbf{D}_x}^T \mathbf{W}_{\mathbf{D}_x})$$
$$+ \sigma_{\mathbf{C}_x} Tr(\mathbf{W}_{\mathbf{C}_x}^T \mathbf{W}_{\mathbf{C}_x}) + \sigma_{\mathbf{g}_x} Tr(\mathbf{W}_{\mathbf{g}_x}^T \mathbf{W}_{\mathbf{g}_x})]$$

and we have proved that $\mathbf{s}$, $\tilde{\mathbf{W}}_{\mathbf{D}_x}$, $\tilde{\mathbf{W}}_{\mathbf{C}_x}$ and $\tilde{\mathbf{W}}_{\mathbf{g}_x}$ are uniformly ultimately bounded. On the other hand, differential inequality (11) implies

$$V(t) \leq e^{-\alpha(t-t_0)}V(t_0) + \frac{1}{2\alpha\lambda_{\min}(\mathbf{K}_d)} \sup_{t_0 < \tau < t} \|\boldsymbol{\varepsilon}_1(\tau)\|^2$$
$$+ \frac{1}{2\alpha}[\sigma_{\mathbf{D}_x} Tr(\mathbf{W}_{\mathbf{D}_x}^T \mathbf{W}_{\mathbf{D}_x}) + \sigma_{\mathbf{C}_x} Tr(\mathbf{W}_{\mathbf{C}_x}^T \mathbf{W}_{\mathbf{C}_x}) + \sigma_{\mathbf{g}_x} Tr(\mathbf{W}_{\mathbf{g}_x}^T \mathbf{W}_{\mathbf{g}_x})]$$

By the relationship from (7) as

$$V \geq \frac{1}{2} [\lambda_{\min}(\mathbf{D}_x) \|\mathbf{s}\|^2 + \lambda_{\min}(\mathbf{Q}_{\mathbf{D}_x}) Tr(\tilde{\mathbf{W}}_{\mathbf{D}_x}^T \tilde{\mathbf{W}}_{\mathbf{D}_x})$$

$$+ \lambda_{\min}(\mathbf{Q}_{\mathbf{C}_x}) Tr(\tilde{\mathbf{W}}_{\mathbf{C}_x}^T \tilde{\mathbf{W}}_{\mathbf{C}_x}) + \lambda_{\min}(\mathbf{Q}_{\mathbf{g}_x}) Tr(\tilde{\mathbf{W}}_{\mathbf{g}_x}^T \tilde{\mathbf{W}}_{\mathbf{g}_x})]$$

we may derive the bound for $\mathbf{s}$ as

$$\|\mathbf{s}(t)\| \leq \sqrt{\frac{2V(t_0)}{\lambda_{\min}(\mathbf{D}_x)}} e^{-\frac{\alpha}{2}(t-t_0)} + \frac{1}{\sqrt{\alpha \lambda_{\min}(\mathbf{D}_x) \lambda_{\min}(\mathbf{K}_d)}} \sup_{t_0 < \tau < t} \|\varepsilon_1(\tau)\|$$

$$+ \frac{1}{\sqrt{\alpha \lambda_{\min}(\mathbf{D}_x)}} [\sigma_{\mathbf{D}_x} Tr(\mathbf{W}_{\mathbf{D}_x}^T \mathbf{W}_{\mathbf{D}_x}) + \sigma_{\mathbf{C}_x} Tr(\mathbf{W}_{\mathbf{C}_x}^T \mathbf{W}_{\mathbf{C}_x})$$

$$+ \sigma_{\mathbf{g}_x} Tr(\mathbf{W}_{\mathbf{g}_x}^T \mathbf{W}_{\mathbf{g}_x})]^{\frac{1}{2}}$$

Remark 3: If a sufficient number of basis functions are employed in the function approximation so that $\varepsilon_1 \approx 0$, the σ-modification terms in (9) can be eliminated and (10) becomes $\dot{V} = -\mathbf{s}^T \mathbf{K}_d \mathbf{s}$ which implies $\mathbf{s} \in L_\infty^n \cap L_2^n$. It is straightforward to prove $\dot{\mathbf{s}}$ to be uniformly bounded, and hence convergence of $\mathbf{s}$ can be concluded by Barbalat's lemma.

Remark 4: Suppose ε_1 cannot be ignored and there exist a positive number δ such that $\|\varepsilon_1\| \leq \delta$ for all $t \geq 0$, then, instead of (1), a new controller can be constructed as

$$\boldsymbol{\tau} = \mathbf{J}_a^T (\mathbf{F}_{ext} + \hat{\mathbf{g}}_x + \hat{\mathbf{D}}_x \dot{\mathbf{v}} + \hat{\mathbf{C}}_x \mathbf{v} - \mathbf{K}_d \mathbf{s} + \boldsymbol{\tau}_{robust}) \tag{12}$$

where $\boldsymbol{\tau}_{robust}$ is the robust term to be designed. Let us consider the Lyapunov function candidate (7) and the update law (9) again but with $\sigma_{(\cdot)} = 0$. The time derivative of V can be computed as

$$\dot{V} \leq -\mathbf{s}^T \mathbf{K}_d \mathbf{s} + \delta \|\mathbf{s}\| + \mathbf{s}^T \boldsymbol{\tau}_{robust}$$

By picking $\boldsymbol{\tau}_{robust} = -\delta [\text{sgn}(s_1) \quad \cdots \quad \text{sgn}(s_n)]^T$, where $s_i, i = 1, \ldots, n$ are the i-th element of the vector $\mathbf{s}$, we may have $\dot{V} \leq 0$, and asymptotic convergence of $\mathbf{s}$ can be concluded by Barbalat's lemma.

Remark 5: The regressor-free adaptive design introduced in this section is also free from the acceleration information and the singularity problem in inertia matrix estimation. In addition, the closed loop system will converge to the actual target impedance provided all parameters are properly estimated.

Table 5.1 summarizes the adaptive impedance control laws derived in this section. Two columns are arranged to present the regressor based and regressor-free designs respectively according to their controller forms, adaptive laws and implementation issues.

Table 5.1 Summary of adaptive impedance control for RR

	Rigid robot interacting with environment $\mathbf{D}_x(\mathbf{x})\ddot{\mathbf{x}} + \mathbf{C}_x(\mathbf{x},\dot{\mathbf{x}})\dot{\mathbf{x}} + \mathbf{g}_x(\mathbf{x}) = \mathbf{J}_a^{-T}\boldsymbol{\tau} - \mathbf{F}_{ext}$ (5.2-2a)	
	Regressor-based	Regressor-free
Controller	$\boldsymbol{\tau} = \mathbf{J}_a^T(\hat{\mathbf{g}}_x + \hat{\mathbf{D}}_x\dot{\mathbf{v}} + \hat{\mathbf{C}}_x\mathbf{v}$ $+\mathbf{F}_{ext} - \mathbf{K}_d\mathbf{s})$ (5.3-3)	$\boldsymbol{\tau} = \mathbf{J}_a^T(\hat{\mathbf{g}}_x + \hat{\mathbf{D}}_x\dot{\mathbf{v}} + \hat{\mathbf{C}}_x\mathbf{v}$ $+\mathbf{F}_{ext} - \mathbf{K}_d\mathbf{s})$ (5.4-1)
Adaptive Law	$\dot{\hat{\mathbf{p}}}_x = -\boldsymbol{\Gamma}^{-1}\mathbf{Y}^T(\mathbf{x},\dot{\mathbf{x}},\mathbf{v},\dot{\mathbf{v}})\mathbf{s}$ (5.3-8)	$\dot{\hat{\mathbf{W}}}_{\mathbf{D}_x} = -\mathbf{Q}_{\mathbf{D}_x}^{-1}(\mathbf{Z}_{\mathbf{D}_x}\dot{\mathbf{v}}\mathbf{s}^T + \sigma_{\mathbf{D}_x}\hat{\mathbf{W}}_{\mathbf{D}_x})$ $\dot{\hat{\mathbf{W}}}_{\mathbf{C}_x} = -\mathbf{Q}_{\mathbf{C}_x}^{-1}(\mathbf{Z}_{\mathbf{C}_x}\mathbf{v}\mathbf{s}^T + \sigma_{\mathbf{C}_x}\hat{\mathbf{W}}_{\mathbf{C}_x})$ $\dot{\hat{\mathbf{W}}}_{\mathbf{g}_x} = -\mathbf{Q}_{\mathbf{g}_x}^{-1}(\mathbf{z}_{\mathbf{g}_x}\mathbf{s}^T + \sigma_{\mathbf{g}_x}\hat{\mathbf{W}}_{\mathbf{g}_x})$ (5.4-9)
Realization	Need regressor matrix	Does not need regressor matrix

Example 5.1:

The 2-DOF planar robot (3.3-4) is considered in this example to verify the efficacy of the strategy developed in this section by computer simulation. Actual values of system parameters are the same as those in example 4.1, i.e. $m_1=m_2=0.5(kg)$, $l_1=l_2=0.75(m)$, $l_{c1}=l_{c2}=0.375(m)$, and $I_1=I_2=0.0234(kg\text{-}m^2)$. The endpoint starts from $\mathbf{x}(0) = \mathbf{x}_i(0) = [0.8m \quad 0.75m \quad 0 \quad 0]^T$ to track a $0.2m$ radius circle centered at $(0.8m, 1.0m)$ in 10 seconds without knowing its precise model. The constraint surface is smooth and can be modeled as a linear spring (Spong and Vidyasagr 1989) $f_{ext}=k_w(x-x_w)$ where f_{ext} is the force acting on the surface, $k_w=5000N/m$ is the environmental stiffness, x is the coordinate of the end-point in the X direction, and $x_w=0.95m$ is the position of the surface. Hence, the external force vector becomes $\mathbf{F}_{ext} = [f_{ext} \quad 0]^T$. Since the surface is away from the desired initial endpoint position $(0.8m, 0,8m)$, different phases of operations can be observed. The controller in (1) is applied with the gain matrices

$$\mathbf{K}_d = \begin{bmatrix} 50 & 0 \\ 0 & 50 \end{bmatrix}, \text{ and } \mathbf{\Lambda} = \begin{bmatrix} 20 & 0 \\ 0 & 20 \end{bmatrix}.$$

Parameter matrices in the target impedance are selected as

$$\mathbf{M}_i = \begin{bmatrix} 0.5 & 0 \\ 0 & 0.5 \end{bmatrix}, \ \mathbf{B}_i = \begin{bmatrix} 100 & 0 \\ 0 & 100 \end{bmatrix} \text{ and } \mathbf{K}_i = \begin{bmatrix} 1500 & 0 \\ 0 & 1500 \end{bmatrix}.$$

The 11-term Fourier series is selected as the basis function for the approximation of entries in $\mathbf{D}_x$, $\mathbf{C}_x$ and $\mathbf{g}_x$. Therefore, $\hat{\mathbf{W}}_{\mathbf{D}}$ and $\hat{\mathbf{W}}_{\mathbf{C}}$ are in $\mathfrak{R}^{44 \times 2}$, and $\hat{\mathbf{W}}_{\mathbf{g}}$ is in $\mathfrak{R}^{22 \times 2}$. The initial weighting vectors for the entries are assigned to be

$$\hat{\mathbf{w}}_{D_{x11}}(0) = [0.05 \quad 0 \quad \cdots \quad 0]^T \in \mathfrak{R}^{11 \times 1}$$

$$\hat{\mathbf{w}}_{D_{x12}}(0) = \hat{\mathbf{w}}_{D_{x21}}(0) = [-0.05 \quad 0 \quad \cdots \quad 0]^T \in \mathfrak{R}^{11 \times 1}$$

$$\hat{\mathbf{w}}_{D_{x22}}(0) = [0.1 \quad 0 \quad \cdots \quad 0]^T \in \mathfrak{R}^{11 \times 1}$$

$$\hat{\mathbf{w}}_{C_{x11}}(0) = [0.05 \quad 0 \quad \cdots \quad 0]^T \in \mathfrak{R}^{11 \times 1}$$

$$\hat{\mathbf{w}}_{C_{x12}}(0) = \hat{\mathbf{w}}_{C_{x21}}(0) = [-0.05 \quad 0 \quad \cdots \quad 0]^T \in \mathfrak{R}^{11 \times 1}$$

$$\hat{\mathbf{w}}_{C_{x22}}(0) = [0.1 \quad 0 \quad \cdots \quad 0]^T \in \mathfrak{R}^{11 \times 1}$$

$$\hat{\mathbf{w}}_{g_{x1}}(0) = \hat{\mathbf{w}}_{g_{x2}}(0) = [0 \quad 0 \quad \cdots \quad 0]^T \in \mathfrak{R}^{11 \times 1}.$$

The gain matrices in the update laws (9) are designed as

$$\mathbf{Q}_{\mathbf{D}_x}^{-1} = 0.1\mathbf{I}_{44}, \ \mathbf{Q}_{\mathbf{C}_x}^{-1} = 0.1\mathbf{I}_{44} \text{ and } \mathbf{Q}_{\mathbf{g}_x}^{-1} = 50\mathbf{I}_{22}.$$

In this simulation, we assume that the approximation error can be neglected, and hence the σ-modification parameters are chosen as $\sigma_{(\cdot)} = 0$. The simulation results are shown in Figure 5.1 to 5.7. Figure 5.1 shows the robot endpoint tracking performance in the Cartesian space. It can be seen that after some transient response the endpoint converges to the desired trajectory in the free space nicely. Afterwards, the endpoint contacts with the constraint surface at $x_w=0.95(m)$ compliantly. When entering the free space again, the endpoint follows the desired trajectory with very small tracking error regardless of the system uncertainties. Computation of the complex regressor is avoided in this strategy which greatly simplifies the design and implementation of the control law. Figure 5.2 presents the time history of the joint space tracking performance. The transient states converge very fast without unwanted oscillations. The joint space trajectory in the constraint motion phase is smooth. The control efforts to the two joints are reasonable that can be verified in Figure 5.3. The external forces exerted on the endpoint during the constraint motion phase are shown in Figure 5.4. Figure 5.5 to 5.7 are the performance of function approximation. Although most parameters do not converge to their actual values, they still remain bounded as desired.

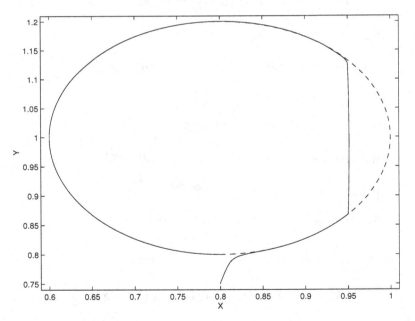

Figure 5.1 Robot endpoint tracking performance in the Cartesian space. After some transient the endpoint converges to the desired trajectory in the free space nicely. Afterwards, the endpoint contacts with the constraint surface compliantly. When entering the free space again, the endpoint follows the desired trajectory with very small tracking error regardless of the system uncertainties

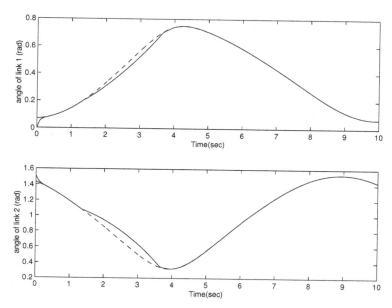

Figure 5.2 The joint space tracking performance. The transient is very fast and the constraint motion phase is smooth

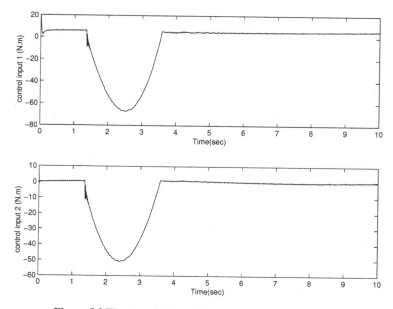

Figure 5.3 The control efforts for both joints are all reasonable

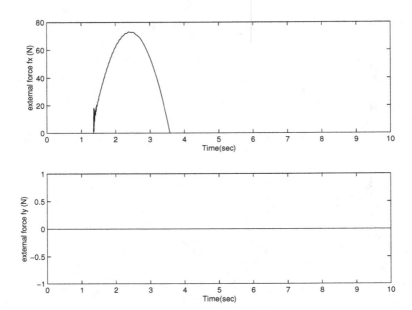

Figure 5.4 Time histories of the external forces in the Cartesian space

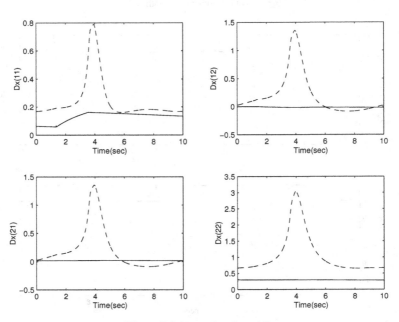

Figure 5.5 Approximation of $\mathbf{D}_x$

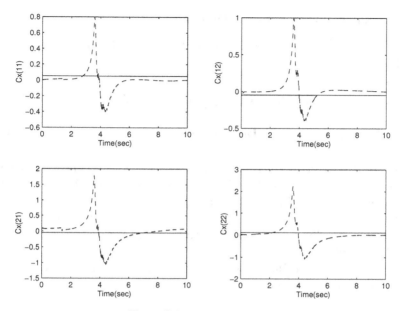

Figure 5.6 Approximation of $\mathbf{C}_x$

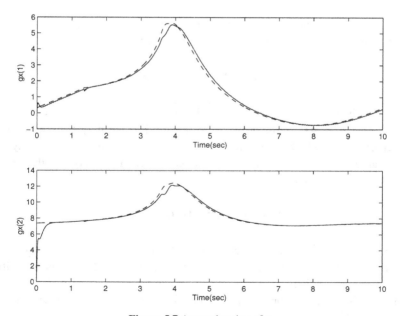

Figure 5.7 Approximation of $\mathbf{g}_x$

5.5 Consideration of Actuator Dynamics

When the actuator dynamics is included, equation (5.2-1) has to be modified to

$$\mathbf{D}(\mathbf{q})\ddot{\mathbf{q}} + \mathbf{C}(\mathbf{q},\dot{\mathbf{q}})\dot{\mathbf{q}} + \mathbf{g}(\mathbf{q}) = \mathbf{Hi} - \mathbf{J}_a^T \mathbf{F}_{ext} \tag{1a}$$

$$\mathbf{L}\dot{\mathbf{i}} + \mathbf{R}\mathbf{i} + \mathbf{K}_b\dot{\mathbf{q}} = \mathbf{u} \tag{1b}$$

In this section, we firstly consider the case when all parameters in (1) are known and a controller is developed so that the closed loop system behaves like the target impedance in (5.2-3). Then, a regressor-based adaptive controller is designed for (1) under the assumption that the system parameters contain uncertainties. Finally, we will derive a regressor-free adaptive controller for the impedance control of system (1).

With the same definitions of $\mathbf{s}$ and $\mathbf{v}$ in (5.3-2), equation (1a) can be represented in the Cartesian space as

$$\mathbf{D}_x\dot{\mathbf{s}} + \mathbf{C}_x\mathbf{s} + \mathbf{g}_x + \mathbf{D}_x\dot{\mathbf{v}} + \mathbf{C}_x\mathbf{v} = \mathbf{J}_a^{-T}\mathbf{Hi} - \mathbf{F}_{ext} \tag{2}$$

Suppose $\mathbf{D}_x$, $\mathbf{C}_x$ and $\mathbf{g}_x$ are known, and we may design a proper control law such that motor armature current $\mathbf{i}$ follows the trajectory

$$\mathbf{i} = \mathbf{H}^{-1}\mathbf{J}_a^T(\mathbf{g}_x + \mathbf{D}_x\dot{\mathbf{v}} + \mathbf{C}_x\mathbf{v} - \mathbf{K}_d\mathbf{s} + \mathbf{F}_{ext}) \tag{3}$$

where $\mathbf{K}_d$ is a positive definite matrix. Then the closed loop dynamics becomes $\mathbf{D}_x\dot{\mathbf{s}} + \mathbf{C}_x\mathbf{s} + \mathbf{K}_d\mathbf{s} = \mathbf{0}$. With proper selection of $\mathbf{K}_d$, we may have asymptotic convergence of $\mathbf{s}$ which further implies convergence of the closed loop system to the target impedance. To make the actual current $\mathbf{i}$ converge to the perfect current in (3), let us select the control input in (1b) as

$$\mathbf{u} = \mathbf{L}\dot{\mathbf{i}}_d + \mathbf{R}\mathbf{i} + \mathbf{K}_b\dot{\mathbf{q}} - \mathbf{K}_c\mathbf{e}_i \tag{4}$$

where $\mathbf{e}_i = \mathbf{i} - \mathbf{i}_d$ is the current error, $\mathbf{i}_d \in \mathfrak{R}^n$ is the desired current which is equivalent to the perfect current trajectory (3), and $\mathbf{K}_c \in \mathfrak{R}^{n \times n}$ is a positive definite matrix. Substituting (4) into (1b), we may have $\mathbf{L}\dot{\mathbf{e}}_i + \mathbf{K}_c\mathbf{e}_i = \mathbf{0}$. Therefore, it is easy to prove that $\mathbf{i} \to \mathbf{i}_d$ as $t \to \infty$ with proper selection of gain matrix $\mathbf{K}_c$.

In summary, if all parameters in the rigid-link electrically-driven robot (1) are available, the controller (4) can give asymptotic convergence of the closed loop system behavior to the target impedance.

5.5.1 Regressor-based adaptive controller

Consider the robot model in (2) and (1b), and assume that $\mathbf{D}_x$, $\mathbf{C}_x$, $\mathbf{g}_x$, $\mathbf{L}$, $\mathbf{R}$ and $\mathbf{K}_b$ are uncertain matrices. Therefore, controller (4) and perfect current trajectory (3) are not realizable. Let us modify (3) to

$$
\begin{aligned}
\mathbf{i}_d &= \mathbf{H}^{-1}\mathbf{J}_a^T(\hat{\mathbf{g}}_x + \hat{\mathbf{D}}_x\dot{\mathbf{v}} + \hat{\mathbf{C}}_x\mathbf{v} - \mathbf{K}_d\mathbf{s} + \mathbf{F}_{ext}) \\
&= \mathbf{H}^{-1}\mathbf{J}_a^T[\mathbf{Y}(\mathbf{x}, \dot{\mathbf{x}}, \mathbf{v}, \dot{\mathbf{v}})\hat{\mathbf{p}}_x - \mathbf{K}_d\mathbf{s} + \mathbf{F}_{ext}]
\end{aligned} \tag{5}
$$

where quantities with hats are respectively the estimated values, and regressor matrix $\mathbf{Y}$ is defined similarly to the one in (5.2-8). The output tracking loop dynamics can be obtained from (2) and (5) as

$$
\begin{aligned}
\mathbf{D}_x\dot{\mathbf{s}} + \mathbf{C}_x\mathbf{s} + \mathbf{K}_d\mathbf{s} &= -\tilde{\mathbf{D}}_x\dot{\mathbf{v}} - \tilde{\mathbf{C}}_x\mathbf{v} - \tilde{\mathbf{g}}_x + \mathbf{J}_a^{-T}\mathbf{H}(\mathbf{i} - \mathbf{i}_d) \\
&= -\mathbf{Y}(\mathbf{x}, \dot{\mathbf{x}}, \mathbf{v}, \dot{\mathbf{v}})\tilde{\mathbf{p}}_x + \mathbf{J}_a^{-T}\mathbf{H}(\mathbf{i} - \mathbf{i}_d)
\end{aligned} \tag{6}
$$

If controller $\mathbf{u}$ can be designed such that $\mathbf{i} \to \mathbf{i}_d$, and update law is selected to have $\hat{\mathbf{p}}_x \to \mathbf{p}_x$, then (6) reduces to $\mathbf{D}_x\dot{\mathbf{s}} + \mathbf{C}_x\mathbf{s} + \mathbf{K}_d\mathbf{s} = \mathbf{0}$, and convergence of the system dynamics to the target impedance can be obtained. A modified controller from (4) is designed as

$$
\begin{aligned}
\mathbf{u} &= \hat{\mathbf{L}}\dot{\mathbf{i}}_d + \hat{\mathbf{R}}\mathbf{i} + \hat{\mathbf{K}}_b\dot{\mathbf{q}} - \mathbf{K}_c\mathbf{e}_i \\
&= \hat{\mathbf{p}}_i^T\boldsymbol{\varphi} - \mathbf{K}_c\mathbf{e}_i
\end{aligned} \tag{7}
$$

where $\boldsymbol{\varphi} = [\dot{\mathbf{i}}_d^T \quad \mathbf{i}^T \quad \dot{\mathbf{q}}^T]^T \in \mathfrak{R}^{3n}$ and $\hat{\mathbf{p}}_i$ is an estimate of the parametric matrix $\mathbf{p}_i = [\mathbf{L}^T \quad \mathbf{R}^T \quad \mathbf{K}_b^T]^T \in \mathfrak{R}^{3n \times n}$. The current tracking loop dynamics with the controller in (7) can be represented in the form

$$
\mathbf{L}\dot{\mathbf{e}}_i + \mathbf{K}_c\mathbf{e}_i = -\tilde{\mathbf{p}}_i^T\boldsymbol{\varphi} \tag{8}
$$

where $\tilde{\mathbf{p}}_i = \mathbf{p}_i - \hat{\mathbf{p}}_i$. A Lyapunov-like function candidate can be found as

$$
V(\mathbf{s}, \mathbf{e}_i, \tilde{\mathbf{p}}_x, \tilde{\mathbf{p}}_i) = \frac{1}{2}\mathbf{s}^T\mathbf{D}_x\mathbf{s} + \frac{1}{2}\mathbf{e}_i^T\mathbf{L}\mathbf{e}_i + \frac{1}{2}\tilde{\mathbf{p}}_x^T\boldsymbol{\Gamma}\tilde{\mathbf{p}}_x + Tr(\tilde{\mathbf{p}}_i^T\boldsymbol{\Gamma}_i\tilde{\mathbf{p}}_i) \tag{9}
$$

where $\mathbf{\Gamma} \in \Re^{r \times r}$ and $\mathbf{\Gamma}_i \in \Re^{3n \times 3n}$ are positive definite matrices. Along the trajectories of (6) and (8), the time derivative of V is computed as

$$\dot{V} = -\mathbf{s}^T \mathbf{K}_d \mathbf{s} + \mathbf{s}^T \mathbf{J}_a^{-T} \mathbf{H} \mathbf{e}_i - \mathbf{e}_i^T \mathbf{K}_c \mathbf{e}_i$$
$$-\tilde{\mathbf{p}}_x^T (\mathbf{\Gamma} \dot{\hat{\mathbf{p}}}_x + \mathbf{Y}^T \mathbf{s}) - Tr[\tilde{\mathbf{p}}_i^T (\mathbf{\Gamma}_i \dot{\hat{\mathbf{p}}}_i + \varphi \mathbf{e}_i^T)] \tag{10}$$

The update laws are selected as

$$\dot{\hat{\mathbf{p}}}_x = -\mathbf{\Gamma}^{-1} \mathbf{Y}^T \mathbf{s}$$
$$\dot{\hat{\mathbf{p}}}_i = -\mathbf{\Gamma}_i^{-1} \varphi \mathbf{e}_i^T \tag{11}$$

and (10) becomes

$$\dot{V} = -[\mathbf{s}^T \quad \mathbf{e}_i^T] \mathbf{Q} \begin{bmatrix} \mathbf{s} \\ \mathbf{e}_i \end{bmatrix} \leq 0 \tag{12}$$

where $\mathbf{Q} = \begin{bmatrix} \mathbf{K}_d & -\dfrac{1}{2} \mathbf{J}_a^{-T} \mathbf{H} \\ -\dfrac{1}{2} \mathbf{J}_a^{-T} \mathbf{H} & \mathbf{K}_c \end{bmatrix}$ is positive definite by proper selection

of $\mathbf{K}_d$ and $\mathbf{K}_c$. Equation (12) implies that $\mathbf{s}$ and $\mathbf{e}_i$ are uniformly bounded and square integrable. It can also be proved that $\dot{\mathbf{s}}$ and $\dot{\mathbf{e}}_i$ are uniformly bounded. Hence, asymptotic convergence in the current tracking loop and output tracking loop can be concluded from Barbalat's lemma.

Remark 6: Realization of controller (7) and update laws in (11) require the time derivative of $\mathbf{i}_d$ in (5) which implies the needs for $\ddot{\mathbf{x}}$, $\dot{\mathbf{F}}_{ext}$ and $\dot{\mathbf{Y}}$. Availability of all of these quantities is impractical in general; therefore, some more feasible controllers are to be developed.

5.5.2 Regressor-free adaptive controller

Suppose $\mathbf{D}_x$, $\mathbf{C}_x$, $\mathbf{g}_x$, $\mathbf{L}$, $\mathbf{R}$ and $\mathbf{K}_b$ are not available, and we would like to design a regressor-free adaptive controller so that the closed-loop dynamics converges to the target impedance. Besides, realization of the controller designed in this section has to be independent to the acceleration feedback and time derivatives of the external force.

Consider the robot model in (2) and (1b) again. Since most robot parameters are unavailable, controller (4) and perfect current trajectory (3) are not realizable. The perfect current trajectory can be modified similar to (5) as

$$\mathbf{i}_d = \mathbf{H}^{-1}\mathbf{J}_a^T(\mathbf{F}_{ext} + \hat{\mathbf{g}}_x + \hat{\mathbf{D}}_x\dot{\mathbf{v}} + \hat{\mathbf{C}}_x\mathbf{v} - \mathbf{K}_d\mathbf{s}) \tag{13}$$

Substituting (13) into (2), we may obtain the output tracking loop dynamics

$$\mathbf{D}_x\dot{\mathbf{s}} + \mathbf{C}_x\mathbf{s} + \mathbf{K}_d\mathbf{s} = -\tilde{\mathbf{D}}_x\dot{\mathbf{v}} - \tilde{\mathbf{C}}_x\mathbf{v} - \tilde{\mathbf{g}}_x + \mathbf{J}_a^{-T}\mathbf{H}(\mathbf{i} - \mathbf{i}_d) \tag{14}$$

If controller $\mathbf{u}$ can be designed such that $\mathbf{i} \rightarrow \mathbf{i}_d$, and there are some update laws to have $\hat{\mathbf{D}}_x \rightarrow \mathbf{D}_x$, $\hat{\mathbf{C}}_x \rightarrow \mathbf{C}_x$ and $\hat{\mathbf{g}}_x \rightarrow \mathbf{g}_x$, then (14) reduces to $\mathbf{D}_x\dot{\mathbf{s}} + \mathbf{C}_x\mathbf{s} + \mathbf{K}_d\mathbf{s} = \mathbf{0}$, and convergence of the system dynamics to the target impedance can be obtained. According to (7), let us select the control input in (1b) as

$$\mathbf{u} = \hat{\mathbf{f}} - \mathbf{K}_c\mathbf{e}_i \tag{15}$$

where $\hat{\mathbf{f}}$ is an estimate of $\mathbf{f}(\mathbf{i}_d, \mathbf{i}, \dot{\mathbf{q}}) = \mathbf{L}\dot{\mathbf{i}}_d + \mathbf{R}\mathbf{i} + \mathbf{K}_b\dot{\mathbf{q}}$. Substituting this control law into (1b), we may have the dynamics in the current tracking loop

$$\mathbf{L}\dot{\mathbf{e}}_i + \mathbf{K}_c\mathbf{e}_i = \hat{\mathbf{f}} - \mathbf{f} \tag{16}$$

If an appropriate update law for $\hat{\mathbf{f}}$ can be selected, we may have $\mathbf{i} \rightarrow \mathbf{i}_d$. Since $\mathbf{D}_x$, $\mathbf{C}_x$, $\mathbf{g}_x$ and $\mathbf{f}$ are functions of time, traditional adaptive controllers are not directly applicable. To design the update laws, let us apply the function approximation representation

$$\mathbf{D}_x = \mathbf{W}_{\mathbf{D}_x}^T \mathbf{Z}_{\mathbf{D}_x} + \boldsymbol{\varepsilon}_{\mathbf{D}_x} \tag{17a}$$

$$\mathbf{C}_x = \mathbf{W}_{\mathbf{C}_x}^T \mathbf{Z}_{\mathbf{C}_x} + \boldsymbol{\varepsilon}_{\mathbf{C}_x} \tag{17b}$$

$$\mathbf{g}_x = \mathbf{W}_{\mathbf{g}_x}^T \mathbf{z}_{\mathbf{g}_x} + \boldsymbol{\varepsilon}_{\mathbf{g}_x} \tag{17c}$$

$$\mathbf{f} = \mathbf{W}_{\mathbf{f}}^T \mathbf{z}_{\mathbf{f}} + \boldsymbol{\varepsilon}_{\mathbf{f}} \tag{17d}$$

where $\mathbf{W}_{\mathbf{D}_x} \in \mathfrak{R}^{n^2\beta_D \times n}$, $\mathbf{W}_{\mathbf{C}_x} \in \mathfrak{R}^{n^2\beta_C \times n}$, $\mathbf{W}_{\mathbf{g}_x} \in \mathfrak{R}^{n\beta_g \times n}$ and $\mathbf{W}_{\mathbf{f}} \in \mathfrak{R}^{n\beta_f \times n}$ are weighting matrices, $\mathbf{Z}_{\mathbf{D}_x} \in \mathfrak{R}^{n^2\beta_D \times n}$, $\mathbf{Z}_{\mathbf{C}_x} \in \mathfrak{R}^{n^2\beta_C \times n}$, $\mathbf{z}_{\mathbf{g}_x} \in \mathfrak{R}^{n\beta_g \times 1}$ and $\mathbf{z}_{\mathbf{f}} \in \mathfrak{R}^{n\beta_f \times 1}$ are matrices of basis functions, and $\boldsymbol{\varepsilon}_{(.)}$ are approximation error matrices. The number $\beta_{(\cdot)}$ represents the number of basis functions used. Using the same set of basis functions, the corresponding estimates can also be represented as

$$\hat{\mathbf{D}}_x = \hat{\mathbf{W}}_{\mathbf{D}_x}^T \mathbf{Z}_{\mathbf{D}_x} \tag{17e}$$

$$\hat{\mathbf{C}}_x = \hat{\mathbf{W}}_{\mathbf{C}_x}^T \mathbf{Z}_{\mathbf{C}_x} \tag{17f}$$

$$\hat{\mathbf{g}}_x = \hat{\mathbf{W}}_{\mathbf{g}_x}^T \mathbf{z}_{\mathbf{g}_x} \tag{17g}$$

$$\hat{\mathbf{f}} = \hat{\mathbf{W}}_{\mathbf{f}}^T \mathbf{z}_{\mathbf{f}} \tag{17h}$$

Define $\tilde{\mathbf{W}}_{(\cdot)} = \hat{\mathbf{W}}_{(\cdot)} - \mathbf{W}_{(\cdot)}$, then equation (14) and (16) become

$$\mathbf{D}_x \dot{\mathbf{s}} + \mathbf{C}_x \mathbf{s} + \mathbf{K}_d \mathbf{s} = \mathbf{J}_a^{-T} \mathbf{H}(\mathbf{i} - \mathbf{i}_d) - \tilde{\mathbf{W}}_{\mathbf{D}_x}^T \mathbf{Z}_{\mathbf{D}_x} \dot{\mathbf{v}}$$
$$- \tilde{\mathbf{W}}_{\mathbf{C}_x}^T \mathbf{Z}_{\mathbf{C}_x} \mathbf{v} - \tilde{\mathbf{W}}_{\mathbf{g}_x}^T \mathbf{z}_{\mathbf{g}_x} + \varepsilon_1 \tag{18a}$$

$$\mathbf{L} \dot{\mathbf{e}}_i + \mathbf{K}_c \mathbf{e}_i = -\tilde{\mathbf{W}}_{\mathbf{f}}^T \mathbf{z}_{\mathbf{f}} + \varepsilon_2 \tag{18b}$$

where $\varepsilon_1 = \varepsilon_1(\varepsilon_{\mathbf{D}}, \varepsilon_{\mathbf{C}}, \varepsilon_{\mathbf{g}}, \mathbf{s}, \ddot{\mathbf{x}}_i)$ and $\varepsilon_2 = \varepsilon_2(\varepsilon_f, \mathbf{e}_i)$ are lumped approximation errors. Since $\mathbf{W}_{(\cdot)}$ are constant matrices, their update laws can be easily found by proper selection of the Lyapunov-like function. Let us consider a candidate

$$V(\mathbf{s}, \mathbf{e}_i, \tilde{\mathbf{W}}_{\mathbf{D}_x}, \tilde{\mathbf{W}}_{\mathbf{C}_x}, \tilde{\mathbf{W}}_{\mathbf{g}_x}, \tilde{\mathbf{W}}_{\mathbf{f}}) = \frac{1}{2}\mathbf{s}^T \mathbf{D}_x \mathbf{s} + \frac{1}{2}\mathbf{e}_i^T \mathbf{L} \mathbf{e}_i + \frac{1}{2} Tr(\tilde{\mathbf{W}}_{\mathbf{D}_x}^T \mathbf{Q}_{\mathbf{D}_x} \tilde{\mathbf{W}}_{\mathbf{D}_x}$$
$$+ \tilde{\mathbf{W}}_{\mathbf{C}_x}^T \mathbf{Q}_{\mathbf{C}_x} \tilde{\mathbf{W}}_{\mathbf{C}_x} + \tilde{\mathbf{W}}_{\mathbf{g}_x}^T \mathbf{Q}_{\mathbf{g}_x} \tilde{\mathbf{W}}_{\mathbf{g}_x} + \tilde{\mathbf{W}}_{\mathbf{f}}^T \mathbf{Q}_{\mathbf{f}} \tilde{\mathbf{W}}_{\mathbf{f}}) \tag{19}$$

The matrices $\mathbf{Q}_{\mathbf{D}_x} \in \Re^{n^2 \beta_{\mathbf{D}} \times n^2 \beta_{\mathbf{D}}}$, $\mathbf{Q}_{\mathbf{C}_x} \in \Re^{n^2 \beta_{\mathbf{C}} \times n^2 \beta_{\mathbf{C}}}$, $\mathbf{Q}_{\mathbf{g}_x} \in \Re^{n \beta_{\mathbf{g}} \times n \beta_{\mathbf{g}}}$ and $\mathbf{Q}_{\mathbf{f}} \in \Re^{n \beta_f \times n \beta_f}$ are all positive definite. The time derivative of V along the trajectory of (18) can be computed as

$$\dot{V} = -\mathbf{s}^T \mathbf{K}_d \mathbf{s} + \mathbf{s}^T \mathbf{J}_a^{-T} \mathbf{H} \mathbf{e}_i - \mathbf{e}_i^T \mathbf{K}_c \mathbf{e}_i + \mathbf{s}^T \varepsilon_1 + \mathbf{e}_i^T \varepsilon_2$$
$$- Tr[\tilde{\mathbf{W}}_{\mathbf{D}_x}^T (\mathbf{Z}_{\mathbf{D}_x} \dot{\mathbf{v}} \mathbf{s}^T + \mathbf{Q}_{\mathbf{D}_x} \dot{\hat{\mathbf{W}}}_{\mathbf{D}_x}) + \tilde{\mathbf{W}}_{\mathbf{C}_x}^T (\mathbf{Z}_{\mathbf{C}_x} \mathbf{v} \mathbf{s}^T + \mathbf{Q}_{\mathbf{C}_x} \dot{\hat{\mathbf{W}}}_{\mathbf{C}_x})$$
$$+ \tilde{\mathbf{W}}_{\mathbf{g}_x}^T (\mathbf{z}_{\mathbf{g}_x} \mathbf{s}^T + \mathbf{Q}_{\mathbf{g}_x} \dot{\hat{\mathbf{W}}}_{\mathbf{g}_x}) + \tilde{\mathbf{W}}_{\mathbf{f}}^T (\mathbf{z}_{\mathbf{f}} \mathbf{e}_i^T + \mathbf{Q}_{\mathbf{f}} \dot{\hat{\mathbf{W}}}_{\mathbf{f}})] \tag{20}$$

The update laws can thus be selected as

$$\dot{\hat{\mathbf{W}}}_{\mathbf{D}_x} = -\mathbf{Q}_{\mathbf{D}_x}^{-1}(\mathbf{Z}_{\mathbf{D}_x}\dot{\mathbf{v}}\mathbf{s}^T + \sigma_{\mathbf{D}_x}\hat{\mathbf{W}}_{\mathbf{D}_x}) \tag{21a}$$

$$\dot{\hat{\mathbf{W}}}_{\mathbf{C}_x} = -\mathbf{Q}_{\mathbf{C}_x}^{-1}(\mathbf{Z}_{\mathbf{C}_x}\mathbf{v}\mathbf{s}^T + \sigma_{\mathbf{C}_x}\hat{\mathbf{W}}_{\mathbf{C}_x}) \tag{21b}$$

$$\dot{\hat{\mathbf{W}}}_{\mathbf{g}_x} = -\mathbf{Q}_{\mathbf{g}_x}^{-1}(\mathbf{z}_{\mathbf{g}_x}\mathbf{s}^T + \sigma_{\mathbf{g}_x}\hat{\mathbf{W}}_{\mathbf{g}_x}) \tag{21c}$$

$$\dot{\hat{\mathbf{W}}}_{\mathbf{f}} = -\mathbf{Q}_{\mathbf{f}}^{-1}(\mathbf{z}_{\mathbf{f}}\mathbf{e}_i^T + \sigma_{\mathbf{f}}\hat{\mathbf{W}}_{\mathbf{f}}) \tag{21d}$$

and (20) becomes

$$\dot{V} = -[\mathbf{s}^T \quad \mathbf{e}_i^T]\mathbf{Q}\begin{bmatrix}\mathbf{s}\\\mathbf{e}_i\end{bmatrix} + [\mathbf{s}^T \quad \mathbf{e}_i^T]\begin{bmatrix}\boldsymbol{\varepsilon}_1\\\boldsymbol{\varepsilon}_2\end{bmatrix} + \sigma_{\mathbf{D}_x}Tr(\tilde{\mathbf{W}}_{\mathbf{D}_x}^T\hat{\mathbf{W}}_{\mathbf{D}_x})$$
$$+\sigma_{\mathbf{C}_x}Tr(\tilde{\mathbf{W}}_{\mathbf{C}_x}^T\hat{\mathbf{W}}_{\mathbf{C}_x}) + \sigma_{\mathbf{g}_x}Tr(\tilde{\mathbf{W}}_{\mathbf{g}_x}^T\hat{\mathbf{W}}_{\mathbf{g}_x}) + \sigma_{\mathbf{f}}Tr(\tilde{\mathbf{W}}_{\mathbf{f}}^T\hat{\mathbf{W}}_{\mathbf{f}}) \tag{22}$$

where $\mathbf{Q} = \begin{bmatrix} \mathbf{K}_d & -\dfrac{1}{2}\mathbf{J}_a^{-T}\mathbf{H} \\ -\dfrac{1}{2}\mathbf{J}_a^{-T}\mathbf{H} & \mathbf{K}_c \end{bmatrix}$ is positive definite by proper selection of

$\mathbf{K}_d$ and $\mathbf{K}_c$. Owing to the existence of ε_1 and ε_2 in (22), definiteness of $\dot{V}$ cannot be determined. Let us proceed by considering the upper bound of V in (19) as

$$V \leq \frac{1}{2}\lambda_{\max}(\mathbf{A})\left\|\begin{bmatrix}\mathbf{s}\\\mathbf{e}_i\end{bmatrix}\right\|^2 + \frac{1}{2}[\lambda_{\max}(\mathbf{Q}_{\mathbf{D}_x})Tr(\tilde{\mathbf{W}}_{\mathbf{D}_x}^T\tilde{\mathbf{W}}_{\mathbf{D}_x})$$
$$+\lambda_{\max}(\mathbf{Q}_{\mathbf{C}_x})Tr(\tilde{\mathbf{W}}_{\mathbf{C}_x}^T\tilde{\mathbf{W}}_{\mathbf{C}_x}) + \lambda_{\max}(\mathbf{Q}_{\mathbf{g}_x})Tr(\tilde{\mathbf{W}}_{\mathbf{g}_x}^T\tilde{\mathbf{W}}_{\mathbf{g}_x})$$
$$+\lambda_{\max}(\mathbf{Q}_{\mathbf{f}})Tr(\tilde{\mathbf{W}}_{\mathbf{f}}^T\tilde{\mathbf{W}}_{\mathbf{f}})]$$

where $\mathbf{A} = \begin{bmatrix}\mathbf{D}_x & \mathbf{0}\\\mathbf{0} & \mathbf{L}\end{bmatrix}$, and the inequalities

$$-[\mathbf{s}^T \quad \mathbf{e}_i^T]\mathbf{Q}\begin{bmatrix}\mathbf{s}\\\mathbf{e}_i\end{bmatrix} + [\mathbf{s}^T \quad \mathbf{e}_i^T]\begin{bmatrix}\boldsymbol{\varepsilon}_1\\\boldsymbol{\varepsilon}_2\end{bmatrix}$$
$$\leq -\frac{1}{2}\left(\lambda_{\min}(\mathbf{Q})\left\|\begin{bmatrix}\mathbf{s}\\\mathbf{e}_i\end{bmatrix}\right\|^2 - \frac{1}{\lambda_{\min}(\mathbf{Q})}\left\|\begin{bmatrix}\boldsymbol{\varepsilon}_1\\\boldsymbol{\varepsilon}_2\end{bmatrix}\right\|^2\right)$$

$$Tr(\tilde{\mathbf{W}}_{(\cdot)}^T \hat{\mathbf{W}}_{(\cdot)}) \leq \frac{1}{2} Tr(\mathbf{W}_{(\cdot)}^T \mathbf{W}_{(\cdot)}) - \frac{1}{2} Tr(\tilde{\mathbf{W}}_{(\cdot)}^T \tilde{\mathbf{W}}_{(\cdot)})$$

then we may rewrite (22) into

$$\dot{V} \leq -\alpha V + \frac{1}{2}[\alpha\lambda_{max}(\mathbf{A}) - \lambda_{min}(\mathbf{Q})] \left\| \begin{bmatrix} \mathbf{s} \\ \mathbf{e}_i \end{bmatrix} \right\|^2 + \frac{1}{2}\{[\alpha\lambda_{max}(\mathbf{Q}_{\mathbf{D}_x}) - \sigma_{\mathbf{D}_x}]Tr(\tilde{\mathbf{W}}_{\mathbf{D}_x}^T \tilde{\mathbf{W}}_{\mathbf{D}_x})$$

$$+[\alpha\lambda_{max}(\mathbf{Q}_{\mathbf{C}_x}) - \sigma_{\mathbf{C}_x}]Tr(\tilde{\mathbf{W}}_{\mathbf{C}_x}^T \tilde{\mathbf{W}}_{\mathbf{C}_x}) +[\alpha\lambda_{max}(\mathbf{Q}_{\mathbf{g}_x}) - \sigma_{\mathbf{g}_x}]Tr(\tilde{\mathbf{W}}_{\mathbf{g}_x}^T \tilde{\mathbf{W}}_{\mathbf{g}_x})$$

$$+[\alpha\lambda_{max}(\mathbf{Q}_{\mathbf{f}}) - \sigma_{\mathbf{f}}]Tr(\tilde{\mathbf{W}}_{\mathbf{f}}^T \tilde{\mathbf{W}}_{\mathbf{f}})\} + \frac{1}{2\lambda_{min}(\mathbf{Q})} \left\| \begin{bmatrix} \varepsilon_1 \\ \varepsilon_2 \end{bmatrix} \right\|^2 + \frac{1}{2}\{\sigma_{\mathbf{D}_x}Tr(\mathbf{W}_{\mathbf{D}_x}^T \mathbf{W}_{\mathbf{D}_x})$$

$$+\sigma_{\mathbf{C}_x}Tr(\mathbf{W}_{\mathbf{C}_x}^T \mathbf{W}_{\mathbf{C}_x}) +\sigma_{\mathbf{g}_x}Tr(\mathbf{W}_{\mathbf{g}_x}^T \mathbf{W}_{\mathbf{g}_x}) + \sigma_{\mathbf{f}}Tr(\mathbf{W}_{\mathbf{f}}^T \mathbf{W}_{\mathbf{f}})\}$$

where α is selected to satisfy

$$\alpha \leq \min\left\{\frac{\lambda_{min}(\mathbf{Q})}{\lambda_{max}(\mathbf{A})}, \frac{\sigma_{\mathbf{D}_x}}{\lambda_{max}(\mathbf{Q}_{\mathbf{D}_x})}, \frac{\sigma_{\mathbf{C}_x}}{\lambda_{max}(\mathbf{Q}_{\mathbf{C}_x})}, \frac{\sigma_{\mathbf{g}_x}}{\lambda_{max}(\mathbf{Q}_{\mathbf{g}_x})}, \frac{\sigma_{\mathbf{f}}}{\lambda_{max}(\mathbf{Q}_{\mathbf{f}})}\right\}.$$

With this selection, we may further have

$$\dot{V} \leq -\alpha V + \frac{1}{2\lambda_{min}(\mathbf{Q})} \left\| \begin{bmatrix} \varepsilon_1 \\ \varepsilon_2 \end{bmatrix} \right\|^2 + \frac{1}{2}[\sigma_{\mathbf{D}_x}Tr(\mathbf{W}_{\mathbf{D}_x}^T \mathbf{W}_{\mathbf{D}_x})$$

$$+ \sigma_{\mathbf{C}_x}Tr(\mathbf{W}_{\mathbf{C}_x}^T \mathbf{W}_{\mathbf{C}_x}) + \sigma_{\mathbf{g}_x}Tr(\mathbf{W}_{\mathbf{g}_x}^T \mathbf{W}_{\mathbf{g}_x}) + \sigma_{\mathbf{f}}Tr(\mathbf{W}_{\mathbf{f}}^T \mathbf{W}_{\mathbf{f}})] \qquad (23)$$

So, $\dot{V} < 0$ can be concluded whenever

$$V > \frac{1}{2\alpha\lambda_{min}(\mathbf{Q})} \sup_{\tau \geq t_0} \left\| \begin{bmatrix} \varepsilon_1(\tau) \\ \varepsilon_2(\tau) \end{bmatrix} \right\|^2 + \frac{1}{2\alpha}[\sigma_{\mathbf{D}_x}Tr(\mathbf{W}_{\mathbf{D}_x}^T \mathbf{W}_{\mathbf{D}_x})$$

$$+\sigma_{\mathbf{C}_x}Tr(\mathbf{W}_{\mathbf{C}_x}^T \mathbf{W}_{\mathbf{C}_x}) + \sigma_{\mathbf{g}_x}Tr(\mathbf{W}_{\mathbf{g}_x}^T \mathbf{W}_{\mathbf{g}_x}) + \sigma_{\mathbf{f}}Tr(\mathbf{W}_{\mathbf{f}}^T \mathbf{W}_{\mathbf{f}})] \qquad (24)$$

Hence, we have proved that $\mathbf{s}$, $\mathbf{e}_i$, $\tilde{\mathbf{W}}_{\mathbf{D}_x}$, $\tilde{\mathbf{W}}_{\mathbf{C}_x}$, $\tilde{\mathbf{W}}_{\mathbf{g}_x}$ and $\tilde{\mathbf{W}}_{\mathbf{f}}$ are uniformly ultimately bounded. On the other hand, (23) also implies

$$V(t) \le e^{-\alpha(t-t_0)}V(t_0) + \frac{1}{2\alpha\lambda_{\min}(\mathbf{Q})} \sup_{t_0 < \tau < t} \left\| \begin{bmatrix} \varepsilon_1(\tau) \\ \varepsilon_2(\tau) \end{bmatrix} \right\|^2$$

$$+ \frac{1}{2\alpha}[\sigma_{\mathbf{D}_x} Tr(\mathbf{W}_{\mathbf{D}_x}^T \mathbf{W}_{\mathbf{D}_x}) + \sigma_{\mathbf{C}_x} Tr(\mathbf{W}_{\mathbf{C}_x}^T \mathbf{W}_{\mathbf{C}_x})$$

$$+ \sigma_{\mathbf{g}_x} Tr(\mathbf{W}_{\mathbf{g}_x}^T \mathbf{W}_{\mathbf{g}_x}) + \sigma_{\mathbf{f}} Tr(\mathbf{W}_{\mathbf{f}}^T \mathbf{W}_{\mathbf{f}})] \tag{25}$$

Consider the lower bound of V in (19) as

$$V \ge \frac{1}{2}\lambda_{\min}(\mathbf{A}) \left\| \begin{bmatrix} \mathbf{s} \\ \mathbf{e}_i \end{bmatrix} \right\|^2 + \frac{1}{2}[\lambda_{\min}(\mathbf{Q}_{\mathbf{D}_x})Tr(\tilde{\mathbf{W}}_{\mathbf{D}_x}^T \tilde{\mathbf{W}}_{\mathbf{D}_x})$$

$$+ \lambda_{\min}(\mathbf{Q}_{\mathbf{C}_x})Tr(\tilde{\mathbf{W}}_{\mathbf{C}_x}^T \tilde{\mathbf{W}}_{\mathbf{C}_x}) + \lambda_{\min}(\mathbf{Q}_{\mathbf{g}_x})Tr(\tilde{\mathbf{W}}_{\mathbf{g}_x}^T \tilde{\mathbf{W}}_{\mathbf{g}_x}) + \lambda_{\min}(\mathbf{Q}_{\mathbf{f}})Tr(\tilde{\mathbf{W}}_{\mathbf{f}}^T \tilde{\mathbf{W}}_{\mathbf{f}})]$$

This implies $\left\| \begin{bmatrix} \mathbf{s} \\ \mathbf{e}_i \end{bmatrix} \right\| \le \sqrt{\dfrac{2V}{\lambda_{\min}(\mathbf{A})}}$. Together with (25), we have the bound for the error signal vector as

$$\left\| \begin{bmatrix} \mathbf{s} \\ \mathbf{e}_i \end{bmatrix} \right\| \le \sqrt{\frac{2V(t_0)}{\lambda_{\min}(\mathbf{A})}} e^{-\frac{\alpha}{2}(t-t_0)} + \frac{1}{\sqrt{2\alpha\lambda_{\min}(\mathbf{A})\lambda_{\min}(\mathbf{Q})}} \sup_{t_0 < \tau < t} \left\| \begin{bmatrix} \varepsilon_1(\tau) \\ \varepsilon_2(\tau) \end{bmatrix} \right\|$$

$$+ \frac{1}{\sqrt{2\alpha\lambda_{\min}(\mathbf{A})}}[\sigma_{\mathbf{D}_x} Tr(\mathbf{W}_{\mathbf{D}_x}^T \mathbf{W}_{\mathbf{D}_x}) + \sigma_{\mathbf{C}_x} Tr(\mathbf{W}_{\mathbf{C}_x}^T \mathbf{W}_{\mathbf{C}_x})$$

$$+ \sigma_{\mathbf{g}_x} Tr(\mathbf{W}_{\mathbf{g}_x}^T \mathbf{W}_{\mathbf{g}_x}) + \sigma_{\mathbf{f}} Tr(\mathbf{W}_{\mathbf{f}}^T \mathbf{W}_{\mathbf{f}})]^{\frac{1}{2}}$$

Remark 7: If the number of basis functions is chosen to be sufficiently large such that $\varepsilon_1 \approx 0$ and $\varepsilon_2 \approx 0$, then (22) becomes

$$\dot{V} = -[\mathbf{s}^T \quad \mathbf{e}_i^T]\mathbf{Q}\begin{bmatrix} \mathbf{s} \\ \mathbf{e}_i \end{bmatrix} \le 0$$

This implies that $\mathbf{s}$ and $\mathbf{e}_i$ are uniformly bounded and square integrable. It is also easy to prove that $\dot{\mathbf{s}}$ and $\dot{\mathbf{e}}_i$ are uniformly bounded; as a result, asymptotic convergence of $\mathbf{s}$ and $\mathbf{e}_i$ can be concluded by Barbalat's lemma. This further implies that $\mathbf{i} \rightarrow \mathbf{i}_d$ and $\mathbf{q} \rightarrow \mathbf{q}_d$, even though the robot model contains uncertainties.

Remark 8: Suppose ε_1 and ε_2 cannot be ignored but their variation bounds are available, i.e. there exists positive constants $\delta_1, \delta_2 > 0$ such that $\| \varepsilon_1 \| \le \delta_1$ and $\| \varepsilon_2 \| \le \delta_2$ for all $t \ge 0$. To cover the effect of these bounded approximation errors, the desired current (13), and the control input (15) are modified to be

$$\mathbf{i}_d = \mathbf{H}^{-1}\mathbf{J}_a^T (\mathbf{F}_{ext} + \hat{\mathbf{g}}_x + \hat{\mathbf{D}}_x \dot{\mathbf{v}} + \hat{\mathbf{C}}_x \mathbf{v} - \mathbf{K}_d \mathbf{s} + \boldsymbol{\tau}_{robust1})$$

$$\mathbf{u} = \hat{\mathbf{f}} - \mathbf{K}_c \mathbf{e}_i + \boldsymbol{\tau}_{robust2}$$

where $\boldsymbol{\tau}_{robust1}$ and $\boldsymbol{\tau}_{robust2}$ are robust terms to be designed. Let us consider the Lyapunov-like function candidate (19) and the update law (21) without σ-modification again. The time derivative of V can be computed as

$$\dot{V} = -[\mathbf{s}^T \quad \mathbf{e}_i^T]\mathbf{Q}\begin{bmatrix} \mathbf{s} \\ \mathbf{e}_i \end{bmatrix} + \delta_1\|\mathbf{s}\| + \delta_2\|\mathbf{e}_i\| + \mathbf{s}^T\boldsymbol{\tau}_{robust1} + \mathbf{e}_i^T\boldsymbol{\tau}_{robust2}$$

By picking $\boldsymbol{\tau}_{robust1} = -\delta_1[\mathrm{sgn}(s_1) \quad \cdots \quad \mathrm{sgn}(s_n)]^T$, where s_i, $i=1,\ldots,n$ is the i-th element of the vector $\mathbf{s}$ and $\boldsymbol{\tau}_{robust2} = -\delta_2[\mathrm{sgn}(e_{i_1}) \quad \cdots \quad \mathrm{sgn}(e_{i_n})]^T$, where e_{i_k}, $k=1,\ldots,n$ is the k-th element of the vector $\mathbf{e}_i$, we may have $\dot{V} \le 0$, and asymptotic convergence of the state error can be concluded by Barbalat's lemma.

Remark 9: Realization of the desired current (13), control law (15) and update laws (21) does not need the information of the regressor matrix, joint accelerations, or time derivatives of the external force, which largely simplified the implementation.

The adaptive impedance control of EDRR is summarized in Table 5.2. Both the regressor-based and regressor-free approaches are listed for comparison. It should be noted that, in the actual implementation, the former needs to know the regressor and its derivative, and the knowledge of the joint acceleration as well as the time derivative of the external force.

Table 5.2 Summary of adaptive impedance control for EDRR

	Electrically driven rigid robot interacting with environment $$\mathbf{D}_x\dot{\mathbf{s}} + \mathbf{C}_x\mathbf{s} + \mathbf{g}_x + \mathbf{D}_x\dot{\mathbf{v}} + \mathbf{C}_x\mathbf{v} = \mathbf{J}_a^{-T}\mathbf{Hi} - \mathbf{F}_{ext}$$ $$\mathbf{Li} + \mathbf{Ri} + \mathbf{K}_b\dot{\mathbf{q}} = \mathbf{u}$$ (5.5-1), (5.5-2)	
	Regressor-based	Regressor-free
Controller	$$\mathbf{i}_d = \mathbf{H}^{-1}\mathbf{J}_a^T(\hat{\mathbf{g}}_x + \hat{\mathbf{D}}_x\dot{\mathbf{v}} + \hat{\mathbf{C}}_x\mathbf{v}$$ $$-\mathbf{K}_d\mathbf{s} + \mathbf{F}_{ext})$$ $$= \mathbf{H}^{-1}\mathbf{J}_a^T[\mathbf{Y}(\mathbf{x},\dot{\mathbf{x}},\mathbf{v},\dot{\mathbf{v}})\hat{\mathbf{p}}_x$$ $$-\mathbf{K}_d\mathbf{s} + \mathbf{F}_{ext}]$$ $$\mathbf{u} = \hat{\mathbf{L}}\mathbf{i}_d + \hat{\mathbf{R}}\mathbf{i} + \hat{\mathbf{K}}_b\dot{\mathbf{q}} - \mathbf{K}_c\mathbf{e}_i$$ $$= \hat{\mathbf{p}}_i^T\boldsymbol{\varphi} - \mathbf{K}_c\mathbf{e}_i$$ (5.5-5), (5.5-7)	$$\mathbf{i}_d = \mathbf{H}^{-1}\mathbf{J}_a^T(\mathbf{F}_{ext} + \hat{\mathbf{g}}_x + \hat{\mathbf{D}}_x\dot{\mathbf{v}}$$ $$+\hat{\mathbf{C}}_x\mathbf{v} - \mathbf{K}_d\mathbf{s})$$ $$\mathbf{u} = \hat{\mathbf{f}} - \mathbf{K}_c\mathbf{e}_i$$ (5.5-13), (5.5-15)
Adaptive Law	$$\dot{\hat{\mathbf{p}}}_x = -\boldsymbol{\Gamma}^{-1}\mathbf{Y}^T\mathbf{s}$$ $$\dot{\hat{\mathbf{p}}}_i = -\boldsymbol{\Gamma}_i^{-1}\boldsymbol{\varphi}\mathbf{e}_i^T$$ (5.5-11)	$$\dot{\hat{\mathbf{W}}}_{\mathbf{D}_x} = -\mathbf{Q}_{\mathbf{D}_x}^{-1}(\mathbf{Z}_{\mathbf{D}_x}\dot{\mathbf{v}}\mathbf{s}^T + \sigma_{\mathbf{D}_x}\hat{\mathbf{W}}_{\mathbf{D}_x})$$ $$\dot{\hat{\mathbf{W}}}_{\mathbf{C}_x} = -\mathbf{Q}_{\mathbf{C}_x}^{-1}(\mathbf{Z}_{\mathbf{C}_x}\mathbf{v}\mathbf{s}^T + \sigma_{\mathbf{C}_x}\hat{\mathbf{W}}_{\mathbf{C}_x})$$ $$\dot{\hat{\mathbf{W}}}_{\mathbf{g}_x} = -\mathbf{Q}_{\mathbf{g}_x}^{-1}(\mathbf{z}_{\mathbf{g}_x}\mathbf{s}^T + \sigma_{\mathbf{g}_x}\hat{\mathbf{W}}_{\mathbf{g}_x})$$ $$\dot{\hat{\mathbf{W}}}_{\mathbf{f}} = -\mathbf{Q}_{\mathbf{f}}^{-1}(\mathbf{z}_{\mathbf{f}}\mathbf{e}_i^T + \sigma_{\mathbf{f}}\hat{\mathbf{W}}_{\mathbf{f}})$$ (5.5-21)
Realization Issue	Need to know the regressor matrix, the derivative of the regressor matrix, the joint accelerations, and the derivative of the external force.	Does not need the information for the regressor matrix, its derivative, the joint accelerations, or the derivative of the external force.

Example 5.2:

Consider the same 2-DOF planar robot in example 5.1 with the inclusion of the actuator dynamics, and we would like to verify the controller developed in this section by computer simulations. Actual values of link parameters are selected as $m_1 = m_2 = 0.5(kg)$, $l_1 = l_2 = 0.75(m)$, $l_{c1} = l_{c2} = 0.375(m)$, and $I_1 = I_2 = 0.0234(kg\text{-}m^2)$. Parameters related to the actuator dynamics are the same with those used in chapter 4 and are given with $h_1 = h_2 = 10(N\text{-}m/A)$, $L_1 = L_2 = 0.025(H)$, $r_1 = r_2 = 1(\Omega)$, and $k_{b1} = k_{b2} = 1(Vol/rad/sec)$. In order to observe the effect of the actuator dynamics, the endpoint is required to track a $0.2m$ radius circle centered at

$(0.8m, 1.0m)$ in 2 seconds which is much faster than the case in example 5.1. The initial conditions of the generalized coordinate vector is $\mathbf{q}(0) = [0.0022 \quad 1.5019 \quad 0 \quad 0]^T$, i.e., the endpoint is still at $(0.8m, 0,75m)$ initially. The controller gain matrices are selected as

$$\mathbf{K}_d = \begin{bmatrix} 50 & 0 \\ 0 & 50 \end{bmatrix}, \; \mathbf{\Lambda} = \begin{bmatrix} 20 & 0 \\ 0 & 20 \end{bmatrix} \text{ and } \mathbf{K}_c = \begin{bmatrix} 100 & 0 \\ 0 & 100 \end{bmatrix}.$$

The initial value for the desired current can be found by calculation as

$$\mathbf{i}_d(0) = \mathbf{i}(0) = [0.8 \quad 0.1]^T.$$

The matrices in the target impedance are picked as

$$\mathbf{M}_i = \begin{bmatrix} 0.5 & 0 \\ 0 & 0.5 \end{bmatrix}, \; \mathbf{B}_i = \begin{bmatrix} 100 & 0 \\ 0 & 100 \end{bmatrix} \text{ and } \mathbf{K}_i = \begin{bmatrix} 1500 & 0 \\ 0 & 1500 \end{bmatrix}.$$

The 11-term Fourier series is selected as the basis function for the approximation so that $\hat{\mathbf{W}}_{\mathbf{D}_x}$ and $\hat{\mathbf{W}}_{\mathbf{C}_x}$ are in $\mathfrak{R}^{44 \times 2}$, while $\hat{\mathbf{W}}_{\mathbf{g}_x}$ and $\hat{\mathbf{W}}_{\mathbf{f}}$ are in $\mathfrak{R}^{22 \times 2}$. The initial weighting vectors for the entries are assigned to be

$$\hat{\mathbf{w}}_{D_{x11}}(0) = [0.05 \quad 0 \quad \cdots \quad 0]^T \in \mathfrak{R}^{11 \times 1}$$

$$\hat{\mathbf{w}}_{D_{x12}}(0) = \hat{\mathbf{w}}_{D_{x21}}(0) = [-0.05 \quad 0 \quad \cdots \quad 0]^T \in \mathfrak{R}^{11 \times 1}$$

$$\hat{\mathbf{w}}_{D_{x22}}(0) = [0.1 \quad 0 \quad \cdots \quad 0]^T \in \mathfrak{R}^{11 \times 1}$$

$$\hat{\mathbf{w}}_{C_{x11}}(0) = [0.05 \quad 0 \quad \cdots \quad 0]^T \in \mathfrak{R}^{11 \times 1}$$

$$\hat{\mathbf{w}}_{C_{x12}}(0) = \hat{\mathbf{w}}_{C_{x21}}(0) = [-0.05 \quad 0 \quad \cdots \quad 0]^T \in \mathfrak{R}^{11 \times 1}$$

$$\hat{\mathbf{w}}_{C_{x22}}(0) = [0.1 \quad 0 \quad \cdots \quad 0]^T \in \mathfrak{R}^{11 \times 1}$$

$$\hat{\mathbf{w}}_{g_{x1}}(0) = \hat{\mathbf{w}}_{g_{x2}}(0) = [0 \quad 0 \quad \cdots \quad 0]^T \in \mathfrak{R}^{11 \times 1}$$

$$\hat{\mathbf{w}}_{f_1}(0) = \hat{\mathbf{w}}_{f_2}(0) = [0 \quad 0 \quad \cdots \quad 0]^T \in \mathfrak{R}^{11 \times 1}$$

The gain matrices in the update laws are selected as

$$\mathbf{Q}_{\mathbf{D}_x}^{-1} = 0.1\mathbf{I}_{44}, \ \mathbf{Q}_{\mathbf{C}_x}^{-1} = 0.1\mathbf{I}_{44}, \ \mathbf{Q}_{\mathbf{g}_x}^{-1} = 50\mathbf{I}_{22} \ \text{and} \ \mathbf{Q}_{\mathbf{f}}^{-1} = 10000\mathbf{I}_{22}.$$

The approximation error is also assumed to be neglected, and the σ-modification parameters are all zero. The simulation results are shown in Figure 5.8 to 5.16. Figure 5.8 shows the tracking performance of the robot endpoint and its desired trajectory in the Cartesian space. It is observed that the endpoint trajectory converges smoothly to the desired trajectory in the free space tracking and contacts compliantly in the constrained motion phase. Although the initial error is quite large, the transient state takes only about 0.2 seconds which can be justified from the joint space tracking history in Figure 5.9. The performance in the current tracking loop is very good as shown in Figure 5.10. The control efforts to the two joints are reasonable that are presented in Figure 5.11. Figure 4.12 shows the time histories of the external forces. Figure 4.13 to 4.16 are the performance of function approximation. Although most parameters do not converge to their actual values, they still remain bounded as desired.

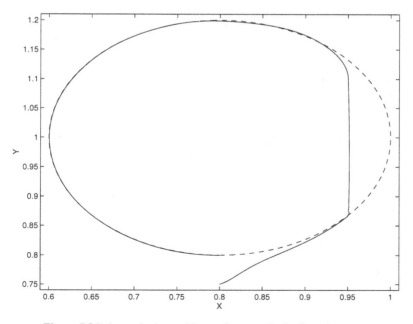

Figure 5.8 Robot endpoint tracking performance in the Cartesian space

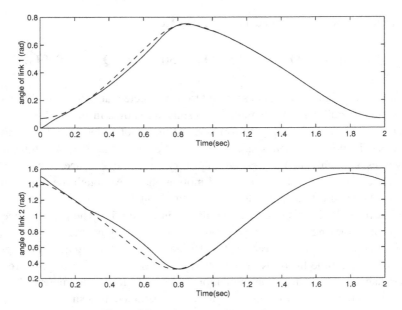

Figure 5.9 Joint space tracking performance

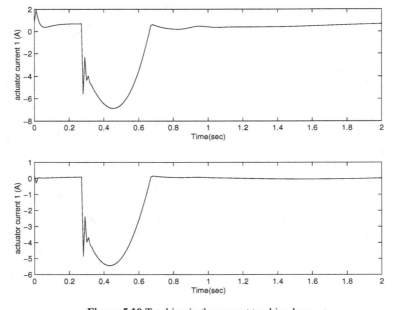

Figure 5.10 Tracking in the current tracking loop

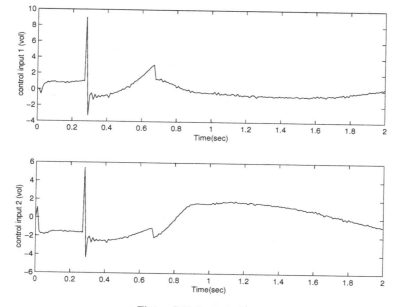

Figure 5.11 Control efforts

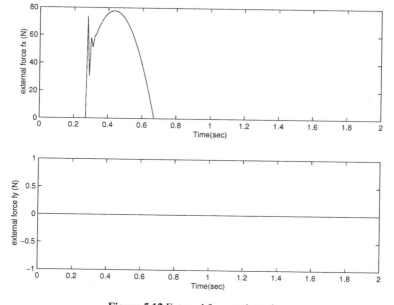

Figure 5.12 External force trajectories

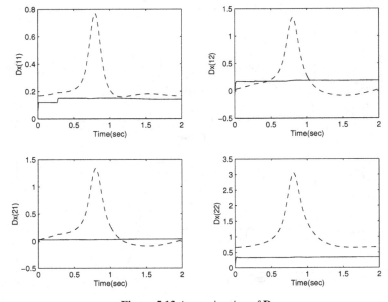

Figure 5.13 Approximation of $\mathbf{D}_x$

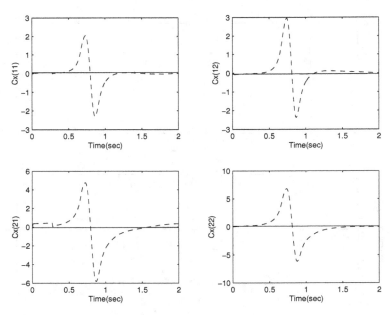

Figure 5.14 Approximation of $\mathbf{C}_x$

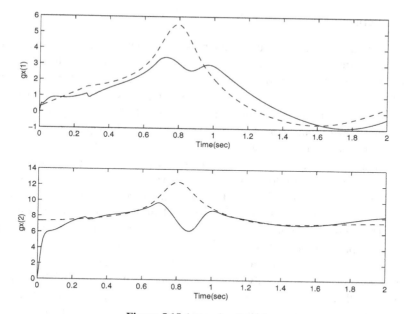

Figure 5.15 Approximation of $\mathbf{g}_x$

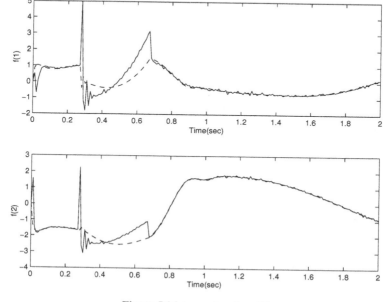

Figure 5.16 Approximation of $\mathbf{f}$

5.6 Conclusions

Compliant interaction between the robot and environment is very important in the industrial applications. In this chapter, we consider the adaptive impedance control of rigid robots where the impedance control enables the robot to have good performance in the free space tracking phase and to behave like the target impedance in the constrained motion phase. In Section 5.2, a regressor-based impedance controller is derived, but the closed loop system may not converge to the target impedance even when all parameters converge to their actual values. In Section 5.3, an adaptive impedance control is constructed by following the design introduced in Section 4.3. However, implementation of this controller requires the knowledge of not only the regressor matrix and its time derivative, but also the joint accelerations and time derivative of the external force. Therefore, it is not feasible for practical applications. A regressor-free adaptive impedance controller is thus designed in Section 5.4 which does not need the availability of the additional information required in the previous section. Finally, the actuator dynamics is considered in Section 5.5. Simulation cases show that the robot can be operated at a much higher speed with good performance for both the free space tracking and compliant motion control.

Chapter 6

Adaptive Control of Flexible-Joint Robots

6.1 Introduction

Most controllers for industrial robots are designed based on the rigid robot assumption. Consideration of the joint flexibility in the controller design is one of the approaches to increase the control performance. For a robot with n links, we need to use $2n$ generalized coordinates to describe its whole dynamic behavior when taking the joint flexibility into account. Therefore, the modeling of the flexible joint robot is far more complex than that of the rigid robot. Since the mathematical model is only an approximation of the real system, the simplified representation of the system behavior will contain model inaccuracies such as parametric uncertainties, unmodeled dynamics and external disturbances. Because these inaccuracies may degrade the performance of the closed-loop system, any practical design should consider their effects. The inherent highly nonlinear coupling and model inaccuracies make the controller design for a flexible joint robot extremely difficult. Spong (1989, 1995) proposed one of the first adaptive controllers for flexible joint robots based on the singular perturbation formulation of the robot dynamics. A simple composite controller was designed such that the joint elastic forces were stabilized by a fast feedback control and link variables were controlled by a slow control law. Ge (1996) suggested a robust adaptive controller based on a new singular perturbation model where the motor tracking error was modeled as the fast variables instead of the joint elastic forces. This innovative approach leads to a controller more robust to the case of load sensor failure, gives a new insight into the control design problem of flexible joint robots and presents an alternative singular perturbation model for controller design. Ott et. al. (2000) verified a singular perturbation based adaptive controller for flexible joint robots experimentally. The tracking quality was improved significantly by the use of the adaptive control law compared to the non-adaptive one. Dixon et. al. (1999) proposed an adaptive partial state feedback controller for flexible joint robots based on the backstepping design with the knowledge of the regression matrix. A backstepping design based output feedback adaptive controller for flexible joint robot was suggested in Yim (2001). Kozlowski and Sauer (1999a, 1999b) suggested an adaptive controller to have semi-global convergence to an arbitrarily small neighborhood of the equilibrium point in the presence of bounded disturbances. Like other adaptive strategies, the uncertain parameters are required to be time-invariant and capable of being collected to form a parameter vector. Huang and Chen (2004a) proposed an adaptive backstepping-like controller based on FAT for single-link flexible-joint robots with mismatched uncertainties. Similar to most backstepping designs, the derivation is too complex to robots with more joints. Chien and Huang (2006b) suggested a FAT-based adaptive controller for general flexible-joint robots without requiring the computation of the regressor matrix. Chien and Huang (2006a) designed an adaptive impedance controller for the flexible-joint robots to give good performance in both the free space tracking and compliant motion phase. Chien and Huang (2007a) included the actuator into consideration in the design of an adaptive controller for flexible-joint robots.

In this chapter, we would like to study the FAT-based adaptive controller designs for n-link flexible-joint robots. The tedious computation of the regressor matrix is avoided. This chapter is organized as following: in Section 6.2, we consider the control of a known flexible-joint robot. Section 6.3 derives the regressor-based adaptive controller and Section 6.4 presents regressor-free adaptive controller. Section 6.5 considers the actuator dynamics.

6.2 Control of Known Flexible-Joint Robots

The rigid-link flexible-joint robot considered in this chapter is shown in (3.6-1) as

$$\mathbf{D}(\mathbf{q})\ddot{\mathbf{q}} + \mathbf{C}(\mathbf{q},\dot{\mathbf{q}})\dot{\mathbf{q}} + \mathbf{g}(\mathbf{q}) = \mathbf{K}(\boldsymbol{\theta} - \mathbf{q}) \tag{1a}$$

$$\mathbf{J}\ddot{\boldsymbol{\theta}} + \mathbf{B}\dot{\boldsymbol{\theta}} + \mathbf{K}(\boldsymbol{\theta} - \mathbf{q}) = \boldsymbol{\tau}_a \tag{1b}$$

Define a transmission torque $\boldsymbol{\tau}_t = \mathbf{K}(\boldsymbol{\theta} - \mathbf{q})$ (Spong 1987, Lin and Goldenberg 1995), then (1) can be rewritten to be

$$\mathbf{D}\ddot{\mathbf{q}} + \mathbf{C}\dot{\mathbf{q}} + \mathbf{g} = \boldsymbol{\tau}_t \tag{2a}$$

$$\mathbf{J}_t\ddot{\boldsymbol{\tau}}_t + \mathbf{B}_t\dot{\boldsymbol{\tau}}_t + \boldsymbol{\tau}_t = \boldsymbol{\tau}_a - \overline{\mathbf{q}}(\dot{\mathbf{q}},\ddot{\mathbf{q}}) \tag{2b}$$

where $\mathbf{J}_t = \mathbf{JK}^{-1}$, $\mathbf{B}_t = \mathbf{BK}^{-1}$, and $\overline{\mathbf{q}}(\dot{\mathbf{q}}, \ddot{\mathbf{q}}) = \mathbf{J}\ddot{\mathbf{q}} + \mathbf{B}\dot{\mathbf{q}}$. Since (2) is in a cascade form connected by the torque $\boldsymbol{\tau}_t$, a backstepping-like design procedure can be employed by regarding $\boldsymbol{\tau}_t$ as a control signal to (2a), and a desired trajectory $\boldsymbol{\tau}_{td}$ is designed for the convergence of $\mathbf{q}$. If a proper $\boldsymbol{\tau}_a$ can be constructed such that $\boldsymbol{\tau}_t \to \boldsymbol{\tau}_{td}$, then we may have $\mathbf{q} \to \mathbf{q}_d$. Define the tracking error $\mathbf{e} = \mathbf{q} - \mathbf{q}_d$ and the error signals $\mathbf{s} = \dot{\mathbf{e}} + \boldsymbol{\Lambda}\mathbf{e}$ and $\mathbf{v} = \dot{\mathbf{q}}_d - \boldsymbol{\Lambda}\mathbf{e}$. Equation (2a) becomes

$$\mathbf{D}\dot{\mathbf{s}} + \mathbf{Cs} + \mathbf{g} + \mathbf{D}\dot{\mathbf{v}} + \mathbf{Cv} = \boldsymbol{\tau}_t \qquad (3)$$

Suppose that all parameters in the system model are available, then the desired torque $\boldsymbol{\tau}_{td}$ can be defined as

$$\boldsymbol{\tau}_{td} = \mathbf{g} + \mathbf{D}\dot{\mathbf{v}} + \mathbf{Cv} - \mathbf{K}_d\mathbf{s} \qquad (4)$$

With this desired torque, (3) gives the output tracking dynamics

$$\mathbf{D}\dot{\mathbf{s}} + \mathbf{Cs} + \mathbf{K}_d\mathbf{s} = \boldsymbol{\tau}_t - \boldsymbol{\tau}_{td} \qquad (5a)$$

If a control torque $\boldsymbol{\tau}_a$ can be designed to have $\boldsymbol{\tau}_t \to \boldsymbol{\tau}_{td}$, then (5a) becomes

$$\mathbf{D}\dot{\mathbf{s}} + \mathbf{Cs} + \mathbf{K}_d\mathbf{s} = 0 \qquad (5b)$$

and convergence of the output error follows. To this end, we would like to employ the model reference control (MRC) rule below. Let us consider a reference model

$$\mathbf{J}_r\ddot{\boldsymbol{\tau}}_r + \mathbf{B}_r\dot{\boldsymbol{\tau}}_r + \mathbf{K}_r\boldsymbol{\tau}_r = \mathbf{J}_r\ddot{\boldsymbol{\tau}}_{td} + \mathbf{B}_r\dot{\boldsymbol{\tau}}_{td} + \mathbf{K}_r\boldsymbol{\tau}_{td} \qquad (6)$$

where $\boldsymbol{\tau}_r \in \Re^n$ is the state vector, and matrices $\mathbf{J}_r \in \Re^{n \times n}$, $\mathbf{B}_r \in \Re^{n \times n}$, and $\mathbf{K}_r \in \Re^{n \times n}$ are selected to give proper dynamics for the convergence of $\boldsymbol{\tau}_r$ to $\boldsymbol{\tau}_{td}$. Define $\overline{\boldsymbol{\tau}}_{td}(\dot{\boldsymbol{\tau}}_{td}, \ddot{\boldsymbol{\tau}}_{td}) = \mathbf{K}_r^{-1}(\mathbf{B}_r\dot{\boldsymbol{\tau}}_{td} + \mathbf{J}_r\ddot{\boldsymbol{\tau}}_{td})$, and rewrite (2b) and (6) into the state space form

$$\dot{\mathbf{x}}_p = \mathbf{A}_p\mathbf{x}_p + \mathbf{B}_p\boldsymbol{\tau}_a - \mathbf{B}_p\overline{\mathbf{q}} \qquad (7a)$$

$$\dot{\mathbf{x}}_m = \mathbf{A}_m\mathbf{x}_m + \mathbf{B}_m(\boldsymbol{\tau}_{td} + \overline{\boldsymbol{\tau}}_{td}) \qquad (7b)$$

where $\mathbf{x}_p = [\boldsymbol{\tau}_t^T \quad \dot{\boldsymbol{\tau}}_t^T]^T \in \mathfrak{R}^{2n}$ and $\mathbf{x}_m = [\boldsymbol{\tau}_r^T \quad \dot{\boldsymbol{\tau}}_r^T]^T \in \mathfrak{R}^{2n}$ are augmented state

vectors. $\mathbf{A}_p = \begin{bmatrix} \mathbf{0} & \mathbf{I}_{n \times n} \\ -\mathbf{J}_t^{-1} & -\mathbf{J}_t^{-1}\mathbf{B}_t \end{bmatrix} \in \mathfrak{R}^{2n \times 2n}$ and $\mathbf{A}_m = \begin{bmatrix} \mathbf{0} & \mathbf{I}_{n \times n} \\ -\mathbf{J}_r^{-1}\mathbf{K}_r & -\mathbf{J}_r^{-1}\mathbf{B}_r \end{bmatrix} \in \mathfrak{R}^{2n \times 2n}$

are augmented system matrices, and $\mathbf{B}_p = \begin{bmatrix} \mathbf{0} \\ \mathbf{J}_t^{-1} \end{bmatrix} \in \mathfrak{R}^{2n \times n}$ and

$\mathbf{B}_m = \begin{bmatrix} \mathbf{0} \\ \mathbf{J}_r^{-1}\mathbf{K}_r \end{bmatrix} \in \mathfrak{R}^{2n \times n}$ are augmented input gain matrices. Let $\boldsymbol{\tau}_t = \mathbf{C}_p \mathbf{x}_p$

and $\boldsymbol{\tau}_r = \mathbf{C}_m \mathbf{x}_m$ be respectively the output signal vector for (7a) and (7b), where $\mathbf{C}_p = \mathbf{C}_m = [\mathbf{I}_{n \times n} \quad \mathbf{0}] \in \mathfrak{R}^{n \times 2n}$ are augmented output signal matrices. It is noted that $(\mathbf{A}_m, \mathbf{B}_m)$ is controllable, $(\mathbf{A}_m, \mathbf{C}_m)$ is observable, and the transfer function $\mathbf{C}_m(s\mathbf{I} - \mathbf{A}_m)^{-1}\mathbf{B}_m$ is strictly positive real. According to the MRC rule, the control torque $\boldsymbol{\tau}_a$ is selected as

$$\boldsymbol{\tau}_a = \mathbf{\Theta}\mathbf{x}_p + \mathbf{\Phi}\boldsymbol{\tau}_{td} + \mathbf{h}(\overline{\boldsymbol{\tau}}_{td}, \overline{\mathbf{q}}) \tag{8}$$

where $\mathbf{\Theta} \in \mathfrak{R}^{n \times 2n}$ and $\mathbf{\Phi} \in \mathfrak{R}^{n \times n}$ are matrices satisfying relations $\mathbf{A}_p + \mathbf{B}_p\mathbf{\Theta} = \mathbf{A}_m$ and $\mathbf{B}_p\mathbf{\Phi} = \mathbf{B}_m$, respectively, and $\mathbf{h}(\overline{\boldsymbol{\tau}}_{td}, \overline{\mathbf{q}}) = \mathbf{\Phi}\overline{\boldsymbol{\tau}}_{td} + \overline{\mathbf{q}} \in \mathfrak{R}^n$. Substitute (8) into (7a) to have

$$\dot{\mathbf{x}}_p = \mathbf{A}_m\mathbf{x}_p + \mathbf{B}_m(\boldsymbol{\tau}_{td} + \overline{\boldsymbol{\tau}}_{td}) \tag{9}$$

Define $\mathbf{e}_m = \mathbf{x}_p - \mathbf{x}_m$ and $\mathbf{e}_\tau = \boldsymbol{\tau}_t - \boldsymbol{\tau}_r$ be error vectors, then from (7b) and (9) we may have the *torque tracking loop* dynamics

$$\dot{\mathbf{e}}_m = \mathbf{A}_m\mathbf{e}_m \tag{10a}$$

$$\mathbf{e}_\tau = \mathbf{C}_m\mathbf{e}_m \tag{10b}$$

To prove stability in the output tracking loop (5a) and the torque tracking loop (10), a Lyapunov-like function candidate is designed as

$$V(\mathbf{s}, \mathbf{e}_m) = \frac{1}{2}\mathbf{s}^T\mathbf{D}\mathbf{s} + \mathbf{e}_m^T\mathbf{P}_t\mathbf{e}_m \tag{11}$$

where $\mathbf{P}_t = \mathbf{P}_t^T \in \mathfrak{R}^{2n \times 2n}$ is a positive definite matrix satisfying the Lyapunov equation $\mathbf{A}_m^T\mathbf{P}_t + \mathbf{P}_t\mathbf{A}_m = -\mathbf{C}_m^T\mathbf{C}_m$. Along the trajectory (5a) and (10), the time derivative of V is computed as

$$\dot{V} = -[\mathbf{s}^T \quad \mathbf{e}_\tau^T]\mathbf{Q}\begin{bmatrix} \mathbf{s} \\ \mathbf{e}_\tau \end{bmatrix} \le 0 \tag{12}$$

where $\mathbf{Q} = \begin{bmatrix} \mathbf{K}_d & -\dfrac{1}{2}\mathbf{I}_{n\times n} \\ -\dfrac{1}{2}\mathbf{I}_{n\times n} & \mathbf{I}_{n\times n} \end{bmatrix}$ is positive definite by proper selection of $\mathbf{K}_d$.

Equation (12) implies uniform boundedness and square integrability of $\mathbf{s}$ and $\mathbf{e}_\tau$. Uniform boundedness of $\dot{\mathbf{s}}$ and $\dot{\mathbf{e}}_\tau$ can also be proved easily; therefore, we may conclude $\mathbf{q} \to \mathbf{q}_d$ and $\boldsymbol{\tau}_t \to \boldsymbol{\tau}_{td}$ as $t \to \infty$ by Barbalat's lemma.

Remark 1: Dependence of $\mathbf{h}(\overline{\boldsymbol{\tau}}_{td}, \overline{\mathbf{q}})$ in (8) implies the requirements for the knowledge of joint accelerations or even their higher time derivatives. This greatly restricts the application of the strategy presented here. In next section, a regressor-based adaptive controller will be derived. However, its realization will still be limited due to its dependence on high-order state variable feedbacks. Finally, a regressor-free adaptive control is developed in section 6.4 to ease its realization.

6.3 Regressor-Based Adaptive Control of Flexible-Joint Robots

Let us consider the system described in (6.2-3) and (6.2-2b) as

$$\mathbf{D}\dot{\mathbf{s}} + \mathbf{C}\mathbf{s} + \mathbf{g} + \mathbf{D}\dot{\mathbf{v}} + \mathbf{C}\mathbf{v} = \boldsymbol{\tau}_t \tag{1a}$$

$$\mathbf{J}_t\ddot{\boldsymbol{\tau}}_t + \mathbf{B}_t\dot{\boldsymbol{\tau}}_t + \boldsymbol{\tau}_t = \boldsymbol{\tau}_a - \overline{\mathbf{q}}(\dot{\mathbf{q}}, \ddot{\mathbf{q}}) \tag{1b}$$

where $\mathbf{D}$, $\mathbf{C}$ and $\mathbf{g}$ are assumed to be unavailable here. Hence, (6.2-4) and (6.2-8) are not feasible. A new version of the transmission torque is designed as

$$\begin{aligned} \boldsymbol{\tau}_{td} &= \hat{\mathbf{g}} + \hat{\mathbf{D}}\dot{\mathbf{v}} + \hat{\mathbf{C}}\mathbf{v} - \mathbf{K}_d\mathbf{s} \\ &= \mathbf{Y}(\mathbf{q}, \dot{\mathbf{q}}, \mathbf{v}, \dot{\mathbf{v}})\hat{\mathbf{p}} - \mathbf{K}_d\mathbf{s} \end{aligned} \tag{2}$$

The dynamics of the output tracking loop can then be obtained by plugging (2) into (1a) as

$$\begin{aligned} \mathbf{D}\dot{\mathbf{s}} + \mathbf{C}\mathbf{s} + \mathbf{K}_d\mathbf{s} &= -\tilde{\mathbf{D}}\dot{\mathbf{v}} - \tilde{\mathbf{C}}\mathbf{v} - \tilde{\mathbf{g}} + (\boldsymbol{\tau}_t - \boldsymbol{\tau}_{td}) \\ &= -\mathbf{Y}(\mathbf{q}, \dot{\mathbf{q}}, \mathbf{v}, \dot{\mathbf{v}})\tilde{\mathbf{p}} + (\boldsymbol{\tau}_t - \boldsymbol{\tau}_{td}) \end{aligned} \tag{3}$$

Therefore, if an effective control torque $\boldsymbol{\tau}_a$ can be designed to have $\boldsymbol{\tau}_t \to \boldsymbol{\tau}_{td}$, and a proper update law can be selected so that $\hat{\mathbf{p}} \to \mathbf{p}$, then (3) implies (6.2-5b). The dynamics for the torque tracking loop is exactly the same as those in the previous section with the control law in (6.2-8). Therefore, we may select a Lyapunov-like function candidate

$$V(\mathbf{s}, \mathbf{e}_m, \tilde{\mathbf{p}}) = \frac{1}{2}\mathbf{s}^T \mathbf{Ds} + \mathbf{e}_m^T \mathbf{P}_t \mathbf{e}_m + \frac{1}{2}\tilde{\mathbf{p}}^T \boldsymbol{\Gamma} \tilde{\mathbf{p}} \tag{4}$$

where $\mathbf{P}_t = \mathbf{P}_t^T \in \Re^{2n \times 2n}$ is a positive definite matrix satisfying the Lyapunov equation $\mathbf{A}_m^T \mathbf{P}_t + \mathbf{P}_t \mathbf{A}_m = -\mathbf{C}_m^T \mathbf{C}_m$, and $\boldsymbol{\Gamma} \in \Re^{r \times r}$ is positive definite. Along the trajectory of (3) and (6.2-10), the time derivative of (4) becomes

$$\dot{V} = -\mathbf{s}^T \mathbf{K}_d \mathbf{s} + \mathbf{s}^T \mathbf{e}_\tau - \mathbf{e}_\tau^T \mathbf{e}_\tau - \tilde{\mathbf{p}}^T (\boldsymbol{\Gamma} \dot{\hat{\mathbf{p}}} + \mathbf{Y}^T \mathbf{s}) \tag{5}$$

The update law is picked to be

$$\dot{\hat{\mathbf{p}}} = -\boldsymbol{\Gamma}^{-1} \mathbf{Y}^T \mathbf{s} \tag{6}$$

Hence, (5) becomes exactly (6.2-12), and all stability properties are the same there.

Remark 2: To implement the strategy designed in this section, we do not need the knowledge of $\mathbf{D}$, $\mathbf{C}$ and $\mathbf{g}$. However, it still needs the feedback of joint accelerations and their higher order time derivatives. In addition, computation of the regressor matrix and its time derivatives are necessary here.

6.4 FAT-Based Adaptive Control of Flexible-Joint Robots

Let us consider the system described in (6.3-1) as

$$\mathbf{D}\dot{\mathbf{s}} + \mathbf{Cs} + \mathbf{g} + \mathbf{D}\dot{\mathbf{v}} + \mathbf{Cv} = \boldsymbol{\tau}_t \tag{1a}$$

$$\mathbf{J}_t \ddot{\boldsymbol{\tau}}_t + \mathbf{B}_t \dot{\boldsymbol{\tau}}_t + \boldsymbol{\tau}_t = \boldsymbol{\tau}_a - \overline{\mathbf{q}}(\dot{\mathbf{q}}, \ddot{\mathbf{q}}) \tag{1b}$$

where $\mathbf{D}(\mathbf{q})$, $\mathbf{C}(\mathbf{q}, \dot{\mathbf{q}})$ and $\mathbf{g}(\mathbf{q})$ are not available and their variation bounds are not given. We would like to design a desired transmission torque $\boldsymbol{\tau}_{td}$ so that a proper control torque $\boldsymbol{\tau}_a$ can be constructed to have convergence in the torque tracking loop, i.e., $\boldsymbol{\tau}_t \to \boldsymbol{\tau}_{td}$. Since $\mathbf{D}(\mathbf{q})$, $\mathbf{C}(\mathbf{q}, \dot{\mathbf{q}})$ and $\mathbf{g}(\mathbf{q})$ are not available, the traditional adaptive control and robust control are not easy to be applied here. In the following, we would like to use the function approximation technique to design an adaptive controller for the rigid-link flexible-joint robot without the knowledge of the regressor matrix. Since the adaptive control of flexible-joint robots is much more difficult than that for its rigid-joint counterpart, avoidance of the regressor computation in the FAT-based design largely simplifies the implementation in the real-time environment.

The desired transmission torque τ_d can be designed as

$$\boldsymbol{\tau}_{td} = \hat{\mathbf{g}} + \hat{\mathbf{D}}\dot{\mathbf{v}} + \hat{\mathbf{C}}\mathbf{v} - \mathbf{K}_d\mathbf{s} \tag{2}$$

where $\hat{\mathbf{D}}$, $\hat{\mathbf{C}}$ and $\hat{\mathbf{g}}$ are estimates of $\mathbf{D}(\mathbf{q})$, $\mathbf{C}(\mathbf{q},\dot{\mathbf{q}})$ and $\mathbf{g}(\mathbf{q})$, respectively. Substituting (2) into (1a), we may have the dynamics for the output tracking loop

$$\mathbf{D}\dot{\mathbf{s}} + \mathbf{C}\mathbf{s} + \mathbf{K}_d\mathbf{s} = -\tilde{\mathbf{D}}\dot{\mathbf{v}} - \tilde{\mathbf{C}}\mathbf{v} - \tilde{\mathbf{g}} + (\boldsymbol{\tau}_t - \boldsymbol{\tau}_{td}) \tag{3}$$

where $\tilde{\mathbf{D}} = \mathbf{D} - \hat{\mathbf{D}}$, $\tilde{\mathbf{C}} = \mathbf{C} - \hat{\mathbf{C}}$, and $\tilde{\mathbf{g}} = \mathbf{g} - \hat{\mathbf{g}}$. If a proper controller $\boldsymbol{\tau}_a$ and update laws for $\hat{\mathbf{D}}$, $\hat{\mathbf{C}}$ and $\hat{\mathbf{g}}$ can be designed, we may have $\boldsymbol{\tau}_t \rightarrow \boldsymbol{\tau}_{td}$, $\hat{\mathbf{D}} \rightarrow \mathbf{D}$, $\hat{\mathbf{C}} \rightarrow \mathbf{C}$ and $\hat{\mathbf{g}} \rightarrow \mathbf{g}$ so that (3) can give desired performance. Hence, we would like to derive the dynamics for the torque tracking loop next. To this end, the same MRC scheme used in Section 6.2 is employed here. Consider the reference model in (6.2-6) again

$$\mathbf{J}_r\ddot{\boldsymbol{\tau}}_r + \mathbf{B}_r\dot{\boldsymbol{\tau}}_r + \mathbf{K}_r\boldsymbol{\tau}_r = \mathbf{J}_r\ddot{\boldsymbol{\tau}}_{td} + \mathbf{B}_r\dot{\boldsymbol{\tau}}_{td} + \mathbf{K}_r\boldsymbol{\tau}_{td} \tag{4}$$

With the definition $\overline{\boldsymbol{\tau}}_{td}(\dot{\boldsymbol{\tau}}_{td}, \ddot{\boldsymbol{\tau}}_{td}) = \mathbf{K}_r^{-1}(\mathbf{B}_r\dot{\boldsymbol{\tau}}_{td} + \mathbf{J}_r\ddot{\boldsymbol{\tau}}_{td})$, we may represent (1b) and (4) into the state space representation as

$$\dot{\mathbf{x}}_p = \mathbf{A}_p\mathbf{x}_p + \mathbf{B}_p\boldsymbol{\tau}_a - \mathbf{B}_p\overline{\mathbf{q}} \tag{5a}$$

$$\dot{\mathbf{x}}_m = \mathbf{A}_m\mathbf{x}_m + \mathbf{B}_m(\boldsymbol{\tau}_{td} + \overline{\boldsymbol{\tau}}_{td}) \tag{5b}$$

where $\mathbf{x}_p$, $\mathbf{x}_m$, $\mathbf{A}_p$, $\mathbf{A}_m$, $\mathbf{B}_p$, and $\mathbf{B}_m$ are defined in Section 6.2. The pair $(\mathbf{A}_m, \mathbf{B}_m)$ is controllable, $(\mathbf{A}_m, \mathbf{C}_m)$ is observable, and the transfer function $\mathbf{C}_m(s\mathbf{I} - \mathbf{A}_m)^{-1}\mathbf{B}_m$ is SPR, where $\mathbf{C}_m$ is also defined in Section 6.2. Since system (1) contains uncertainties, the control torque in (6.2-8) is not feasible. A new one is constructed as

$$\boldsymbol{\tau}_a = \boldsymbol{\Theta}\mathbf{x}_p + \boldsymbol{\Phi}\boldsymbol{\tau}_{td} + \hat{\mathbf{h}} \tag{6}$$

where $\hat{\mathbf{h}}$ is the estimate of $\mathbf{h}(\bar{\boldsymbol{\tau}}_{td},\bar{\mathbf{q}}) = \boldsymbol{\Phi}\bar{\boldsymbol{\tau}}_{td} + \bar{\mathbf{q}}$, $\boldsymbol{\Theta} \in \mathfrak{R}^{n \times 2n}$ and $\boldsymbol{\Phi} \in \mathfrak{R}^{n \times n}$ are matrices satisfying $\mathbf{A}_p + \mathbf{B}_p\boldsymbol{\Theta} = \mathbf{A}_m$ and $\mathbf{B}_p\boldsymbol{\Phi} = \mathbf{B}_m$ respectively. Plugging (6) into (5a), we have the dynamics

$$\dot{\mathbf{x}}_p = \mathbf{A}_m\mathbf{x}_p + \mathbf{B}_m(\boldsymbol{\tau}_{td} + \bar{\boldsymbol{\tau}}_{td}) + \mathbf{B}_p(\hat{\mathbf{h}} - \mathbf{h}) \tag{7}$$

With the definition of $\mathbf{e}_m = \mathbf{x}_p - \mathbf{x}_m$ and $\mathbf{e}_\tau = \boldsymbol{\tau}_t - \boldsymbol{\tau}_r$, we may have the dynamics in the torque tracking loop

$$\dot{\mathbf{e}}_m = \mathbf{A}_m\mathbf{e}_m + \mathbf{B}_p(\hat{\mathbf{h}} - \mathbf{h}) \tag{8a}$$

$$\mathbf{e}_\tau = \mathbf{C}_m\mathbf{e}_m \tag{8b}$$

If we may design an appropriate update law such that $\hat{\mathbf{h}} \to \mathbf{h}$, then (8) implies $\mathbf{e}_m \to 0$ as $t \to \infty$. To proceed further, let us apply the function approximation representation

$$\mathbf{D} = \mathbf{W}_\mathbf{D}^T\mathbf{Z}_\mathbf{D} + \boldsymbol{\varepsilon}_\mathbf{D}$$
$$\mathbf{C} = \mathbf{W}_\mathbf{C}^T\mathbf{Z}_\mathbf{C} + \boldsymbol{\varepsilon}_\mathbf{C}$$
$$\mathbf{g} = \mathbf{W}_\mathbf{g}^T\mathbf{z}_\mathbf{g} + \boldsymbol{\varepsilon}_\mathbf{g} \tag{9}$$
$$\mathbf{h} = \mathbf{W}_\mathbf{h}^T\mathbf{z}_\mathbf{h} + \boldsymbol{\varepsilon}_\mathbf{h}$$

where $\mathbf{W}_\mathbf{D} \in \mathfrak{R}^{n^2\beta_D \times n}$, $\mathbf{W}_\mathbf{C} \in \mathfrak{R}^{n^2\beta_C \times n}$, $\mathbf{W}_\mathbf{g} \in \mathfrak{R}^{n\beta_g \times n}$, and $\mathbf{W}_\mathbf{h} \in \mathfrak{R}^{n\beta_h \times n}$ are weighting matrices, $\mathbf{Z}_\mathbf{D} \in \mathfrak{R}^{n^2\beta_D \times n}$, $\mathbf{Z}_\mathbf{C} \in \mathfrak{R}^{n^2\beta_C \times n}$, $\mathbf{z}_\mathbf{g} \in \mathfrak{R}^{n\beta_g \times 1}$, and $\mathbf{z}_\mathbf{h} \in \mathfrak{R}^{n\beta_h \times 1}$ are matrices of basis functions, and $\boldsymbol{\varepsilon}_{(\cdot)}$ are approximation error matrices. Using the same set of basis functions, the corresponding estimates can also be represented as

$$\hat{\mathbf{D}} = \hat{\mathbf{W}}_\mathbf{D}^T\mathbf{Z}_\mathbf{D}$$
$$\hat{\mathbf{C}} = \hat{\mathbf{W}}_\mathbf{C}^T\mathbf{Z}_\mathbf{C}$$
$$\hat{\mathbf{g}} = \hat{\mathbf{W}}_\mathbf{g}^T\mathbf{z}_\mathbf{g} \tag{10}$$
$$\hat{\mathbf{h}} = \hat{\mathbf{W}}_\mathbf{h}^T\mathbf{z}_\mathbf{h}$$

Define $\tilde{\mathbf{W}}_{(\cdot)} = \mathbf{W}_{(\cdot)} - \hat{\mathbf{W}}_{(\cdot)}$, then the output error tracking dynamics (3) and the torque tracking dynamics (8a) become

$$\mathbf{D}\dot{\mathbf{s}} + \mathbf{C}\mathbf{s} + \mathbf{K}_d\mathbf{s} = (\boldsymbol{\tau}_t - \boldsymbol{\tau}_{td}) - \tilde{\mathbf{W}}_{\mathbf{D}}^T\mathbf{Z}_{\mathbf{D}}\dot{\mathbf{v}}$$

$$- \tilde{\mathbf{W}}_{\mathbf{C}}^T\mathbf{Z}_{\mathbf{C}}\mathbf{v} - \tilde{\mathbf{W}}_{\mathbf{g}}^T\mathbf{z}_{\mathbf{g}} + \boldsymbol{\varepsilon}_1 \tag{11a}$$

$$\dot{\mathbf{e}}_m = \mathbf{A}_m\mathbf{e}_m - \mathbf{B}_p\tilde{\mathbf{W}}_{\mathbf{h}}^T\mathbf{z}_{\mathbf{h}} + \mathbf{B}_p\boldsymbol{\varepsilon}_2 \tag{11b}$$

where $\varepsilon_1 = \varepsilon_1(\varepsilon_{\mathbf{D}}, \varepsilon_{\mathbf{C}}, \varepsilon_{\mathbf{g}}, \mathbf{s}, \ddot{\mathbf{q}}_d)$ and $\varepsilon_2 = \varepsilon_2(\varepsilon_{\mathbf{h}}, \mathbf{e}_m)$ are lumped approximation errors. Since $\mathbf{W}_{(\cdot)}$ are constant vectors, their update laws can be easily found by proper selection of the Lyapunov-like function. Let us consider a candidate

$$V(\mathbf{s}, \mathbf{e}_m, \tilde{\mathbf{W}}_{\mathbf{D}}, \tilde{\mathbf{W}}_{\mathbf{C}}, \tilde{\mathbf{W}}_{\mathbf{g}}, \tilde{\mathbf{W}}_{\mathbf{h}}) = \frac{1}{2}\mathbf{s}^T\mathbf{D}\mathbf{s} + \mathbf{e}_m^T\mathbf{P}_t\mathbf{e}_m + \frac{1}{2}Tr(\tilde{\mathbf{W}}_{\mathbf{D}}^T\mathbf{Q}_{\mathbf{D}}\tilde{\mathbf{W}}_{\mathbf{D}}$$

$$+ \tilde{\mathbf{W}}_{\mathbf{C}}^T\mathbf{Q}_{\mathbf{C}}\tilde{\mathbf{W}}_{\mathbf{C}} + \tilde{\mathbf{W}}_{\mathbf{g}}^T\mathbf{Q}_{\mathbf{g}}\tilde{\mathbf{W}}_{\mathbf{g}} + \tilde{\mathbf{W}}_{\mathbf{h}}^T\mathbf{Q}_{\mathbf{h}}\tilde{\mathbf{W}}_{\mathbf{h}}) \tag{12}$$

where $\mathbf{P}_t = \mathbf{P}_t^T \in \mathfrak{R}^{2n \times 2n}$ is a positive definite matrix satisfying the Lyapunov equation $\mathbf{A}_m^T\mathbf{P}_t + \mathbf{P}_t\mathbf{A}_m = -\mathbf{C}_m^T\mathbf{C}_m$. The matrices $\mathbf{Q}_{\mathbf{D}} \in \mathfrak{R}^{n^2\beta_{\mathbf{D}} \times n^2\beta_{\mathbf{D}}}$, $\mathbf{Q}_{\mathbf{C}} \in \mathfrak{R}^{n^2\beta_C \times n^2\beta_C}$, $\mathbf{Q}_{\mathbf{g}} \in \mathfrak{R}^{n\beta_g \times n\beta_g}$ and $\mathbf{Q}_{\mathbf{h}} \in \mathfrak{R}^{n\beta_h \times n\beta_h}$ are positive definite. The time derivative of V along the trajectory of (11) can be computed as

$$\dot{V} = -\mathbf{s}^T\mathbf{K}_d\mathbf{s} + \mathbf{s}^T\mathbf{e}_\tau - \mathbf{e}_\tau^T\mathbf{e}_\tau + \mathbf{s}^T\boldsymbol{\varepsilon}_1 + \mathbf{e}_m^T\mathbf{P}_t\mathbf{B}_p\boldsymbol{\varepsilon}_2$$

$$- Tr[\tilde{\mathbf{W}}_{\mathbf{D}}^T(\mathbf{Z}_{\mathbf{D}}\dot{\mathbf{v}}\mathbf{s}^T + \mathbf{Q}_{\mathbf{D}}\dot{\hat{\mathbf{W}}}_{\mathbf{D}}) + \tilde{\mathbf{W}}_{\mathbf{C}}^T(\mathbf{Z}_{\mathbf{C}}\mathbf{v}\mathbf{s}^T + \mathbf{Q}_{\mathbf{C}}\dot{\hat{\mathbf{W}}}_{\mathbf{C}})]$$

$$- Tr[\tilde{\mathbf{W}}_{\mathbf{g}}^T(\mathbf{z}_{\mathbf{g}}\mathbf{s}^T + \mathbf{Q}_{\mathbf{g}}\dot{\hat{\mathbf{W}}}_{\mathbf{g}}) + \tilde{\mathbf{W}}_{\mathbf{h}}^T(\mathbf{z}_{\mathbf{h}}\mathbf{e}_m^T\mathbf{P}_t\mathbf{B}_p + \mathbf{Q}_{\mathbf{h}}\dot{\hat{\mathbf{W}}}_{\mathbf{h}})] \tag{13}$$

If we design $\mathbf{B}_m = \mathbf{B}_p$ such that $\mathbf{e}_m^T\mathbf{P}_t\mathbf{B}_p = \mathbf{e}_\tau^T$ and select the update laws as

$$\dot{\hat{\mathbf{W}}}_{\mathbf{D}} = -\mathbf{Q}_{\mathbf{D}}^{-1}(\mathbf{Z}_{\mathbf{D}}\dot{\mathbf{v}}\mathbf{s}^T + \sigma_{\mathbf{D}}\hat{\mathbf{W}}_{\mathbf{D}}) \tag{14a}$$

$$\dot{\hat{\mathbf{W}}}_{\mathbf{C}} = -\mathbf{Q}_{\mathbf{C}}^{-1}(\mathbf{Z}_{\mathbf{C}}\mathbf{v}\mathbf{s}^T + \sigma_{\mathbf{C}}\hat{\mathbf{W}}_{\mathbf{C}}) \tag{14b}$$

$$\dot{\hat{\mathbf{W}}}_{\mathbf{g}} = -\mathbf{Q}_{\mathbf{g}}^{-1}(\mathbf{z}_{\mathbf{g}}\mathbf{s}^T + \sigma_{\mathbf{g}}\hat{\mathbf{W}}_{\mathbf{g}}) \tag{14c}$$

$$\dot{\hat{\mathbf{W}}}_{\mathbf{h}} = -\mathbf{Q}_{\mathbf{h}}^{-1}(\mathbf{z}_{\mathbf{h}}\mathbf{e}_\tau^T + \sigma_{\mathbf{h}}\hat{\mathbf{W}}_{\mathbf{h}}) \tag{14d}$$

where $\sigma_{(\cdot)}$ are positive constants, then (13) becomes

$$\dot{V} = -[\mathbf{s}^T \quad \mathbf{e}_\tau^T]\mathbf{Q}\begin{bmatrix} \mathbf{s} \\ \mathbf{e}_\tau \end{bmatrix} + [\mathbf{s}^T \quad \mathbf{e}_\tau^T]\begin{bmatrix} \mathbf{\epsilon}_1 \\ \mathbf{\epsilon}_2 \end{bmatrix} + \sigma_D Tr(\tilde{\mathbf{W}}_D^T\hat{\mathbf{W}}_D)$$
$$+ \sigma_C Tr(\tilde{\mathbf{W}}_C^T\hat{\mathbf{W}}_C) + \sigma_g Tr(\tilde{\mathbf{W}}_g^T\hat{\mathbf{W}}_g) + \sigma_h Tr(\tilde{\mathbf{W}}_h^T\hat{\mathbf{W}}_h) \qquad (15)$$

where $\mathbf{Q} = \begin{bmatrix} \mathbf{K}_d & -\dfrac{1}{2}\mathbf{I}_{n\times n} \\ -\dfrac{1}{2}\mathbf{I}_{n\times n} & \mathbf{I}_{n\times n} \end{bmatrix}$ is positive definite due to the selection of

gain matrix $\mathbf{K}_d$.

Remark 3: If the number of basis functions are chosen to be sufficiently large such that $\mathbf{\epsilon}_1 \approx \mathbf{0}$ and $\mathbf{\epsilon}_2 \approx \mathbf{0}$, then (15) becomes

$$\dot{V} = -[\mathbf{s}^T \quad \mathbf{e}_\tau^T]\mathbf{Q}\begin{bmatrix} \mathbf{s} \\ \mathbf{e}_\tau \end{bmatrix} \le 0$$

Therefore it implies that that $\mathbf{e}_\tau$ and $\mathbf{s}$ are uniformly bounded and square integrable. Furthermore, $\dot{\mathbf{e}}_\tau$ and $\dot{\mathbf{s}}$ can be shown to be uniformly bounded; as a result, asymptotic convergence of $\mathbf{e}_\tau$ and $\mathbf{s}$ can easily be concluded by Barbalat's lemma. This further implies that $\mathbf{\tau} \to \mathbf{\tau}_d$ and $\mathbf{q} \to \mathbf{q}_d$ as $t \to \infty$ even though $\mathbf{D}$, $\mathbf{C}$, $\mathbf{g}$ and $\mathbf{h}$ are all unknown.

Remark 4: Suppose $\mathbf{\epsilon}_1$ and $\mathbf{\epsilon}_2$ cannot be ignored but their variation bounds are available i.e. there exists positive constants $\delta_1, \delta_2 > 0$ such that $\|\mathbf{\epsilon}_1\| \le \delta_1$ and $\|\mathbf{\epsilon}_2\| \le \delta_2$. To cover the effect of these bounded approximation errors, the desired transmission torque (2) and actuator input (6) are modified to be

$$\mathbf{\tau}_{td} = \hat{\mathbf{D}}\dot{\mathbf{v}} + \hat{\mathbf{C}}\mathbf{v} + \hat{\mathbf{g}} - \mathbf{K}_d\mathbf{s} + \mathbf{\tau}_{robust1}$$

$$\mathbf{\tau}_a = \mathbf{\Theta}\mathbf{x}_p + \mathbf{\Phi}\mathbf{\tau}_{td} + \hat{\mathbf{h}} + \mathbf{\tau}_{robust2}$$

where $\mathbf{\tau}_{robust1}$ and $\mathbf{\tau}_{robust2}$ are robust terms to be designed. Let us consider the Lyapunov-like function candidate (12) and update laws (14) without the σ-modification terms again. The time derivative of V can be computed as

$$\dot{V} \le -[\mathbf{s}^T \quad \mathbf{e}_\tau^T]\mathbf{Q}\begin{bmatrix} \mathbf{s} \\ \mathbf{e}_\tau \end{bmatrix} + \delta_1\|\mathbf{s}\| + \delta_2\|\mathbf{e}_\tau\| + \mathbf{s}^T\boldsymbol{\tau}_{robust\,1} + \mathbf{e}_\tau^T\boldsymbol{\tau}_{robust\,2}$$

By picking $\boldsymbol{\tau}_{robust\,1} = -\delta_2[\mathrm{sgn}(s_1) \quad \cdots \quad \mathrm{sgn}(s_n)]^T$, where s_i, $i=1,\ldots,n$ is the i-th element of the vector $\mathbf{s}$, and $\boldsymbol{\tau}_{robust\,2} = -\delta_2[\mathrm{sgn}(e_{\tau_1}) \quad \cdots \quad \mathrm{sgn}(e_{\tau_{2n}})]^T$ where e_{τ_i}, $i=1,\ldots,2n$ is the i-th element of the vector $\mathbf{e}_\tau$, we may have $\dot{V} \le 0$ and asymptotic convergence of the state error can be concluded by Barbalat's lemma.

Owing to the existence of $\boldsymbol{\varepsilon}_1$ and $\boldsymbol{\varepsilon}_2$ in (15), the definiteness of $\dot{V}$ cannot be determined. By using the inequalities similar to those in (4.6-31), we may define $\mathbf{A} = \begin{bmatrix} \mathbf{D} & \mathbf{0} \\ \mathbf{0} & 2\mathbf{C}_m^T\mathbf{P}_t\mathbf{C}_m \end{bmatrix}$ and rewrite (15) into

$$\dot{V} \le -\alpha V + \frac{1}{2}[\alpha\lambda_{\max}(\mathbf{A}) - \lambda_{\min}(\mathbf{Q})]\left\|\begin{bmatrix} \mathbf{s} \\ \mathbf{e}_\tau \end{bmatrix}\right\|^2 + \frac{1}{2\lambda_{\min}(\mathbf{Q})}\left\|\begin{bmatrix} \boldsymbol{\varepsilon}_1 \\ \boldsymbol{\varepsilon}_2 \end{bmatrix}\right\|^2$$

$$+\frac{1}{2}\{[\alpha\lambda_{\max}(\mathbf{Q_D}) - \sigma_D]Tr(\tilde{\mathbf{W}}_D^T\tilde{\mathbf{W}}_D) + [\alpha\lambda_{\max}(\mathbf{Q_C})$$

$$-\sigma_C]Tr(\tilde{\mathbf{W}}_C^T\tilde{\mathbf{W}}_C) + [\alpha\lambda_{\max}(\mathbf{Q_g}) - \sigma_g]Tr(\tilde{\mathbf{W}}_g^T\tilde{\mathbf{W}}_g)$$

$$+[\alpha\lambda_{\max}(\mathbf{Q_h}) - \sigma_h]Tr(\tilde{\mathbf{W}}_h^T\tilde{\mathbf{W}}_h)\} + \frac{1}{2}[\sigma_D Tr(\mathbf{W}_D^T\mathbf{W}_D)$$

$$+\sigma_C Tr(\mathbf{W}_C^T\mathbf{W}_C) + \sigma_g Tr(\mathbf{W}_g^T\mathbf{W}_g) + \sigma_h Tr(\mathbf{W}_h^T\mathbf{W}_h)] \qquad (16)$$

where α is a constant selected to satisfy

$$\alpha \le \min\left\{\frac{\lambda_{\min}(\mathbf{Q})}{\lambda_{\max}(\mathbf{A})}, \frac{\sigma_D}{\lambda_{\max}(\mathbf{Q_D})}, \frac{\sigma_C}{\lambda_{\max}(\mathbf{Q_C})}, \frac{\sigma_g}{\lambda_{\max}(\mathbf{Q_g})}, \frac{\sigma_h}{\lambda_{\max}(\mathbf{Q_h})}\right\}$$

so that we have

$$\dot{V} \le -\alpha V + \frac{1}{2\lambda_{\min}(\mathbf{Q})}\left\|\begin{bmatrix} \boldsymbol{\varepsilon}_1 \\ \boldsymbol{\varepsilon}_2 \end{bmatrix}\right\|^2 + \frac{1}{2}[\sigma_D Tr(\mathbf{W}_D^T\mathbf{W}_D)$$

$$+\sigma_C Tr(\mathbf{W}_C^T\mathbf{W}_C) + \sigma_g Tr(\mathbf{W}_g^T\mathbf{W}_g) + \sigma_h Tr(\mathbf{W}_h^T\mathbf{W}_h)] \qquad (17)$$

Therefore, $\dot{V} < 0$ whenever

$$V > \frac{1}{2\alpha\lambda_{\min}(\mathbf{Q})} \sup_{\tau \geq t_0} \left\| \begin{bmatrix} \varepsilon_1(\tau) \\ \varepsilon_2(\tau) \end{bmatrix} \right\|^2 + \frac{1}{2\alpha} [\sigma_D Tr(\mathbf{W}_D^T \mathbf{W}_D)$$

$$+ \sigma_C Tr(\mathbf{W}_C^T \mathbf{W}_C) + \sigma_g Tr(\mathbf{W}_g^T \mathbf{W}_g) + \sigma_h Tr(\mathbf{W}_h^T \mathbf{W}_h)] \qquad (18)$$

This verifies that $\mathbf{s}$, $\mathbf{e}_\tau$, $\tilde{\mathbf{W}}_D$, $\tilde{\mathbf{W}}_C$, $\tilde{\mathbf{W}}_g$, and $\tilde{\mathbf{W}}_h$ are uniformly ultimately bounded. Differential inequality (17) can be solved to have the upper bound for $V(t)$ as

$$V(t) \leq e^{-\alpha(t-t_0)}V(t_0) + \frac{1}{2\alpha\lambda_{\min}(\mathbf{Q})} \sup_{t_0 < \tau < t} \left\| \begin{bmatrix} \varepsilon_1(\tau) \\ \varepsilon_2(\tau) \end{bmatrix} \right\|^2$$

$$+ \frac{1}{2\alpha}[\sigma_D Tr(\mathbf{W}_D^T \mathbf{W}_D) + \sigma_C Tr(\mathbf{W}_C^T \mathbf{W}_C)$$

$$+ \sigma_g Tr(\mathbf{W}_g^T \mathbf{W}_g) + \sigma_h Tr(\mathbf{W}_h^T \mathbf{W}_h)] \qquad (19)$$

From (12), we may estimate the lower bound for V as $V \geq \frac{1}{2}\lambda_{\min}(\mathbf{A}) \left\| \begin{bmatrix} \mathbf{s} \\ \mathbf{e}_\tau \end{bmatrix} \right\|^2$

which gives the expression

$$\left\| \begin{bmatrix} \mathbf{s} \\ \mathbf{e}_\tau \end{bmatrix} \right\| \leq \sqrt{\frac{2V(t)}{\lambda_{\min}(\mathbf{A})}}$$

Together with (19), we may have the bound for the error signals as

$$\left\| \begin{bmatrix} \mathbf{s} \\ \mathbf{e}_\tau \end{bmatrix} \right\| \leq \sqrt{\frac{2V(t_0)}{\lambda_{\min}(\mathbf{A})}} e^{-\frac{\alpha}{2}(t-t_0)} + \frac{1}{\sqrt{\alpha\lambda_{\min}(\mathbf{A})\lambda_{\min}(\mathbf{Q})}} \sup_{t_0 < \tau < t} \left\| \begin{bmatrix} \varepsilon_1(\tau) \\ \varepsilon_2(\tau) \end{bmatrix} \right\|$$

$$+ \frac{1}{\sqrt{\alpha\lambda_{\min}(\mathbf{A})}}[\sigma_D Tr(\mathbf{W}_D^T \mathbf{W}_D) + \sigma_C Tr(\mathbf{W}_C^T \mathbf{W}_C)$$

$$+ \sigma_g Tr(\mathbf{W}_g^T \mathbf{W}_g) + \sigma_h Tr(\mathbf{W}_h^T \mathbf{W}_h)]^{\frac{1}{2}}$$

Therefore, the bound is a weighted exponential function shifted with a constant.

Remark 5: To implement the desired transmission torque (2), actuator input (6) and update law (14), we do not need the regressor matrix, the knowledge of joint accelerations, or their derivatives. Therefore, it is feasible for realization.

Table 6.1 summarizes the adaptive control laws derived in this section based on their controller forms, update laws and implementation issues.

Table 6.1 Summary of the adaptive control for FJR

	Flexible-Joint Robot $$\mathbf{D}\dot{\mathbf{s}} + \mathbf{Cs} + \mathbf{g} + \mathbf{D}\dot{\mathbf{v}} + \mathbf{Cv} = \boldsymbol{\tau}_t$$ $$\mathbf{J}_t\ddot{\boldsymbol{\tau}}_t + \mathbf{B}_t\dot{\boldsymbol{\tau}}_t + \boldsymbol{\tau}_t = \boldsymbol{\tau}_a - \overline{\mathbf{q}}(\dot{\mathbf{q}}, \ddot{\mathbf{q}})$$ (6.3-1)		
	Regressor-based	Regressor-free
Controller	$\boldsymbol{\tau}_{td} = \hat{\mathbf{g}} + \hat{\mathbf{D}}\dot{\mathbf{v}} + \hat{\mathbf{C}}\mathbf{v} - \mathbf{K}_d\mathbf{s}$ $= \mathbf{Y}(\mathbf{q}, \dot{\mathbf{q}}, \mathbf{v}, \dot{\mathbf{v}})\hat{\mathbf{p}} - \mathbf{K}_d\mathbf{s}$ $\boldsymbol{\tau}_a = \boldsymbol{\Theta}\mathbf{x}_p + \boldsymbol{\Phi}\boldsymbol{\tau}_{td} + \mathbf{h}(\overline{\boldsymbol{\tau}}_{td}, \overline{\mathbf{q}})$ (6.3-2), (6.2-8)	$\boldsymbol{\tau}_{td} = \hat{\mathbf{g}} + \hat{\mathbf{D}}\dot{\mathbf{v}} + \hat{\mathbf{C}}\mathbf{v} - \mathbf{K}_d\mathbf{s}$ $\boldsymbol{\tau}_a = \boldsymbol{\Theta}\mathbf{x}_p + \boldsymbol{\Phi}\boldsymbol{\tau}_{td} + \hat{\mathbf{h}}$ (6.4-2), (6.4-6)	
Adaptive Law	$$\dot{\hat{\mathbf{p}}} = -\boldsymbol{\Gamma}^{-1}\mathbf{Y}^T\mathbf{s}$$ (6.3-6)	$\dot{\hat{\mathbf{W}}}_{\mathbf{D}} = -\mathbf{Q}_{\mathbf{D}}^{-1}(\mathbf{Z}_{\mathbf{D}}\dot{\mathbf{v}}\mathbf{s}^T + \sigma_{\mathbf{D}}\hat{\mathbf{W}}_{\mathbf{D}})$ $\dot{\hat{\mathbf{W}}}_{\mathbf{C}} = -\mathbf{Q}_{\mathbf{C}}^{-1}(\mathbf{Z}_{\mathbf{C}}\mathbf{v}\mathbf{s}^T + \sigma_{\mathbf{C}}\hat{\mathbf{W}}_{\mathbf{C}})$ $\dot{\hat{\mathbf{W}}}_{\mathbf{g}} = -\mathbf{Q}_{\mathbf{g}}^{-1}(\mathbf{z}_{\mathbf{g}}\mathbf{s}^T + \sigma_{\mathbf{g}}\hat{\mathbf{W}}_{\mathbf{g}})$ $\dot{\hat{\mathbf{W}}}_{\mathbf{h}} = -\mathbf{Q}_{\mathbf{h}}^{-1}(\mathbf{z}_{\mathbf{h}}\mathbf{e}_r^T + \sigma_{\mathbf{h}}\hat{\mathbf{W}}_{\mathbf{h}})$ (6.4-14)	
Realization Issue	Need information of joint accelerations and their higher derivatives. Need to compute the regressor matrix and its time derivatives.	Does not need joint accelerations. Does not need to compute the regressor matrix.	

Example 6.1:

Consider the flexible-joint robot in (3.6-3), and we are going to verify the regressor-free adaptive control strategy developed in this section by computer simulations. Actual values of link parameters are selected as $m_1 = m_2 = 0.5(kg)$, $l_1 = l_2 = 0.75(m)$, $l_{c1} = l_{c2} = 0.375(m)$, $I_1 = I_2 = 0.0234(kg\text{-}m^2)$, and $k_1 = k_2 = 100(\text{N-}m/\text{rad})$. Parameters for the actuator part are chosen as $j_1 = 0.02(kg\text{-}m^2)$, $j_2 = 0.01(kg\text{-}m^2)$, $b_1 = 5(\text{N-}m\text{-}sec/rad)$, and $b_2 = 4(\text{N-}m\text{-}sec/rad)$(Chien and Huang 2007a). We would like the endpoint to track a $0.2m$ radius circle centered at $(0.8m, 1.0m)$ in

10 seconds without knowing its precise model. The initial condition for the generalized coordinate is at $\mathbf{q}(0) = \boldsymbol{\theta}(0) = [0.0022 \quad 1.5019 \quad 0 \quad 0]^T$, i.e., the endpoint is initially at $(0.8m, 0,75m)$. It is away form the desired initial endpoint position $(0.8m, 0,8m)$ for observation of the transient. The initial state for the reference model is $\boldsymbol{\tau}_r(0) = [1.8 \quad -2.8 \quad 0 \quad 0]^T$, which is the same as the initial state for the desired torque. The controller in (6) is applied with the gain matrices

$$\mathbf{K}_d = \begin{bmatrix} 5 & 0 \\ 0 & 5 \end{bmatrix}, \text{ and } \boldsymbol{\Lambda} = \begin{bmatrix} 5 & 0 \\ 0 & 5 \end{bmatrix}.$$

The 11-term Fourier series is selected as the basis function for the approximation. Therefore, $\hat{\mathbf{W}}_D$ and $\hat{\mathbf{W}}_C$ are in $\mathfrak{R}^{44 \times 2}$, while $\hat{\mathbf{W}}_g$ and $\hat{\mathbf{W}}_h$ are in $\mathfrak{R}^{22 \times 2}$. The initial weighting vectors for the entries are assigned to be

$$\hat{\mathbf{w}}_{D_{11}}(0) = [0.05 \quad 0 \quad \cdots \quad 0]^T \in \mathfrak{R}^{11 \times 1}$$

$$\hat{\mathbf{w}}_{D_{12}}(0) = \hat{\mathbf{w}}_{D_{21}}(0) = [-0.05 \quad 0 \quad \cdots \quad 0]^T \in \mathfrak{R}^{11 \times 1}$$

$$\hat{\mathbf{w}}_{D_{22}}(0) = [0.1 \quad 0 \quad \cdots \quad 0]^T \in \mathfrak{R}^{11 \times 1}$$

$$\hat{\mathbf{w}}_{C_{11}}(0) = [0.05 \quad 0 \quad \cdots \quad 0]^T \in \mathfrak{R}^{11 \times 1}$$

$$\hat{\mathbf{w}}_{C_{12}}(0) = \hat{\mathbf{w}}_{C_{21}}(0) = [-0.05 \quad 0 \quad \cdots \quad 0]^T \in \mathfrak{R}^{11 \times 1}$$

$$\hat{\mathbf{w}}_{C_{22}}(0) = [0.1 \quad 0 \quad \cdots \quad 0]^T \in \mathfrak{R}^{11 \times 1}$$

$$\hat{\mathbf{w}}_{g_1}(0) = \hat{\mathbf{w}}_{g_2}(0) = [0 \quad 0 \quad \cdots \quad 0]^T \in \mathfrak{R}^{11 \times 1}$$

$$\hat{\mathbf{w}}_{h_1}(0) = \hat{\mathbf{w}}_{h_2}(0) = [0 \quad 0 \quad \cdots \quad 0]^T \in \mathfrak{R}^{11 \times 1}$$

The gain matrices in the update law (14) are selected as

$$\mathbf{Q}_D^{-1} = 2\mathbf{I}_{44}, \quad \mathbf{Q}_C^{-1} = 2\mathbf{I}_{44}, \quad \mathbf{Q}_g^{-1} = 10\mathbf{I}_{22}, \text{ and } \mathbf{Q}_h^{-1} = 10000\mathbf{I}_{22}$$

The approximation error is assumed to be neglected in this simulation, and the σ-modification parameters are chosen as $\sigma_{(\cdot)} = 0$.

The simulation results are shown in Figure 6.1 to 6.8. Figure 6.1 shows the tracking performance of the robot endpoint and its desired trajectory in the Cartesian space. It is observed that the endpoint trajectory converges nicely to the desired trajectory, although the initial position error is quite large. After the transient state, the tracking error is small regardless of the time-varying uncertainties in **D**, **C** and **g**. Computation of the complex regressor is avoided in this strategy which greatly simplifies the design and implementation of the control law. Figure 6.2 presents the time history of the joint space tracking performance. The transient states converge very fast and the tracking errors are small. The control efforts to the two joints are reasonable that can be verified in Figure 6.3. The control torque for both joints can be seen in Figure 6.4. Figure 6.5 to 6.8 are the performance of function approximation. Although most parameters do not converge to their actual values, they still remain bounded as desired.

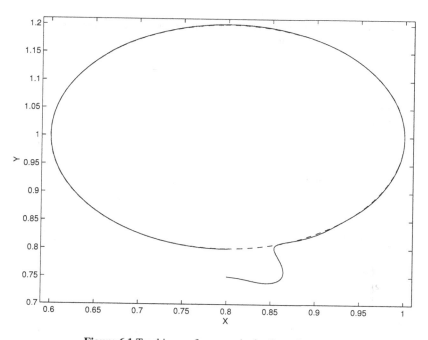

Figure 6.1 Tracking performance in the Cartesian space

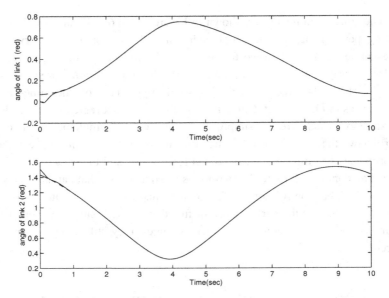

Figure 6.2 Joint space tracking performance. It can be seen that the transient is fast, and the tracking error is very small

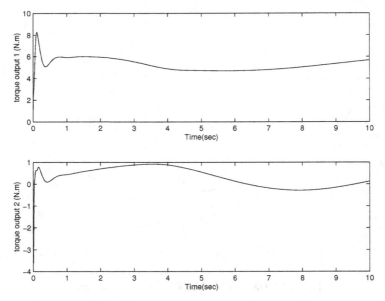

Figure 6.3 Torque tracking performance. It can be seen that the torque errors for both joints are small

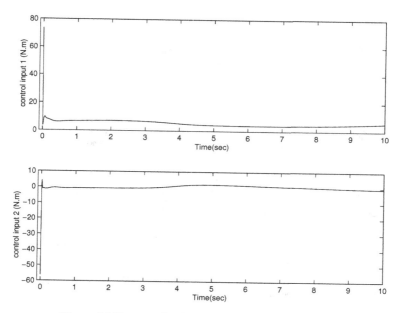

Figure 6.4 The control torques for both joints are reasonable

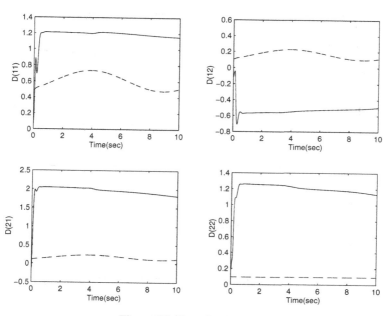

Figure 6.5 Approximation of **D**

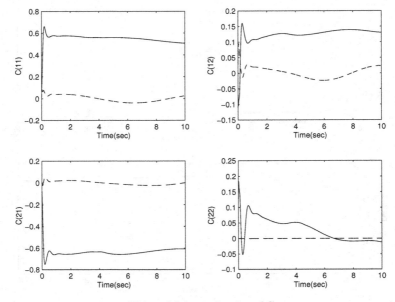

Figure 6.6 Approximation of **C**

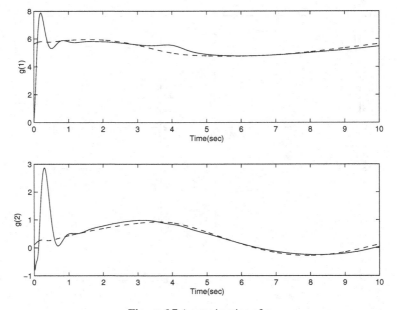

Figure 6.7 Approximation of **g**

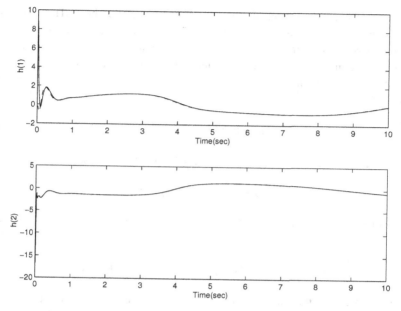

Figure 6.8 Approximation of **h**

6.5 Consideration of Actuator Dynamics

According to (6.4-1) and (3.8-1), the dynamics of a rigid-link flexible-joint electrically-driven robot can be described by

$$\mathbf{D}\dot{\mathbf{s}} + \mathbf{C}\mathbf{s} + \mathbf{g} + \mathbf{D}\dot{\mathbf{v}} + \mathbf{C}\mathbf{v} = \boldsymbol{\tau}_t \tag{1a}$$

$$\mathbf{J}_t\ddot{\boldsymbol{\tau}}_t + \mathbf{B}_t\dot{\boldsymbol{\tau}}_t + \boldsymbol{\tau}_t = \mathbf{H}\mathbf{i} - \overline{\mathbf{q}}(\dot{\mathbf{q}}, \ddot{\mathbf{q}}) \tag{1b}$$

$$\mathbf{L}\dot{\mathbf{i}} + \mathbf{R}\mathbf{i} + \mathbf{K}_b\dot{\mathbf{q}} = \mathbf{u} \tag{1c}$$

This system is in a cascade form with the configuration shown in Figure 6.9.

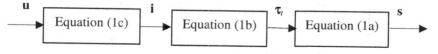

Figure 6.9 Cascade structure in equation (1)

Therefore, the backstepping-like procedure can be applied here. The concept is to design a desired torque trajectory $\boldsymbol{\tau}_{td}$ first for convergence of $\mathbf{s}$ in (1a). A desired current trajectory $\mathbf{i}_d$ can then be found to ensure $\boldsymbol{\tau}_t \to \boldsymbol{\tau}_{td}$ in (1b). Finally, the control effort $\mathbf{u}$ is constructed to have convergence of $\mathbf{i}$ to $\mathbf{i}_d$.

Assuming that all parameters in (1) are known, then the desired torque can be designed as

$$\boldsymbol{\tau}_{td} = \mathbf{g} + \mathbf{D}\dot{\mathbf{v}} + \mathbf{Cv} - \mathbf{K}_d \mathbf{s} \tag{2}$$

Therefore, the dynamics for output error tracking is found to be

$$\mathbf{D}\dot{\mathbf{s}} + \mathbf{Cs} + \mathbf{K}_d \mathbf{s} = \boldsymbol{\tau}_t - \boldsymbol{\tau}_{td} \tag{3}$$

To ensure torque tracking in (1b), the MRC rule is applied with the reference model

$$\mathbf{J}_r \ddot{\boldsymbol{\tau}}_r + \mathbf{B}_r \dot{\boldsymbol{\tau}}_r + \mathbf{K}_r \boldsymbol{\tau}_r = \mathbf{J}_r \ddot{\boldsymbol{\tau}}_{td} + \mathbf{B}_r \dot{\boldsymbol{\tau}}_{td} + \mathbf{K}_r \boldsymbol{\tau}_{td} \tag{4}$$

where $\boldsymbol{\tau}_r \in \mathfrak{R}^n$ is the state vector of the reference model, and $\mathbf{J}_r \in \mathfrak{R}^{n \times n}$, $\mathbf{B}_r \in \mathfrak{R}^{n \times n}$, and $\mathbf{K}_r \in \mathfrak{R}^{n \times n}$ are selected to give convergence of $\boldsymbol{\tau}_r$ to $\boldsymbol{\tau}_{td}$. With the definition of $\overline{\boldsymbol{\tau}}_{td}(\dot{\boldsymbol{\tau}}_{td}, \ddot{\boldsymbol{\tau}}_{td}) = \mathbf{K}_r^{-1}(\mathbf{B}_r \dot{\boldsymbol{\tau}}_{td} + \mathbf{J}_r \ddot{\boldsymbol{\tau}}_{td})$, we may rewrite (1b) and (4) into the state space representation

$$\dot{\mathbf{x}}_p = \mathbf{A}_p \mathbf{x}_p + \mathbf{B}_p \mathbf{Hi} - \mathbf{B}_p \overline{\mathbf{q}} \tag{5a}$$

$$\dot{\mathbf{x}}_m = \mathbf{A}_m \mathbf{x}_m + \mathbf{B}_m (\boldsymbol{\tau}_{td} + \overline{\boldsymbol{\tau}}_{td}) \tag{5b}$$

where $\mathbf{x}_p = [\boldsymbol{\tau}_t^T \;\; \dot{\boldsymbol{\tau}}_t^T]^T \in \mathfrak{R}^{2n}$ and $\mathbf{x}_m = [\boldsymbol{\tau}_r^T \;\; \dot{\boldsymbol{\tau}}_r^T]^T \in \mathfrak{R}^{2n}$ are augmented state vectors. $\mathbf{A}_p = \begin{bmatrix} \mathbf{0} & \mathbf{I}_{n \times n} \\ -\mathbf{J}_t^{-1} & -\mathbf{J}_t^{-1}\mathbf{B}_t \end{bmatrix} \in \mathfrak{R}^{2n \times 2n}$ and $\mathbf{A}_m = \begin{bmatrix} \mathbf{0} & \mathbf{I}_{n \times n} \\ -\mathbf{J}_r^{-1}\mathbf{K}_r & -\mathbf{J}_r^{-1}\mathbf{B}_r \end{bmatrix} \in \mathfrak{R}^{2n \times 2n}$ are augmented system matrices, and $\mathbf{B}_p = \begin{bmatrix} \mathbf{0} \\ \mathbf{J}_t^{-1} \end{bmatrix} \in \mathfrak{R}^{2n \times n}$ and $\mathbf{B}_m = \begin{bmatrix} \mathbf{0} \\ \mathbf{J}_r^{-1}\mathbf{K}_r \end{bmatrix} \in \mathfrak{R}^{2n \times n}$ are augmented input gain matrices. Let $\boldsymbol{\tau}_t = \mathbf{C}_p \mathbf{x}_p$ and $\boldsymbol{\tau}_r = \mathbf{C}_m \mathbf{x}_m$ be respectively the output signal vector for (5a) and (5b), where $\mathbf{C}_p = \mathbf{C}_m = [\mathbf{I}_{n \times n} \quad \mathbf{0}] \in \mathfrak{R}^{n \times 2n}$ are augmented output signal matrices. The pair $(\mathbf{A}_m, \mathbf{B}_m)$ is controllable, $(\mathbf{A}_m, \mathbf{C}_m)$ is observable, and the transfer function $\mathbf{C}_m (s\mathbf{I} - \mathbf{A}_m)^{-1} \mathbf{B}_m$ is SPR. According to the MRC design, the desired current $\mathbf{i}_d$ is selected as

$$\mathbf{i}_d = \mathbf{H}^{-1}[\mathbf{\Theta}\mathbf{x}_p + \mathbf{\Phi}\boldsymbol{\tau}_{td} + \mathbf{h}(\overline{\boldsymbol{\tau}}_{td}, \overline{\mathbf{q}})] \tag{6}$$

where $\mathbf{\Theta}$ and $\mathbf{\Phi}$ are matrices satisfying $\mathbf{A}_p + \mathbf{B}_p\mathbf{\Theta} = \mathbf{A}_m$ and $\mathbf{B}_p\mathbf{\Phi} = \mathbf{B}_m$, respectively, and $\mathbf{h}$ is defined as $\mathbf{h}(\overline{\boldsymbol{\tau}}_{td}, \overline{\mathbf{q}}) = \mathbf{\Phi}\overline{\boldsymbol{\tau}}_{td} + \overline{\mathbf{q}}$. Using (6) and (5a), we may obtain

$$\dot{\mathbf{x}}_p = \mathbf{A}_m\dot{\mathbf{x}}_p + \mathbf{B}_m(\boldsymbol{\tau}_d + \overline{\boldsymbol{\tau}}_{td}) + \mathbf{B}_p\mathbf{H}(\mathbf{i} - \mathbf{i}_d) \tag{7}$$

With the definition $\mathbf{e}_m = \mathbf{x}_p - \mathbf{x}_m$ and $\mathbf{e}_\tau = \boldsymbol{\tau}_t - \boldsymbol{\tau}_r$, the dynamics for the torque tracking loop becomes

$$\dot{\mathbf{e}}_m = \mathbf{A}_m\mathbf{e}_m + \mathbf{B}_p\mathbf{H}(\mathbf{i} - \mathbf{i}_d) \tag{8a}$$

$$\mathbf{e}_\tau = \mathbf{C}_m\mathbf{e}_m \tag{8b}$$

In order to ensure $\boldsymbol{\tau}_t \to \boldsymbol{\tau}_{td}$ and $\mathbf{i} \to \mathbf{i}_d$, the control law in (1c) is designed as

$$\mathbf{u} = \mathbf{L}\dot{\mathbf{i}}_d + \mathbf{R}\mathbf{i} + \mathbf{K}_b\dot{\mathbf{q}} - \mathbf{K}_c\mathbf{e}_i \tag{9}$$

where $\mathbf{e}_i = \mathbf{i} - \mathbf{i}_d$ is the current error vector, and $\mathbf{K}_c$ is a positive definite matrix. Plugging, (9) into (1c), we may have the dynamics for the current tracking loop

$$\mathbf{L}\dot{\mathbf{e}}_i + \mathbf{K}_c\mathbf{e}_i = \mathbf{0} \tag{10}$$

At this stage, we have the output error dynamics in (3), the dynamics of the torque tracking loop in (8) and the dynamics of the current tracking loop in (10). We have to ensure that all of these dynamics be stable. To this end, let us consider a Lyapunov-like function candidate

$$V(\mathbf{s}, \mathbf{e}_m, \mathbf{e}_i) = \frac{1}{2}\mathbf{s}^T\mathbf{D}\mathbf{s} + \mathbf{e}_m^T\mathbf{P}_t\mathbf{e}_m + \frac{1}{2}\mathbf{e}_i^T\mathbf{L}\mathbf{e}_i \tag{11}$$

where $\mathbf{P}_t = \mathbf{P}_t^T \in \mathfrak{R}^{2n \times 2n}$ is a positive definite matrix satisfying the Lyapunov equation $\mathbf{A}_m^T\mathbf{P}_t + \mathbf{P}_t\mathbf{A}_m = -\mathbf{C}_m^T\mathbf{C}_m$. Along the trajectories of (3), (8) and (10), the time derivative of V can be computed as

$$\dot{V} = -\mathbf{s}^T \mathbf{K}_d \mathbf{s} + \frac{1}{2}\mathbf{s}^T (\dot{\mathbf{D}} - 2\mathbf{C})\mathbf{s} + \mathbf{s}^T \mathbf{e}_\tau$$

$$-\mathbf{e}_\tau^T \mathbf{e}_\tau + \mathbf{e}_m^T \mathbf{P}_t \mathbf{B}_p \mathbf{H} \mathbf{e}_i - \mathbf{e}_i^T \mathbf{K}_c \mathbf{e}_i \qquad (12)$$

Selecting $\mathbf{B}_m = \mathbf{B}_p$ and according to the Kalman-Yakubovic lemma, equation (12) becomes

$$\dot{V} = -[\mathbf{s}^T \quad \mathbf{e}_\tau^T \quad \mathbf{e}_i^T]\mathbf{Q}\begin{bmatrix} \mathbf{s} \\ \mathbf{e}_\tau \\ \mathbf{e}_i \end{bmatrix} \leq 0 \qquad (13)$$

where $\mathbf{Q} = \begin{bmatrix} \mathbf{K}_d & -\dfrac{1}{2}\mathbf{I}_{n\times n} & \mathbf{0} \\[2mm] -\dfrac{1}{2}\mathbf{I}_{n\times n} & \mathbf{I}_{n\times n} & -\dfrac{1}{2}\mathbf{H} \\[2mm] \mathbf{0} & -\dfrac{1}{2}\mathbf{H} & \mathbf{K}_c \end{bmatrix}$ is positive definite due to proper

selection of $\mathbf{K}_d$ and $\mathbf{K}_c$. Therefore, we have proved that $\mathbf{s}$, $\mathbf{e}_\tau$ and $\mathbf{e}_i$ are uniformly bounded, and their square integrability can also be proved from (13). Furthermore, uniformly boundedness of $\dot{\mathbf{s}}$, $\dot{\mathbf{e}}_\tau$, and $\dot{\mathbf{e}}_i$ are also easy to be proved, and hence, $\mathbf{q} \to \mathbf{q}_d$, $\boldsymbol{\tau}_t \to \boldsymbol{\tau}_{td}$, and $\mathbf{i} \to \mathbf{i}_d$ follow by Barbalat's lemma.

Remark 6: To implement the control strategy, all system parameters are required to be available, and we need to feedback $\ddot{\mathbf{q}}$ and its higher order derivatives. Therefore, the design introduced in this section is not feasible for practical applications.

6.5.1 Regressor-based adaptive controller design

Consider the system in (1) again, but $\mathbf{D}$, $\mathbf{C}$, $\mathbf{g}$, $\mathbf{L}$, $\mathbf{R}$ and $\mathbf{K}_b$ are unavailable. The desired torque trajectory in (2) is not feasible, and we modify it as

$$\boldsymbol{\tau}_{td} = \hat{\mathbf{g}} + \hat{\mathbf{D}}\dot{\mathbf{v}} + \hat{\mathbf{C}}\mathbf{v} - \mathbf{K}_d \mathbf{s}$$

$$= \mathbf{Y}(\mathbf{q}, \dot{\mathbf{q}}, \mathbf{v}, \dot{\mathbf{v}})\hat{\mathbf{p}} - \mathbf{K}_d \mathbf{s} \qquad (14)$$

where $\hat{\mathbf{D}}$, $\hat{\mathbf{C}}$ and $\hat{\mathbf{g}}$ are estimates of $\mathbf{D}$, $\mathbf{C}$, and $\mathbf{g}$, respectively. Plugging (14) into (1a), and we may obtain the error dynamics for the output tracking loop

$$\mathbf{D}\dot{\mathbf{s}} + \mathbf{C}\mathbf{s} + \mathbf{K}_d\mathbf{s} = -\tilde{\mathbf{D}}\dot{\mathbf{v}} - \tilde{\mathbf{C}}\mathbf{v} - \tilde{\mathbf{g}} + (\boldsymbol{\tau}_t - \boldsymbol{\tau}_{td})$$
$$= -\mathbf{Y}(\mathbf{q}, \dot{\mathbf{q}}, \mathbf{v}, \dot{\mathbf{v}})\tilde{\mathbf{p}} + (\boldsymbol{\tau}_t - \boldsymbol{\tau}_{td}) \tag{15}$$

where $\tilde{\mathbf{D}} = \mathbf{D} - \hat{\mathbf{D}}$, $\tilde{\mathbf{C}} = \mathbf{C} - \hat{\mathbf{C}}$, $\tilde{\mathbf{g}} = \mathbf{g} - \hat{\mathbf{g}}$, and $\tilde{\mathbf{p}} = \mathbf{p} - \hat{\mathbf{p}}$. Therefore, if we may find a control law to drive $\boldsymbol{\tau}_t \to \boldsymbol{\tau}_{td}$ and an update law to have $\hat{\mathbf{p}} \to \mathbf{p}$, then (15) implies convergence of the output error. To this end, we would like to use the MRC rule with the reference model in (4) and the state space representation in (5). The desired current trajectory is designed as the one in (6) to have the dynamics for the torque tracking loop as in (8). Instead of the controller in (9), we use

$$\mathbf{u} = \hat{\mathbf{L}}\mathbf{i}_d + \hat{\mathbf{R}}\mathbf{i} + \hat{\mathbf{K}}_b\dot{\mathbf{q}} - \mathbf{K}_c\mathbf{e}_i$$
$$= \hat{\mathbf{p}}_i^T \boldsymbol{\varphi} - \mathbf{K}_c\mathbf{e}_i \tag{16}$$

where $\boldsymbol{\varphi} = [\mathbf{i}_d^T \ \ \mathbf{i}^T \ \ \dot{\mathbf{q}}^T]^T \in \mathfrak{R}^{3n}$, $\hat{\mathbf{p}}_i = [\hat{\mathbf{L}}^T \ \ \hat{\mathbf{R}}^T \ \ \hat{\mathbf{K}}_b^T]^T \in \mathfrak{R}^{3n\times n}$, and we may have the dynamics for the current tracking loop as

$$\mathbf{L}\dot{\mathbf{e}}_i + \mathbf{K}_c\mathbf{e}_i = -\tilde{\mathbf{p}}_i^T \boldsymbol{\varphi} \tag{17}$$

where $\tilde{\mathbf{p}}_i = \mathbf{p}_i - \hat{\mathbf{p}}_i$. To prove stability, we select the Lyapunov-like function candidate

$$V(\mathbf{s}, \mathbf{e}_m, \mathbf{e}_i, \tilde{\mathbf{p}}, \tilde{\mathbf{p}}_i) = \frac{1}{2}(\mathbf{s}^T\mathbf{D}\mathbf{s} + 2\mathbf{e}_m^T\mathbf{P}_t\mathbf{e}_m + \mathbf{e}_i^T\mathbf{L}\mathbf{e}_i$$
$$+ \tilde{\mathbf{p}}^T\boldsymbol{\Gamma}\tilde{\mathbf{p}}) + Tr(\tilde{\mathbf{p}}_i^T\boldsymbol{\Gamma}_i\tilde{\mathbf{p}}_i) \tag{18}$$

where $\mathbf{P}_t = \mathbf{P}_t^T \in \mathfrak{R}^{2n\times 2n}$ is positive definite satisfying the Lyapunov equation $\mathbf{A}_m^T\mathbf{P}_t + \mathbf{P}_t\mathbf{A}_m = -\mathbf{C}_m^T\mathbf{C}_m$. Taking time derivative of (18) along the trajectories of (8), (15) and (17), we have

$$\dot{V} = -\mathbf{s}^T\mathbf{K}_d\mathbf{s} + \mathbf{s}^T\mathbf{e}_\tau - \mathbf{e}_\tau^T\mathbf{e}_\tau + \mathbf{e}_m^T\mathbf{P}_t\mathbf{B}_p\mathbf{H}\mathbf{e}_i - \mathbf{e}_i^T\mathbf{K}_c\mathbf{e}_i$$
$$- \tilde{\mathbf{p}}^T(\boldsymbol{\Gamma}\dot{\hat{\mathbf{p}}} + \mathbf{Y}^T\mathbf{s}) - Tr[\tilde{\mathbf{p}}_i^T(\boldsymbol{\Gamma}_i\dot{\hat{\mathbf{p}}}_i + \boldsymbol{\varphi}\mathbf{e}_i^T)] \tag{19}$$

Pick $\mathbf{B}_m = \mathbf{B}_p$ to have $\mathbf{e}_m^T \mathbf{P}_t \mathbf{B}_p = \mathbf{e}_\tau^T$, then the update law can be selected to be

$$\dot{\hat{\mathbf{p}}} = -\mathbf{\Gamma}^{-1} \mathbf{Y}^T \mathbf{s} \tag{20a}$$

$$\dot{\hat{\mathbf{p}}}_i = -\mathbf{\Gamma}_i^{-1} \boldsymbol{\varphi} \mathbf{e}_i^T \tag{20b}$$

Thus, (19) becomes (13); therefore, we have proved that $\mathbf{s}$, $\mathbf{e}_\tau$ and $\mathbf{e}_i$ are uniformly bounded, and their square integrability can also be proved from (13). Furthermore, uniformly boundedness of $\dot{\mathbf{s}}$, $\dot{\mathbf{e}}_\tau$, and $\dot{\mathbf{e}}_i$ are also easy to be proved, and hence, $\mathbf{q} \to \mathbf{q}_d$, $\boldsymbol{\tau}_t \to \boldsymbol{\tau}_{td}$, and $\mathbf{i} \to \mathbf{i}_d$ follow by Barbalat's lemma.

Remark 7: To implement the controller strategy, we do not need to have the knowledge of most system parameters, but we have to feedback $\ddot{\mathbf{q}}$ and calculate the regressor matrix and their higher order derivatives. Therefore, the design introduced in this section is not feasible for practical applications, either.

6.5.2 Regressor-free adaptive controller design

Consider the system in (1) again, but $\mathbf{D}$, $\mathbf{C}$, $\mathbf{g}$, $\mathbf{L}$, $\mathbf{R}$ and $\mathbf{K}_b$ are unavailable. The desired torque trajectory in (2) is modified as

$$\boldsymbol{\tau}_{td} = \hat{\mathbf{g}} + \hat{\mathbf{D}}\dot{\mathbf{v}} + \hat{\mathbf{C}}\mathbf{v} - \mathbf{K}_d \mathbf{s} \tag{21}$$

The dynamics for the output tracking loop can thus be written as

$$\mathbf{D}\dot{\mathbf{s}} + \mathbf{C}\mathbf{s} + \mathbf{K}_d \mathbf{s} = -\tilde{\mathbf{D}}\dot{\mathbf{v}} - \tilde{\mathbf{C}}\mathbf{v} - \tilde{\mathbf{g}} + (\boldsymbol{\tau}_t - \boldsymbol{\tau}_{td}) \tag{22}$$

Therefore, if we may find a control law to drive $\boldsymbol{\tau}_t \to \boldsymbol{\tau}_{td}$ and update laws to have $\hat{\mathbf{D}} \to \mathbf{D}$, $\hat{\mathbf{C}} \to \mathbf{C}$, and $\hat{\mathbf{g}} \to \mathbf{g}$, then (22) implies convergence of the output error. To this end, we would like to use the MRC rule with the reference model in (4) and the state space representation in (5). The desired current trajectory is designed according to (6) to be

$$\mathbf{i}_d = \mathbf{H}^{-1}[\boldsymbol{\Theta}\mathbf{x}_p + \boldsymbol{\Phi}\boldsymbol{\tau}_{td} + \hat{\mathbf{h}}] \tag{23}$$

where $\hat{\mathbf{h}}$ is an estimate of $\mathbf{h}(\overline{\boldsymbol{\tau}}_{td}, \overline{\mathbf{q}}) = \boldsymbol{\Phi}\overline{\boldsymbol{\tau}}_{td} + \overline{\mathbf{q}}$. Consequently, the dynamics for the torque tracking loop becomes

$$\dot{\mathbf{e}}_m = \mathbf{A}_m \mathbf{e}_m + \mathbf{B}_p \mathbf{H}(\mathbf{i} - \mathbf{i}_d) + \mathbf{B}_p (\hat{\mathbf{h}} - \mathbf{h}) \tag{24a}$$

$$\mathbf{e}_\tau = \mathbf{C}_m \mathbf{e}_m \tag{24b}$$

Hence, if we may design a control law to ensure $\mathbf{i} \to \mathbf{i}_d$ and an update law to have $\hat{\mathbf{h}} \to \mathbf{h}$, then we may have convergence of the torque tracking loop. The control strategy can be constructed as

$$\mathbf{u} = \hat{\mathbf{f}} - \mathbf{K}_c \mathbf{e}_i \tag{25}$$

where $\mathbf{e}_i = \mathbf{i} - \mathbf{i}_d$ is the current error, $\mathbf{K}_c$ is a positive definite matrix and $\hat{\mathbf{f}}$ is an estimate of $\mathbf{f}(\mathbf{i}_d, \mathbf{i}, \dot{\mathbf{q}}) = \mathbf{L}\dot{\mathbf{i}}_d + \mathbf{R}\mathbf{i} + \mathbf{K}_b \dot{\mathbf{q}}$. With this control law, the dynamics for the current tracking loop can be found as

$$\mathbf{L}\dot{\mathbf{e}}_i + \mathbf{K}_c \mathbf{e}_i = \hat{\mathbf{f}} - \mathbf{f} \tag{26}$$

If we may select a proper update law to have $\hat{\mathbf{f}} \to \mathbf{f}$, (26) ensures convergence in the current tracking loop. Since $\mathbf{D}(\mathbf{q})$, $\mathbf{C}(\mathbf{q}, \dot{\mathbf{q}})$, $\mathbf{g}(\mathbf{q})$, $\mathbf{h}(\overline{\tau}_{td}, \overline{\mathbf{q}})$ and $\mathbf{f}(\mathbf{i}_d, \mathbf{i}, \dot{\mathbf{q}})$ are time-varying functions and their variation bounds are not given, we would like to use their function approximation representations as

$$\mathbf{D} = \mathbf{W}_\mathbf{D}^T \mathbf{Z}_\mathbf{D} + \varepsilon_\mathbf{D} \tag{27a}$$

$$\mathbf{C} = \mathbf{W}_\mathbf{C}^T \mathbf{Z}_\mathbf{C} + \varepsilon_\mathbf{C} \tag{27b}$$

$$\mathbf{g} = \mathbf{W}_\mathbf{g}^T \mathbf{z}_\mathbf{g} + \varepsilon_\mathbf{g} \tag{27c}$$

$$\mathbf{h} = \mathbf{W}_\mathbf{h}^T \mathbf{z}_\mathbf{h} + \varepsilon_\mathbf{h} \tag{27d}$$

$$\mathbf{f} = \mathbf{W}_\mathbf{f}^T \mathbf{z}_\mathbf{f} + \varepsilon_\mathbf{f} \tag{27e}$$

where $\mathbf{W}_\mathbf{D} \in \mathfrak{R}^{n^2 \beta_D \times n}$, $\mathbf{W}_\mathbf{C} \in \mathfrak{R}^{n^2 \beta_C \times n}$, $\mathbf{W}_\mathbf{g} \in \mathfrak{R}^{n \beta_g \times n}$, $\mathbf{W}_\mathbf{h} \in \mathfrak{R}^{n \beta_h \times n}$, and $\mathbf{W}_\mathbf{f} \in \mathfrak{R}^{n \beta_f \times n}$ are weighting matrices for $\mathbf{D}$, $\mathbf{C}$, $\mathbf{g}$, $\mathbf{h}$, and $\mathbf{f}$, respectively, while $\mathbf{Z}_\mathbf{D} \in \mathfrak{R}^{n^2 \beta_D \times n}$, $\mathbf{Z}_\mathbf{C} \in \mathfrak{R}^{n^2 \beta_C \times n}$, $\mathbf{z}_\mathbf{g} \in \mathfrak{R}^{n \beta_g \times 1}$, $\mathbf{z}_\mathbf{h} \in \mathfrak{R}^{n \beta_h \times 1}$, and $\mathbf{z}_\mathbf{f} \in \mathfrak{R}^{n \beta_f \times 1}$ are basis function matrices. Likewise, we have the representations for the estimates as

$$\hat{\mathbf{D}} = \hat{\mathbf{W}}_\mathbf{D}^T \mathbf{Z}_\mathbf{D} \tag{27f}$$

$$\hat{\mathbf{C}} = \hat{\mathbf{W}}_\mathbf{C}^T \mathbf{Z}_\mathbf{C} \tag{27g}$$

$$\hat{\mathbf{g}} = \hat{\mathbf{W}}_{\mathbf{g}}^T \mathbf{z_g} \tag{27h}$$

$$\hat{\mathbf{h}} = \hat{\mathbf{W}}_{\mathbf{h}}^T \mathbf{z_h} \tag{27i}$$

$$\hat{\mathbf{f}} = \hat{\mathbf{W}}_{\mathbf{f}}^T \mathbf{z_f} \tag{27j}$$

Thus the output error dynamics (22), torque tracking error dynamics (24a), and current tracking error dynamics (26) can be rewritten as

$$\mathbf{D}\dot{\mathbf{s}} + \mathbf{C}\mathbf{s} + \mathbf{K}_d\mathbf{s} = (\boldsymbol{\tau}_t - \boldsymbol{\tau}_{td}) - \tilde{\mathbf{W}}_{\mathbf{D}}^T \mathbf{Z_D}\dot{\mathbf{v}} - \tilde{\mathbf{W}}_{\mathbf{C}}^T \mathbf{Z_C}\mathbf{v} - \tilde{\mathbf{W}}_{\mathbf{g}}^T \mathbf{z_g} + \boldsymbol{\varepsilon}_1 \tag{28a}$$

$$\dot{\mathbf{e}}_m = \mathbf{A}_m \mathbf{e}_m - \mathbf{B}_p \tilde{\mathbf{W}}_{\mathbf{h}}^T \mathbf{z_h} + \mathbf{B}_p \mathbf{H}\mathbf{e}_i + \mathbf{B}_p \boldsymbol{\varepsilon}_2 \tag{28b}$$

$$\mathbf{L}\dot{\mathbf{e}}_i + \mathbf{K}_c \mathbf{e}_i = -\tilde{\mathbf{W}}_{\mathbf{f}}^T \mathbf{z_f} + \boldsymbol{\varepsilon}_3 \tag{28c}$$

where $\boldsymbol{\varepsilon}_1 = \boldsymbol{\varepsilon}_1(\boldsymbol{\varepsilon}_{\mathbf{D}}, \boldsymbol{\varepsilon}_{\mathbf{C}}, \boldsymbol{\varepsilon}_{\mathbf{g}}, \mathbf{s}, \ddot{\mathbf{q}}_d)$, $\boldsymbol{\varepsilon}_2 = \boldsymbol{\varepsilon}_2(\boldsymbol{\varepsilon}_{\mathbf{h}}, \mathbf{e}_m)$, and $\boldsymbol{\varepsilon}_3 = \boldsymbol{\varepsilon}_3(\boldsymbol{\varepsilon}_{\mathbf{f}}, \mathbf{e}_i)$ are lumped approximation error vectors. Define the Lyapunov-like function candidate as

$$V(\mathbf{s}, \mathbf{e}_m, \mathbf{e}_i, \tilde{\mathbf{W}}_{\mathbf{D}}, \tilde{\mathbf{W}}_{\mathbf{C}}, \tilde{\mathbf{W}}_{\mathbf{g}}, \tilde{\mathbf{W}}_{\mathbf{h}}, \tilde{\mathbf{W}}_{\mathbf{f}}) = \frac{1}{2}\mathbf{s}^T \mathbf{D}\mathbf{s} + \mathbf{e}_m^T \mathbf{P}_t \mathbf{e}_m$$

$$+ \frac{1}{2}\mathbf{e}_i^T \mathbf{L}\mathbf{e}_i + \frac{1}{2}Tr(\tilde{\mathbf{W}}_{\mathbf{D}}^T \mathbf{Q_D}\tilde{\mathbf{W}}_{\mathbf{D}} + \tilde{\mathbf{W}}_{\mathbf{C}}^T \mathbf{Q_C}\tilde{\mathbf{W}}_{\mathbf{C}}$$

$$+ \tilde{\mathbf{W}}_{\mathbf{g}}^T \mathbf{Q_g}\tilde{\mathbf{W}}_{\mathbf{g}} + \tilde{\mathbf{W}}_{\mathbf{h}}^T \mathbf{Q_h}\tilde{\mathbf{W}}_{\mathbf{h}} + \tilde{\mathbf{W}}_{\mathbf{f}}^T \mathbf{Q_f}\tilde{\mathbf{W}}_{\mathbf{f}}) \tag{29}$$

where matrices $\mathbf{Q_D} \in \Re^{n^2\beta_D \times n^2\beta_D}$, $\mathbf{Q_C} \in \Re^{n^2\beta_C \times n^2\beta_C}$, $\mathbf{Q_g} \in \Re^{n\beta_g \times n\beta_g}$, $\mathbf{Q_h} \in \Re^{n\beta_h \times n\beta_h}$, and $\mathbf{Q_f} \in \Re^{n\beta_f \times n\beta_f}$ are positive definite. Along the trajectory of (28), we may compute the time derivative of V as

$$\dot{V} = -\mathbf{s}^T \mathbf{K}_d\mathbf{s} + \mathbf{s}^T \mathbf{e}_\tau - \mathbf{e}_\tau^T \mathbf{e}_\tau + \mathbf{e}_m^T \mathbf{P}_t \mathbf{B}_p \mathbf{H}\mathbf{e}_i - \mathbf{e}_i^T \mathbf{K}_c \mathbf{e}_i + \mathbf{s}^T \boldsymbol{\varepsilon}_1$$

$$+ \mathbf{e}_m^T \mathbf{P}_t \mathbf{B}_p \boldsymbol{\varepsilon}_2 + \mathbf{e}_i^T \boldsymbol{\varepsilon}_3 - Tr[\tilde{\mathbf{W}}_{\mathbf{D}}^T (\mathbf{Z_D}\dot{\mathbf{v}}\mathbf{s}^T + \mathbf{Q_D}\dot{\hat{\mathbf{W}}}_{\mathbf{D}})$$

$$+ \tilde{\mathbf{W}}_{\mathbf{C}}^T (\mathbf{Z_C}\mathbf{v}\mathbf{s}^T + \mathbf{Q_C}\dot{\hat{\mathbf{W}}}_{\mathbf{C}})] - Tr[\tilde{\mathbf{W}}_{\mathbf{g}}^T (\mathbf{z_g}\mathbf{s}^T + \mathbf{Q_g}\dot{\hat{\mathbf{W}}}_{\mathbf{g}})$$

$$+ \tilde{\mathbf{W}}_{\mathbf{h}}^T (\mathbf{z_h}\mathbf{e}_m^T \mathbf{P}_t \mathbf{B}_p + \mathbf{Q_h}\dot{\hat{\mathbf{W}}}_{\mathbf{h}})] - Tr[\tilde{\mathbf{W}}_{\mathbf{f}}^T (\mathbf{z_f}\mathbf{e}_i^T + \mathbf{Q_f}\dot{\hat{\mathbf{W}}}_{\mathbf{f}})] \tag{30}$$

If we select $\mathbf{B}_m = \mathbf{B}_p$ so that $\mathbf{e}_m^T \mathbf{P}_t \mathbf{B}_p = \mathbf{e}_\tau^T$, and if we pick the update laws as

$$\dot{\hat{\mathbf{W}}}_{\mathbf{D}} = -\mathbf{Q}_{\mathbf{D}}^{-1}(\mathbf{Z}_{\mathbf{D}}\dot{\mathbf{v}}\mathbf{s}^T + \sigma_{\mathbf{D}}\hat{\mathbf{W}}_{\mathbf{D}}) \tag{31a}$$

$$\dot{\hat{\mathbf{W}}}_{\mathbf{C}} = -\mathbf{Q}_{\mathbf{C}}^{-1}(\mathbf{Z}_{\mathbf{C}}\mathbf{v}\mathbf{s}^T + \sigma_{\mathbf{C}}\hat{\mathbf{W}}_{\mathbf{C}}) \tag{31b}$$

$$\dot{\hat{\mathbf{W}}}_{\mathbf{g}} = -\mathbf{Q}_{\mathbf{g}}^{-1}(\mathbf{z}_{\mathbf{g}}\mathbf{s}^T + \sigma_{\mathbf{g}}\hat{\mathbf{W}}_{\mathbf{g}}) \tag{31c}$$

$$\dot{\hat{\mathbf{W}}}_{\mathbf{h}} = -\mathbf{Q}_{\mathbf{h}}^{-1}(\mathbf{z}_{\mathbf{h}}\mathbf{e}_\tau^T + \sigma_{\mathbf{h}}\hat{\mathbf{W}}_{\mathbf{h}}) \tag{31d}$$

$$\dot{\hat{\mathbf{W}}}_{\mathbf{f}} = -\mathbf{Q}_{\mathbf{f}}^{-1}(\mathbf{z}_{\mathbf{f}}\mathbf{e}_i^T + \sigma_{\mathbf{f}}\hat{\mathbf{W}}_{\mathbf{f}}) \tag{31e}$$

then (30) becomes

$$\dot{V} = -[\mathbf{s}^T \quad \mathbf{e}_\tau^T \quad \mathbf{e}_i^T]\mathbf{Q}\begin{bmatrix} \mathbf{s} \\ \mathbf{e}_\tau \\ \mathbf{e}_i \end{bmatrix} + [\mathbf{s}^T \quad \mathbf{e}_\tau^T \quad \mathbf{e}_i^T]\begin{bmatrix} \varepsilon_1 \\ \varepsilon_2 \\ \varepsilon_3 \end{bmatrix}$$
$$+ \sigma_{\mathbf{D}}Tr(\tilde{\mathbf{W}}_{\mathbf{D}}^T\hat{\mathbf{W}}_{\mathbf{D}}) + \sigma_{\mathbf{C}}Tr(\tilde{\mathbf{W}}_{\mathbf{C}}^T\hat{\mathbf{W}}_{\mathbf{C}}) + \sigma_{\mathbf{g}}Tr(\tilde{\mathbf{W}}_{\mathbf{g}}^T\hat{\mathbf{W}}_{\mathbf{g}})$$
$$+ \sigma_{\mathbf{h}}Tr(\tilde{\mathbf{W}}_{\mathbf{h}}^T\hat{\mathbf{W}}_{\mathbf{h}}) + \sigma_{\mathbf{f}}Tr(\tilde{\mathbf{W}}_{\mathbf{f}}^T\hat{\mathbf{W}}_{\mathbf{f}}) \tag{32}$$

where $\mathbf{Q} = \begin{bmatrix} \mathbf{K}_d & -\dfrac{1}{2}\mathbf{I}_{n\times n} & \mathbf{0} \\ -\dfrac{1}{2}\mathbf{I}_{n\times n} & \mathbf{I}_{n\times n} & -\dfrac{1}{2}\mathbf{H} \\ \mathbf{0} & -\dfrac{1}{2}\mathbf{H} & \mathbf{K}_c \end{bmatrix}$ is positive definite due to proper

selection of gain matrices $\mathbf{K}_d$ and $\mathbf{K}_c$.

Remark 8: Realization of the control law (25) and update laws (31) does not need the information of joint accelerations, regressor matrix, or their higher order derivatives, which largely simplified its implementation.

Remark 9: Suppose a sufficient number of basis functions are used and the approximation error can be ignored, then it is not necessary to include the

σ-modification terms in (31). Hence, (32) can be reduced to (13), and convergence of $\mathbf{s}$, $\mathbf{e}_\tau$ and $\mathbf{e}_i$ can be further proved by Barbalat's lemma.

Remark 10: If the approximation error cannot be ignored, but we can find positive numbers δ_1, δ_2 and δ_3 such that $\|\varepsilon_i\| \le \delta_i$, $i=1,2,3$, then robust terms $\tau_{robust\,1}$, $\tau_{robust\,2}$ and $\tau_{robust\,3}$ can be included into (21), (23) and (25) to have

$$\tau_{td} = \hat{\mathbf{g}} + \hat{\mathbf{D}}\dot{\mathbf{v}} + \hat{\mathbf{C}}\mathbf{v} - \mathbf{K}_d \mathbf{s} + \tau_{robust\,1}$$

$$\mathbf{i}_d = \mathbf{H}^{-1}[\boldsymbol{\Theta}\mathbf{x}_p + \boldsymbol{\Phi}\tau_{td} + \hat{\mathbf{h}} + \tau_{robust\,2}]$$

$$\mathbf{u} = \hat{\mathbf{f}} - \mathbf{K}_c \mathbf{e}_i + \tau_{robust\,3}$$

Consider the Lyapunov-like function candidate (29) again, and the update law (31) without σ-modification; then the time derivative of V becomes

$$\dot{V} = -[\mathbf{s}^T \quad \mathbf{e}_\tau^T \quad \mathbf{e}_i^T]\mathbf{Q}\begin{bmatrix} \mathbf{s} \\ \mathbf{e}_\tau \\ \mathbf{e}_i \end{bmatrix} + \delta_1\|\mathbf{s}\| + \delta_2\|\mathbf{e}_\tau\| + \delta_3\|\mathbf{e}_i\|$$

$$+\mathbf{s}^T\tau_{robust\,1} + \mathbf{e}_\tau^T\tau_{robust\,2} + \mathbf{e}_i^T\tau_{robust\,3}$$

If we select $\tau_{robust\,1} = -\delta_1[\text{sgn}(s_1) \quad \text{sgn}(s_2) \quad \cdots \quad \text{sgn}(s_n)]^T$ where s_i, $i=1,\ldots,n$ is the i-th entry in $\mathbf{s}$, $\tau_{robust\,2} = -\delta_2[\text{sgn}(e_{\tau_1}) \quad \cdots \quad \text{sgn}(e_{\tau_{2n}})]^T$ where e_{τ_j}, $j=1,\ldots,n$ is the j-th entry in $\mathbf{e}_\tau$, and $\tau_{robust\,3} = -\delta_3[\text{sgn}(e_{i_1}) \quad \cdots \quad \text{sgn}(e_{i_n})]^T$ where e_{i_k}, $k=1,\ldots,n$ is the k-th element in $\mathbf{e}_i$, then we may have (13) again. This will further give convergence of the output error by Barbalat's lemma.

Owing to the existence of the approximation errors, the definiteness of $\dot{V}$ cannot be determined. By considering the inequality

$$V \le \frac{1}{2}\lambda_{\max}(\mathbf{A})\left\|\begin{bmatrix} \mathbf{s} \\ \mathbf{e}_\tau \\ \mathbf{e}_i \end{bmatrix}\right\|^2 + \frac{1}{2}[\lambda_{\max}(\mathbf{Q_D})Tr(\tilde{\mathbf{W}}_D^T\tilde{\mathbf{W}}_D) + \lambda_{\max}(\mathbf{Q_C})Tr(\tilde{\mathbf{W}}_C^T\tilde{\mathbf{W}}_C)$$

$$+ \lambda_{\max}(\mathbf{Q_g})Tr(\tilde{\mathbf{W}}_g^T\tilde{\mathbf{W}}_g) + \lambda_{\max}(\mathbf{Q_h})Tr(\tilde{\mathbf{W}}_h^T\tilde{\mathbf{W}}_h) + \lambda_{\max}(\mathbf{Q_f})Tr(\tilde{\mathbf{W}}_f^T\tilde{\mathbf{W}}_f)]$$

where $\mathbf{A} = \begin{bmatrix} \mathbf{D} & \mathbf{0} & \mathbf{0} \\ \mathbf{0} & 2\mathbf{C}_m^T\mathbf{P}_t\mathbf{C}_m & \mathbf{0} \\ \mathbf{0} & \mathbf{0} & \mathbf{L} \end{bmatrix}$, we may rewrite (32) as

$$\dot{V} \leq -\alpha V + \frac{1}{2}[\alpha\lambda_{\max}(\mathbf{A}) - \lambda_{\min}(\mathbf{Q})]\left\|\begin{bmatrix} \mathbf{s} \\ \mathbf{e}_\tau \\ \mathbf{e}_i \end{bmatrix}\right\|^2 + \frac{1}{2\lambda_{\min}(\mathbf{Q})}\left\|\begin{bmatrix} \boldsymbol{\varepsilon}_1 \\ \boldsymbol{\varepsilon}_2 \\ \boldsymbol{\varepsilon}_3 \end{bmatrix}\right\|^2$$

$$+\frac{1}{2}\{[\alpha\lambda_{\max}(\mathbf{Q_D}) - \sigma_{\mathbf{D}}]Tr(\tilde{\mathbf{W}}_{\mathbf{D}}^T\tilde{\mathbf{W}}_{\mathbf{D}}) + [\alpha\lambda_{\max}(\mathbf{Q_C}) - \sigma_{\mathbf{C}}]Tr(\tilde{\mathbf{W}}_{\mathbf{C}}^T\tilde{\mathbf{W}}_{\mathbf{C}})$$

$$+[\alpha\lambda_{\max}(\mathbf{Q_g}) - \sigma_{\mathbf{g}}]Tr(\tilde{\mathbf{W}}_{\mathbf{g}}^T\tilde{\mathbf{W}}_{\mathbf{g}}) + [\alpha\lambda_{\max}(\mathbf{Q_h}) - \sigma_{\mathbf{h}}]Tr(\tilde{\mathbf{W}}_{\mathbf{h}}^T\tilde{\mathbf{W}}_{\mathbf{h}})$$

$$+[\alpha\lambda_{\max}(\mathbf{Q_f}) - \sigma_{\mathbf{f}}]Tr(\tilde{\mathbf{W}}_{\mathbf{f}}^T\tilde{\mathbf{W}}_{\mathbf{f}})\} + \frac{1}{2}[\sigma_{\mathbf{D}}Tr(\mathbf{W}_{\mathbf{D}}^T\mathbf{W}_{\mathbf{D}}) + \sigma_{\mathbf{C}}Tr(\mathbf{W}_{\mathbf{C}}^T\mathbf{W}_{\mathbf{C}})$$

$$+\sigma_{\mathbf{g}}Tr(\mathbf{W}_{\mathbf{g}}^T\mathbf{W}_{\mathbf{g}}) + \sigma_{\mathbf{h}}Tr(\mathbf{W}_{\mathbf{h}}^T\mathbf{W}_{\mathbf{h}}) + \sigma_{\mathbf{f}}Tr(\mathbf{W}_{\mathbf{f}}^T\mathbf{W}_{\mathbf{f}})]$$

where α is selected to satisfy

$$\alpha \leq \min\left\{\frac{\lambda_{\min}(\mathbf{Q})}{\lambda_{\max}(\mathbf{A})}, \frac{\sigma_{\mathbf{D}}}{\lambda_{\max}(\mathbf{Q_D})}, \frac{\sigma_{\mathbf{C}}}{\lambda_{\max}(\mathbf{Q_C})}, \frac{\sigma_{\mathbf{g}}}{\lambda_{\max}(\mathbf{Q_g})}, \frac{\sigma_{\mathbf{h}}}{\lambda_{\max}(\mathbf{Q_h})}, \frac{\sigma_{\mathbf{f}}}{\lambda_{\max}(\mathbf{Q_f})}\right\}$$

such that we may further have

$$\dot{V} \leq -\alpha V + \frac{1}{2\lambda_{\min}(\mathbf{Q})}\left\|\begin{bmatrix} \boldsymbol{\varepsilon}_1 \\ \boldsymbol{\varepsilon}_2 \\ \boldsymbol{\varepsilon}_3 \end{bmatrix}\right\|^2 + \frac{1}{2}[\sigma_{\mathbf{D}}Tr(\mathbf{W}_{\mathbf{D}}^T\mathbf{W}_{\mathbf{D}})$$

$$+\sigma_{\mathbf{C}}Tr(\mathbf{W}_{\mathbf{C}}^T\mathbf{W}_{\mathbf{C}}) + \sigma_{\mathbf{g}}Tr(\mathbf{W}_{\mathbf{g}}^T\mathbf{W}_{\mathbf{g}})$$

$$+\sigma_{\mathbf{h}}Tr(\mathbf{W}_{\mathbf{h}}^T\mathbf{W}_{\mathbf{h}}) + \sigma_{\mathbf{f}}Tr(\mathbf{W}_{\mathbf{f}}^T\mathbf{W}_{\mathbf{f}})] \tag{33}$$

Therefore, we have proved that $\dot{V} < 0$ whenever

$$V > \frac{1}{2\alpha\lambda_{\min}(\mathbf{Q})}\sup_{\tau \geq t_0}\left\|\begin{bmatrix} \boldsymbol{\varepsilon}_1(\tau) \\ \boldsymbol{\varepsilon}_2(\tau) \\ \boldsymbol{\varepsilon}_3(\tau) \end{bmatrix}\right\|^2 + \frac{1}{2\alpha}[\sigma_{\mathbf{D}}Tr(\mathbf{W}_{\mathbf{D}}^T\mathbf{W}_{\mathbf{D}})$$

$$+\sigma_{\mathbf{C}}Tr(\mathbf{W}_{\mathbf{C}}^T\mathbf{W}_{\mathbf{C}}) + \sigma_{\mathbf{g}}Tr(\mathbf{W}_{\mathbf{g}}^T\mathbf{W}_{\mathbf{g}}) + \sigma_{\mathbf{h}}Tr(\mathbf{W}_{\mathbf{h}}^T\mathbf{W}_{\mathbf{h}}) + \sigma_{\mathbf{f}}Tr(\mathbf{W}_{\mathbf{f}}^T\mathbf{W}_{\mathbf{f}})]$$

i.e., $\mathbf{s}$, $\mathbf{e}_\tau$, $\mathbf{e}_i$, $\tilde{\mathbf{W}}_{\mathbf{D}}$, $\tilde{\mathbf{W}}_{\mathbf{C}}$, $\tilde{\mathbf{W}}_{\mathbf{g}}$, $\tilde{\mathbf{W}}_{\mathbf{h}}$, and $\tilde{\mathbf{W}}_{\mathbf{f}}$ are uniformly ultimately bounded. In addition, (33) implies

$$V(t) \le e^{-\alpha(t-t_0)} V(t_0) + \frac{1}{2\alpha\lambda_{\min}(\mathbf{Q})} \sup_{t_0 < \tau < t} \left\| \begin{bmatrix} \boldsymbol{\varepsilon}_1(\tau) \\ \boldsymbol{\varepsilon}_2(\tau) \\ \boldsymbol{\varepsilon}_3(\tau) \end{bmatrix} \right\|^2$$

$$+ \frac{1}{2\alpha} [\sigma_{\mathbf{D}} Tr(\mathbf{W}_{\mathbf{D}}^T \mathbf{W}_{\mathbf{D}}) + \sigma_{\mathbf{C}} Tr(\mathbf{W}_{\mathbf{C}}^T \mathbf{W}_{\mathbf{C}})$$

$$+ \sigma_{\mathbf{g}} Tr(\mathbf{W}_{\mathbf{g}}^T \mathbf{W}_{\mathbf{g}}) + \sigma_{\mathbf{h}} Tr(\mathbf{W}_{\mathbf{h}}^T \mathbf{W}_{\mathbf{h}}) + \sigma_{\mathbf{f}} Tr(\mathbf{W}_{\mathbf{f}}^T \mathbf{W}_{\mathbf{f}})] \tag{34}$$

Together with the inequality

$$V \ge \frac{1}{2} \lambda_{\min}(\mathbf{A}) \left\| \begin{bmatrix} \mathbf{s} \\ \mathbf{e}_\tau \\ \mathbf{e}_i \end{bmatrix} \right\|^2 + \frac{1}{2} [\lambda_{\min}(\mathbf{Q}_{\mathbf{D}}) Tr(\tilde{\mathbf{W}}_{\mathbf{D}}^T \tilde{\mathbf{W}}_{\mathbf{D}}) + \lambda_{\min}(\mathbf{Q}_{\mathbf{C}}) Tr(\tilde{\mathbf{W}}_{\mathbf{C}}^T \tilde{\mathbf{W}}_{\mathbf{C}})$$

$$+ \lambda_{\min}(\mathbf{Q}_{\mathbf{g}}) Tr(\tilde{\mathbf{W}}_{\mathbf{g}}^T \tilde{\mathbf{W}}_{\mathbf{g}}) + \lambda_{\min}(\mathbf{Q}_{\mathbf{h}}) Tr(\tilde{\mathbf{W}}_{\mathbf{h}}^T \tilde{\mathbf{W}}_{\mathbf{h}}) + \lambda_{\min}(\mathbf{Q}_{\mathbf{f}}) Tr(\tilde{\mathbf{W}}_{\mathbf{f}}^T \tilde{\mathbf{W}}_{\mathbf{f}})]$$

we may find the error bound

$$\left\| \begin{bmatrix} \mathbf{s} \\ \mathbf{e}_\tau \\ \mathbf{e}_i \end{bmatrix} \right\| \le \sqrt{\frac{2V(t_0)}{\lambda_{\min}(\mathbf{A})}} e^{-\frac{\alpha}{2}(t-t_0)} + \frac{1}{\sqrt{\alpha\lambda_{\min}(\mathbf{A})\lambda_{\min}(\mathbf{Q})}} \sup_{t_0 < \tau < t} \left\| \begin{bmatrix} \boldsymbol{\varepsilon}_1(\tau) \\ \boldsymbol{\varepsilon}_2(\tau) \\ \boldsymbol{\varepsilon}_3(\tau) \end{bmatrix} \right\|$$

$$+ \frac{1}{\sqrt{\alpha\lambda_{\min}(\mathbf{A})}} [\sigma_{\mathbf{D}} Tr(\mathbf{W}_{\mathbf{D}}^T \mathbf{W}_{\mathbf{D}}) + \sigma_{\mathbf{C}} Tr(\mathbf{W}_{\mathbf{C}}^T \mathbf{W}_{\mathbf{C}})$$

$$+ \sigma_{\mathbf{g}} Tr(\mathbf{W}_{\mathbf{g}}^T \mathbf{W}_{\mathbf{g}}) + \sigma_{\mathbf{h}} Tr(\mathbf{W}_{\mathbf{h}}^T \mathbf{W}_{\mathbf{h}}) + \sigma_{\mathbf{f}} Tr(\mathbf{W}_{\mathbf{f}}^T \mathbf{W}_{\mathbf{f}})]^{\frac{1}{2}}$$

Table 6.2 summarizes the adaptive control for EDFJR derived in this chapter in terms of the controller forms, update laws and implementation issues.

Table 6.2 Summary of the adaptive control for EDFJR

Electrically driven flexible-joint robots
$\mathbf{D}\dot{\mathbf{s}} + \mathbf{C}\mathbf{s} + \mathbf{g} + \mathbf{D}\dot{\mathbf{v}} + \mathbf{C}\mathbf{v} = \boldsymbol{\tau}_t$
$\mathbf{J}_t\ddot{\boldsymbol{\tau}}_t + \mathbf{B}_t\dot{\boldsymbol{\tau}}_t + \boldsymbol{\tau}_t = \mathbf{H}\mathbf{i} - \overline{\mathbf{q}}(\dot{\mathbf{q}}, \ddot{\mathbf{q}})$ (6.5-1)
$\mathbf{L}\dot{\mathbf{i}} + \mathbf{R}\mathbf{i} + \mathbf{K}_b\dot{\mathbf{q}} = \mathbf{u}$

	Regressor-based	Regressor-free
Controller	$\boldsymbol{\tau}_{td} = \hat{\mathbf{g}} + \hat{\mathbf{D}}\dot{\mathbf{v}} + \hat{\mathbf{C}}\mathbf{v} - \mathbf{K}_d\mathbf{s}$ $= \mathbf{Y}(\mathbf{q},\dot{\mathbf{q}},\mathbf{v},\dot{\mathbf{v}})\hat{\mathbf{p}} - \mathbf{K}_d\mathbf{s}$ $\mathbf{i}_d = \mathbf{H}^{-1}[\boldsymbol{\Theta}\mathbf{x}_p + \boldsymbol{\Phi}\boldsymbol{\tau}_{td} + \mathbf{h}(\boldsymbol{\tau}_{td},\overline{\mathbf{q}})]$ $\mathbf{u} = \hat{\mathbf{L}}\mathbf{i}_d + \hat{\mathbf{R}}\mathbf{i} + \hat{\mathbf{K}}_b\dot{\mathbf{q}} - \mathbf{K}_c\mathbf{e}_i$ $= \hat{\mathbf{p}}_i^T\boldsymbol{\varphi} - \mathbf{K}_c\mathbf{e}_i$ (6.5-14), (6.5-6), (6.5-16)	$\boldsymbol{\tau}_{td} = \hat{\mathbf{g}} + \hat{\mathbf{D}}\dot{\mathbf{v}} + \hat{\mathbf{C}}\mathbf{v} - \mathbf{K}_d\mathbf{s}$ $\mathbf{i}_d = \mathbf{H}^{-1}[\boldsymbol{\Theta}\mathbf{x}_p + \boldsymbol{\Phi}\boldsymbol{\tau}_{td} + \hat{\mathbf{h}}]$ $\mathbf{u} = \hat{\mathbf{f}} - \mathbf{K}_c\mathbf{e}_i$ (6.5-21), (6.5-23), (6.5-25)
Adaptive Law	$\dot{\hat{\mathbf{p}}} = -\boldsymbol{\Gamma}^{-1}\mathbf{Y}^T\mathbf{s}$ $\dot{\hat{\mathbf{p}}}_i = -\boldsymbol{\Gamma}_i^{-1}\boldsymbol{\varphi}\mathbf{e}_i^T$ (6.5-20)	$\dot{\hat{\mathbf{W}}}_{\mathbf{D}} = -\mathbf{Q}_{\mathbf{D}}^{-1}(\mathbf{Z}_{\mathbf{D}}\dot{\mathbf{v}}\mathbf{s}^T + \sigma_{\mathbf{D}}\hat{\mathbf{W}}_{\mathbf{D}})$ $\dot{\hat{\mathbf{W}}}_{\mathbf{C}} = -\mathbf{Q}_{\mathbf{C}}^{-1}(\mathbf{Z}_{\mathbf{C}}\mathbf{v}\mathbf{s}^T + \sigma_{\mathbf{C}}\hat{\mathbf{W}}_{\mathbf{C}})$ $\dot{\hat{\mathbf{W}}}_{\mathbf{g}} = -\mathbf{Q}_{\mathbf{g}}^{-1}(\mathbf{z}_{\mathbf{g}}\mathbf{s}^T + \sigma_{\mathbf{g}}\hat{\mathbf{W}}_{\mathbf{g}})$ $\dot{\hat{\mathbf{W}}}_{\mathbf{h}} = -\mathbf{Q}_{\mathbf{h}}^{-1}(\mathbf{z}_{\mathbf{h}}\mathbf{e}_\tau^T + \sigma_{\mathbf{h}}\hat{\mathbf{W}}_{\mathbf{h}})$ $\dot{\hat{\mathbf{W}}}_{\mathbf{f}} = -\mathbf{Q}_{\mathbf{f}}^{-1}(\mathbf{z}_{\mathbf{f}}\mathbf{e}_i^T + \sigma_{\mathbf{f}}\hat{\mathbf{W}}_{\mathbf{f}})$ (6.5-31)
Realization Issue	Need to know the regressor matrix, joint accelerations and their higher derivatives.	Does not need the information for the regressor matrix, joint accelerations, or their derivatives.

Example 6.2:

Consider the flexible-joint robot in example 6.1 but with consideration of the motor dynamics. Actual values of link parameters are selected as $m_1=m_2=0.5(kg)$, $l_1=l_2=0.75(m)$, $l_{c1}=l_{c2}=0.375(m)$, $I_1=I_2=0.0234(kg\text{-}m^2)$, and $k_1=k_2=100$(N-m/rad). Parameters for the actuator part are chosen as $j_1=0.02(kg\text{-}m^2)$, $j_2=0.01(kg\text{-}m^2)$, $b_1=5$(N-m-sec/rad), $b_2=4$(N-m-sec/rad), and $h_1=h_2=10$(N-m/A). Electrical parameter for the actuator are $L_1=L_2=0.025$(H), $r_1=r_2=1(\Omega)$, $k_{b1}=k_{b2}=1$ (Vol/rad/sec) (Chien and Huang 2007a). In order to observe the effect of the actuator dynamics, the endpoint is required to track a $0.2m$ radius circle centered at $(0.8m, 1.0m)$ in 2 seconds which is much faster than the case in example 6.1. The initial condition for the generalized coordinate vector is at $\mathbf{q}(0) = \boldsymbol{\theta}(0) = [0.0022 \quad 1.5019 \quad 0 \quad 0]^T$, i.e., the endpoint is initially at

$(0.8m, 0,75m)$. It is away from the desired initial endpoint position $(0.8m, 0,8m)$ for observation of the transient. The initial state for the reference model is $\boldsymbol{\tau}_r(0) = [15.5 \quad -33.6 \quad 0 \quad 0]^T$, which is the same as the initial state for the desired torque. The controller gain matrices are selected as

$$\mathbf{K}_d = \begin{bmatrix} 20 & 0 \\ 0 & 20 \end{bmatrix}, \quad \mathbf{\Lambda} = \begin{bmatrix} 10 & 0 \\ 0 & 10 \end{bmatrix}, \text{ and } \mathbf{K}_c = \begin{bmatrix} 50 & 0 \\ 0 & 50 \end{bmatrix}.$$

The initial value for the desired current can be found by calculation as

$$\mathbf{i}(0) = [77.5 \quad -83.9]^T.$$

The 11-term Fourier series is selected as the basis function for the approximation. Therefore, $\hat{\mathbf{W}}_{\mathbf{D}}$ and $\hat{\mathbf{W}}_{\mathbf{C}}$ are in $\mathfrak{R}^{44\times2}$, while $\hat{\mathbf{W}}_{\mathbf{g}}$, $\hat{\mathbf{W}}_{\mathbf{h}}$, and $\hat{\mathbf{W}}_{\mathbf{f}}$ are in $\mathfrak{R}^{22\times2}$. The initial weighting vectors for the entries are assigned to be

$$\hat{\mathbf{w}}_{D_{11}}(0) = [0.05 \quad 0 \quad \cdots \quad 0]^T \in \mathfrak{R}^{11\times1}$$

$$\hat{\mathbf{w}}_{D_{12}}(0) = \hat{\mathbf{w}}_{D_{21}}(0) = [-0.05 \quad 0 \quad \cdots \quad 0]^T \in \mathfrak{R}^{11\times1}$$

$$\hat{\mathbf{w}}_{D_{22}}(0) = [0.1 \quad 0 \quad \cdots \quad 0]^T \in \mathfrak{R}^{11\times1}$$

$$\hat{\mathbf{w}}_{C_{11}}(0) = [0.05 \quad 0 \quad \cdots \quad 0]^T \in \mathfrak{R}^{11\times1}$$

$$\hat{\mathbf{w}}_{C_{12}}(0) = \hat{\mathbf{w}}_{C_{21}}(0) = [-0.05 \quad 0 \quad \cdots \quad 0]^T \in \mathfrak{R}^{11\times1}$$

$$\hat{\mathbf{w}}_{C_{22}}(0) = [0.1 \quad 0 \quad \cdots \quad 0]^T \in \mathfrak{R}^{11\times1}$$

$$\hat{\mathbf{w}}_{g_1}(0) = \hat{\mathbf{w}}_{g_2}(0) = [0 \quad 0 \quad \cdots \quad 0]^T \in \mathfrak{R}^{11\times1}$$

$$\hat{\mathbf{w}}_{h_1}(0) = \hat{\mathbf{w}}_{h_2}(0) = [0 \quad 0 \quad \cdots \quad 0]^T \in \mathfrak{R}^{11\times1}$$

$$\hat{\mathbf{w}}_{f_1}(0) = \hat{\mathbf{w}}_{f_2}(0) = [0 \quad 0 \quad \cdots \quad 0]^T \in \mathfrak{R}^{11\times1}$$

The gain matrices in the update law (31) are selected as $\mathbf{Q}_{\mathbf{D}}^{-1} = 0.1\mathbf{I}_{44}$, $\mathbf{Q}_{\mathbf{C}}^{-1} = 0.1\mathbf{I}_{44}$, $\mathbf{Q}_{\mathbf{g}}^{-1} = 50\mathbf{I}_{22}$, $\mathbf{Q}_{\mathbf{h}}^{-1} = 1000\mathbf{I}_{22}$, and $\mathbf{Q}_{\mathbf{f}}^{-1} = 10000\mathbf{I}_{22}$. The approximation error is assumed to be neglected in this simulation, and the σ-modification parameters are chosen as $\sigma_{(\cdot)} = 0$.

The simulation results are shown in Figure 6.10 to 6.19. Figure 6.10 shows the tracking performance of the robot endpoint and its desired trajectory in the Cartesian space. It is observed that the endpoint trajectory converges nicely to the desired trajectory, although the initial position error is quite large. After the transient state, the tracking error is small regardless of the time-varying uncertainties in $\mathbf{D}$, $\mathbf{C}$, $\mathbf{g}$, $\mathbf{h}$, and $\mathbf{f}$. Computation of the complex regressor is avoided in this strategy which greatly simplifies the design and implementation of the control law. Figure 6.11 presents the time history of the joint space tracking performance. The torque tracking performance is shown in Figure 6.12. It can be seen that the torque errors for both joints are small. Figure 6.13 presents the current tracking performance. It is observed that the strategy can give very small current error. The control voltages to the two motors are reasonable that can be verified in Figure 6.14. Figure 6.15 to 6.19 are the performance of function approximation. Although most parameters do not converge to their actual values, they still remain bounded as desired.

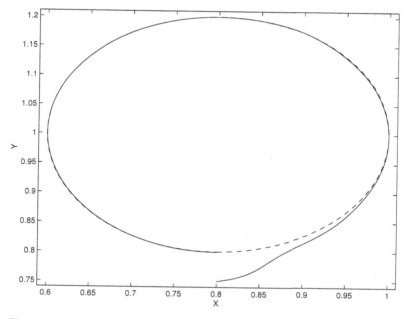

Figure 6.10 Robot endpoint tracking performance in the Cartesian space. After some transient, the endpoint converges to the desired trajectory nicely regardless of the uncertainties in the robot model

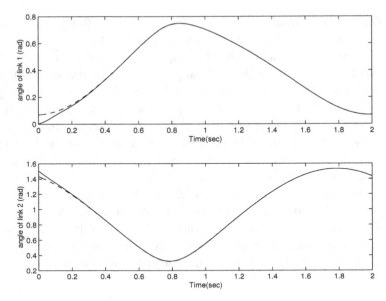

Figure 6.11 Joint space tracking performance. It can be seen that the transient is fast, and the tracking error is very small

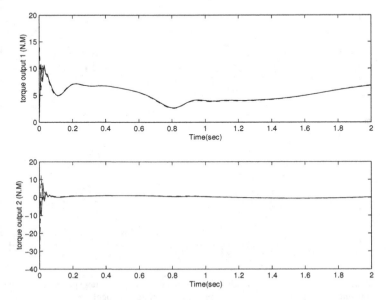

Figure 6.12 Torque tracking performance. It can be seen that the torque errors for both joints are small

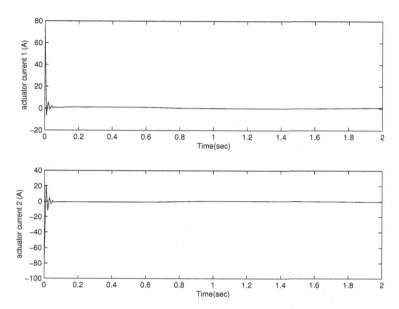

Figure 6.13 Current tracking performance. It can be seen that the current errors for both joints are small

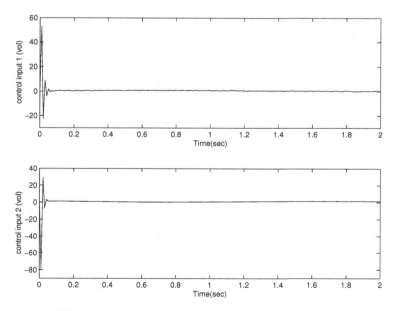

Figure 6.14 The control voltages for both joints are reasonable

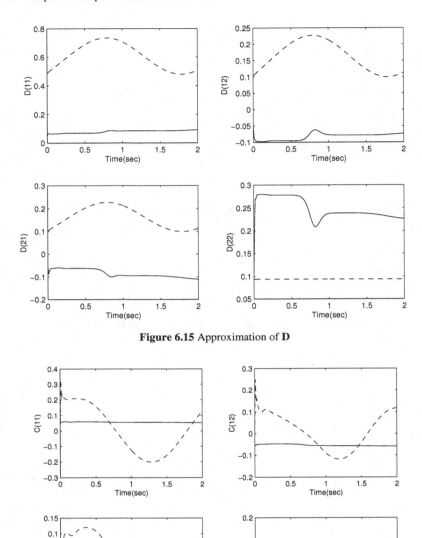

Figure 6.15 Approximation of **D**

Figure 6.16 Approximation of **C**

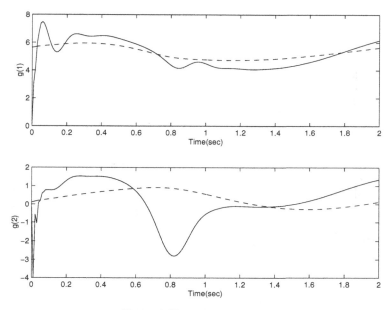

Figure 6.17 Approximation of **g**

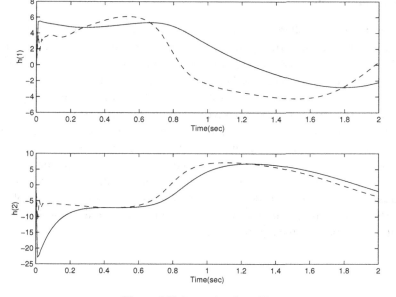

Figure 6.18 Approximation of **h**

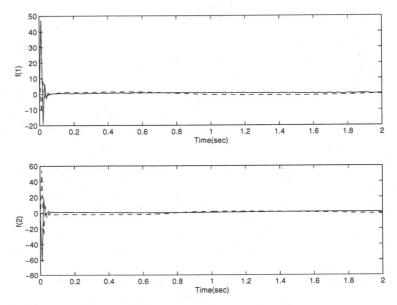

Figure 6.19 Approximation of **f**

6.6 Conclusions

Adaptive controllers for flexible-joint robots in the free space are derived in this chapter. The MRC rule is utilized in Section 6.2 for the control of a known FJR. In Section 6.3, a regressor based adaptive controller is derived. However, its implementation requires the knowledge of joint accelerations, the regressor matrix, and their higher order derivatives. Therefore, the control strategy is not practical. The regressor-free adaptive controller based on FAT is designed in Section 6.4 whose realization do not need the joint accelerations, the regressor matrix, or their higher order derivatives. The actuator dynamics is considered in Section 6.5. A regreesor-based adaptive controller is developed for EDFJR in Section 6.5.1. However, its realization still needs the joint accelerations, regressor matrix, and their derivatives. A regreesor-free adaptive controller for EDFJR is then introduced in Section 6.5.2 and it is free from the information for joint accelerations or the regressor matrix.

Chapter 7

Adaptive Impedance Control of Flexible-Joint Robots

7.1 Introduction

Many practical operations of industrial robots such as grinding and assembling involve contact problems between the end-effector and the environment. To control the robot to interact compliantly with the environment, several approaches have been proposed. Two major strategies are hybrid control presented by Raibert and Craig (1981), and impedance control proposed by Hogan (1985). These two approaches are based on the same assumption that the robot is constructed with rigid links and joints. However, a lot of industrial robot manipulators are designed with harmonic drives to gain high torque with reduced motor speed. To achieve better output performance, the joint flexibility due to the harmonic drives should be carefully considered.

Using the singular perturbation formulation and the concept of integral manifold, Spong (1989) derived a control approach for force/impedance control of flexible-joint robot. Jankowski and ElMaraghy (1991) proposed a nonlinear decoupling and linearizing feedback control based on inverse dynamics. Ahmad (1993) addressed the problem of hybrid force/position control utilizing the constraint formulation developed by Yoshikawa (1986). Based on the solution of the acceleration level inverse dynamic equations, Ider (2000) presented a hybrid force and motion trajectory tracking control law. Ott et. al. (2003) proposed an impedance controller based on an internal torque controller in a cascade structure. Schaffer et. al. (2003) implemented the impedance controllers based on three different kinds of nullspace projections for the realization of nullspace stiffness.

Since the mathematical model is only an approximation of the real system, the simplified representation of the system behavior will contain model inaccuracies such as parametric uncertainties, and unmodeled dynamics.

201

Because these inaccuracies may degrade the performance of the closed-loop system, any practical design should consider their effects. Lian et. al. (1991) presented an adaptive force tracking control scheme for a single-link mechanism with flexible joint based on a two-stage controller. Lin and Goldenberg (1995, 1996, 1997) proposed a combined adaptive and robust control approach to control the motion, internal force, contact force and joint torque simultaneously. Colbaugh et. al. (1997) presented two adaptive schemes for flexible-joint robots based on impedance control and position/force control. In addition, it was shown that the schemes ensure semiglobal uniform boundedness of all signals, and that the ultimate size of the system errors can be made arbitrarily small. Hu and Vukovich (2001) developed a position and force control scheme for flexible-joint robots based on the concept of the integral manifold. However, most of the adaptive constrained motion control approaches need the information of time derivative of the external force which is rarely available precisely in practical applications. Therefore, an adaptive compliant control strategy for flexible joint robot without requiring the time rate of the force feedback is imperative. Moreover, in most adaptive control strategies for robot manipulators, the uncertainties should be linearly parameterizable into the regressor form. It is well-known that derivation of the regressor matrix of a high DOF rigid robot is generally tedious. For the flexible-joint robot, its dynamics is much more complex than that of its rigid-joint counterpart. Hence, the computation of the regressor matrix becomes extremely difficult. Chien and Huang (2006a) proposed a regressor-free adaptive controller for the impedance control of a flexible-joint robot. In this chapter, we would like to consider the impedance control problem for a flexible-joint robot. In Section 7.2, an impedance controller is designed for a known flexible-joint robot. A regressor-based adaptive impedance controller for a flexible-joint robot is derived in Section 7.3. In Section 7.4, a regressor-free impedance controller is constructed with consideration of the joint flexibility. Finally, in Section 7.5, the actuator dynamics is include in the system equation of a flexible-joint robot, resulting in the most complex case in this book.

7.2 Impedance Control of Known Flexible-Joint Robots

The dynamics of an n-link flexible-joint robot interacting with the environment is described in (3.7-1) which can be transformed in the form below by using the same technique in (6.2-2)

$$\mathbf{D}_x\ddot{\mathbf{x}} + \mathbf{C}_x\dot{\mathbf{x}} + \mathbf{g}_x = \mathbf{J}_a^{-T}\boldsymbol{\tau}_t - \mathbf{F}_{ext} \tag{1a}$$

$$\mathbf{J}_t\ddot{\boldsymbol{\tau}}_t + \mathbf{B}_t\dot{\boldsymbol{\tau}}_t + \boldsymbol{\tau}_t = \boldsymbol{\tau}_a - \overline{\mathbf{q}}(\dot{\mathbf{q}}, \ddot{\mathbf{q}}) \tag{1b}$$

where $\mathbf{x} \in \mathfrak{R}^n$, $\mathbf{J}_t = \mathbf{J}\mathbf{K}^{-1}$, $\mathbf{B}_t = \mathbf{B}\mathbf{K}^{-1}$, $\overline{\mathbf{q}}(\dot{\mathbf{q}}, \ddot{\mathbf{q}}) = \mathbf{J}\ddot{\mathbf{q}} + \mathbf{B}\dot{\mathbf{q}}$ and $\boldsymbol{\tau}_t = \mathbf{K}(\boldsymbol{\theta} \text{-} \mathbf{q})$. Suppose all system parameters are known and a controller is to be designed so that the closed-loop system behaves like the target impedance

$$\mathbf{M}_i(\ddot{\mathbf{x}} - \ddot{\mathbf{x}}_d) + \mathbf{B}_i(\dot{\mathbf{x}} - \dot{\mathbf{x}}_d) + \mathbf{K}_i(\mathbf{x} - \mathbf{x}_d) = -\mathbf{F}_{ext} \tag{2}$$

where $\mathbf{x}_d \in \mathfrak{R}^n$ is the desired trajectory, and $\mathbf{M}_i \in \mathfrak{R}^{n \times n}$, $\mathbf{B}_i \in \mathfrak{R}^{n \times n}$, and $\mathbf{K}_i \in \mathfrak{R}^{n \times n}$ are diagonal matrices representing the desired apparent inertia, damping, and stiffness, respectively. The strategy is to regard the impedance controller as a model reference controller, and the target impedance in (2) plays the role of the reference model. Instead of direct utilization of (2), we consider the new target impedance

$$\mathbf{M}_i(\ddot{\mathbf{x}}_i - \ddot{\mathbf{x}}_d) + \mathbf{B}_i(\dot{\mathbf{x}}_i - \dot{\mathbf{x}}_d) + \mathbf{K}_i(\mathbf{x}_i - \mathbf{x}_d) = -\mathbf{F}_{ext} \tag{3}$$

where $\mathbf{x}_i \in \mathfrak{R}^n$ is the state vector of (3). If we can design a controller such that $\mathbf{x}$ converges asymptotically to $\mathbf{x}_i$ then the two target impedances are equivalent.

Define $\mathbf{e} = \mathbf{x} - \mathbf{x}_i$, $\mathbf{s} = \dot{\mathbf{e}} + \Lambda\mathbf{e}$, and $\mathbf{v} = \dot{\mathbf{x}}_i - \Lambda\mathbf{e}$, then we may rewrite (1a) into

$$\mathbf{D}_x\dot{\mathbf{s}} + \mathbf{C}_x\mathbf{s} + \mathbf{g}_x + \mathbf{D}_x\dot{\mathbf{v}} + \mathbf{C}_x\mathbf{v} = \mathbf{J}_a^{-T}\boldsymbol{\tau}_t - \mathbf{F}_{ext} \tag{4}$$

We would like to regard $\boldsymbol{\tau}_t$ as the control torque to drive the system in (4) so that the output tracking error will converge. This can be done by considering the desired torque trajectory

$$\boldsymbol{\tau}_{td} = \mathbf{J}_a^T(\mathbf{F}_{ext} + \mathbf{g}_x + \mathbf{D}_x\dot{\mathbf{v}} + \mathbf{C}_x\mathbf{v} - \mathbf{K}_d\mathbf{s}) \tag{5}$$

Plugging (5) into (4), we have the dynamics for the output tracking error

$$\mathbf{D}_x\dot{\mathbf{s}} + \mathbf{C}_x\mathbf{s} + \mathbf{K}_d\mathbf{s} = \mathbf{J}_a^{-T}(\boldsymbol{\tau}_t - \boldsymbol{\tau}_{td}) \tag{6}$$

Therefore, if we may construct a proper controller $\boldsymbol{\tau}_a$ in (1b) to have $\boldsymbol{\tau}_t \to \boldsymbol{\tau}_{td}$, then (6) implies convergence of $\mathbf{x}$ to $\mathbf{x}_i$. This further implies convergence of the closed-loop system dynamics to the target impedance. To this end, we are going to employ the MRC rule with the reference model

$$\mathbf{J}_r \ddot{\boldsymbol{\tau}}_r + \mathbf{B}_r \dot{\boldsymbol{\tau}}_r + \mathbf{K}_r \boldsymbol{\tau}_r = \mathbf{J}_r \ddot{\boldsymbol{\tau}}_{td} + \mathbf{B}_r \dot{\boldsymbol{\tau}}_{td} + \mathbf{K}_r \boldsymbol{\tau}_{td} \tag{7}$$

where $\boldsymbol{\tau}_r \in \mathfrak{R}^n$ is the state vector of the reference model. Matrices $\mathbf{J}_r \in \mathfrak{R}^{n \times n}$, $\mathbf{B}_r \in \mathfrak{R}^{n \times n}$, and $\mathbf{K}_r \in \mathfrak{R}^{n \times n}$ are selected for the convergence of $\boldsymbol{\tau}_r$ to $\boldsymbol{\tau}_{td}$. Define $\overline{\boldsymbol{\tau}}_{td}(\dot{\boldsymbol{\tau}}_{td}, \ddot{\boldsymbol{\tau}}_{td}) = \mathbf{K}_r^{-1}(\mathbf{B}_r \dot{\boldsymbol{\tau}}_{td} + \mathbf{J}_r \ddot{\boldsymbol{\tau}}_{td})$, and we may represent (1b) and (7) into their state space representations

$$\dot{\mathbf{x}}_p = \mathbf{A}_p \mathbf{x}_p + \mathbf{B}_p \boldsymbol{\tau}_a - \mathbf{B}_p \overline{\mathbf{q}} \tag{8a}$$

$$\dot{\mathbf{x}}_m = \mathbf{A}_m \mathbf{x}_m + \mathbf{B}_m(\boldsymbol{\tau}_{td} + \overline{\boldsymbol{\tau}}_{td}) \tag{8b}$$

where $\mathbf{x}_p = [\boldsymbol{\tau}_t^T \quad \dot{\boldsymbol{\tau}}_t^T]^T \in \mathfrak{R}^{2n}$ and $\mathbf{x}_m = [\boldsymbol{\tau}_r^T \quad \dot{\boldsymbol{\tau}}_r^T]^T \in \mathfrak{R}^{2n}$ are augmented state vectors, $\mathbf{A}_p = \begin{bmatrix} \mathbf{0} & \mathbf{I}_{n \times n} \\ -\mathbf{J}_t^{-1} & -\mathbf{J}_t^{-1}\mathbf{B}_t \end{bmatrix} \in \mathfrak{R}^{2n \times 2n}$ and $\mathbf{A}_m = \begin{bmatrix} \mathbf{0} & \mathbf{I}_{n \times n} \\ -\mathbf{J}_r^{-1}\mathbf{K}_r & -\mathbf{J}_r^{-1}\mathbf{B}_r \end{bmatrix} \in \mathfrak{R}^{2n \times 2n}$ are augmented system matrices, and $\mathbf{B}_p = \begin{bmatrix} \mathbf{0} \\ \mathbf{J}_t^{-1} \end{bmatrix} \in \mathfrak{R}^{2n \times n}$ and $\mathbf{B}_m = \begin{bmatrix} \mathbf{0} \\ \mathbf{J}_r^{-1}\mathbf{K}_r \end{bmatrix} \in \mathfrak{R}^{2n \times n}$ are augmented input gain matrices. The pair $(\mathbf{A}_m, \mathbf{B}_m)$ is controllable, and $(\mathbf{A}_m, \mathbf{C}_m)$ is observable, where $\mathbf{C}_p = \mathbf{C}_m = [\mathbf{I}_{n \times n} \quad \mathbf{0}] \in \mathfrak{R}^{n \times 2n}$ are augmented output matrices characterizing the output signals $\boldsymbol{\tau}_t = \mathbf{C}_p \mathbf{x}_p$ and $\boldsymbol{\tau}_r = \mathbf{C}_m \mathbf{x}_m$, respectively. The transfer function $\mathbf{C}_m(s\mathbf{I} - \mathbf{A}_m)^{-1}\mathbf{B}_m$ is SPR due to proper selection of the matrices $\mathbf{J}_r$, $\mathbf{B}_r$ and $\mathbf{K}_r$. The MRC rule can thus be designed as

$$\boldsymbol{\tau}_a = \boldsymbol{\Theta}\mathbf{x}_p + \boldsymbol{\Phi}\boldsymbol{\tau}_{td} + \mathbf{h}(\overline{\boldsymbol{\tau}}_{td}, \overline{\mathbf{q}}) \tag{9}$$

where $\boldsymbol{\Theta} \in \mathfrak{R}^{n \times 2n}$ and $\boldsymbol{\Phi} \in \mathfrak{R}^{n \times n}$ are matrices satisfying $\mathbf{A}_p + \mathbf{B}_p \boldsymbol{\Theta} = \mathbf{A}_m$ and $\mathbf{B}_p \boldsymbol{\Phi} = \mathbf{B}_m$. The vector $\mathbf{h} \in \mathfrak{R}^n$ is defined as $\mathbf{h}(\overline{\boldsymbol{\tau}}_{td}, \overline{\mathbf{q}}) = \boldsymbol{\Phi}\overline{\boldsymbol{\tau}}_{td} + \overline{\mathbf{q}}$. Substituting (9) into (8a), we may have the dynamics after some manipulation

$$\dot{\mathbf{x}}_p = \mathbf{A}_m \mathbf{x}_p + \mathbf{B}_m(\boldsymbol{\tau}_{td} + \overline{\boldsymbol{\tau}}_{td}) \tag{10}$$

Define $\mathbf{e}_m = \mathbf{x}_p - \mathbf{x}_m$ and $\mathbf{e}_\tau = \mathbf{\tau}_t - \mathbf{\tau}_r$, then using (8b) and (10) we may obtain the error dynamics

$$\dot{\mathbf{e}}_m = \mathbf{A}_m \mathbf{e}_m \tag{11a}$$

$$\mathbf{e}_\tau = \mathbf{C}_m \mathbf{e}_m \tag{11b}$$

Let us consider the Lyapunov-like function candidate

$$V(\mathbf{s}, \mathbf{e}_m) = \frac{1}{2} \mathbf{s}^T \mathbf{D}_x \mathbf{s} + \mathbf{e}_m^T \mathbf{P}_t \mathbf{e}_m \tag{12}$$

where $\mathbf{P}_t = \mathbf{P}_t^T \in \Re^{2n \times 2n}$ is a positive definite matrix satisfying the Lyapunov equation $\mathbf{A}_m^T \mathbf{P}_t + \mathbf{P}_t \mathbf{A}_m = -\mathbf{C}_m^T \mathbf{C}_m$. Along the trajectory of (6) and (11), we may compute the time derivative of V as

$$\dot{V} = -[\mathbf{s}^T \quad \mathbf{e}_\tau^T] \mathbf{Q} \begin{bmatrix} \mathbf{s} \\ \mathbf{e}_\tau \end{bmatrix} \leq 0 \tag{13}$$

where $\mathbf{Q} = \begin{bmatrix} \mathbf{K}_d & -\frac{1}{2} \mathbf{J}_a^{-T} \\ -\frac{1}{2} \mathbf{J}_a^{-T} & \mathbf{I}_{n \times n} \end{bmatrix}$ is positive definite with proper selection of

$\mathbf{K}_d$ and $\mathbf{K}_c$. Therefore, $\mathbf{s}$ and $\mathbf{e}_m$ are uniformly bounded and square integrable. It is also easy to prove that $\dot{\mathbf{s}}$ and $\dot{\mathbf{e}}_\tau$ are uniformly bounded. Hence, convergence of $\mathbf{s}$ and $\mathbf{e}_m$ can be concluded by Barbalat's lemma. This implies that the closed loop system converges asymptotically to the target impedance.

Remark 1: This strategy is feasible only when all system parameters are available. In addition, we need to feedback accelerations and external forces as well as their higher order derivatives that largely restrict the practical applications.

7.3 Regressor-Based Adaptive Impedance Control of Flexible-Joint Robots

In this section, we are concerned with the case when the system parameters are not available. Consider the system equation (7.2-4) and (7.2-1b) again

$$\mathbf{D}_x \dot{\mathbf{s}} + \mathbf{C}_x \mathbf{s} + \mathbf{g}_x + \mathbf{D}_x \dot{\mathbf{v}} + \mathbf{C}_x \mathbf{v} = \mathbf{J}_a^{-T} \boldsymbol{\tau}_t - \mathbf{F}_{ext} \tag{1a}$$

$$\mathbf{J}_t \ddot{\boldsymbol{\tau}}_t + \mathbf{B}_t \dot{\boldsymbol{\tau}}_t + \boldsymbol{\tau}_t = \boldsymbol{\tau}_a - \overline{\mathbf{q}}(\dot{\mathbf{q}}, \ddot{\mathbf{q}}) \tag{1b}$$

We would like to design a control torque $\boldsymbol{\tau}_a$ such that the closed-loop system converges to the target impedance (7.2-3)

$$\mathbf{M}_i(\ddot{\mathbf{x}}_i - \ddot{\mathbf{x}}_d) + \mathbf{B}_i(\dot{\mathbf{x}}_i - \dot{\mathbf{x}}_d) + \mathbf{K}_i(\mathbf{x}_i - \mathbf{x}_d) = -\mathbf{F}_{ext} \tag{2}$$

Since $\mathbf{D}_x$, $\mathbf{C}_x$, and $\mathbf{g}_x$ are not known, desired transmission torque $\boldsymbol{\tau}_{td}$ in (7.2-5) is not realizable. Let us consider a modified version

$$\begin{aligned}
\boldsymbol{\tau}_{td} &= \mathbf{J}_a^T (\mathbf{F}_{ext} + \hat{\mathbf{g}}_x + \hat{\mathbf{D}}_x \dot{\mathbf{v}} + \hat{\mathbf{C}}_x \mathbf{v} - \mathbf{K}_d \mathbf{s}) \\
&= \mathbf{J}_a^T [\mathbf{F}_{ext} + \mathbf{Y}(\mathbf{x}, \dot{\mathbf{x}}, \mathbf{v}, \dot{\mathbf{v}})\hat{\mathbf{p}}_x - \mathbf{K}_d \mathbf{s}]
\end{aligned} \tag{3}$$

where $\hat{\mathbf{D}}_x$, $\hat{\mathbf{C}}_x$, $\hat{\mathbf{g}}_x$, and $\hat{\mathbf{p}}_x$ are estimates of $\mathbf{D}_x$, $\mathbf{C}_x$, $\mathbf{g}_x$, and $\mathbf{p}_x$, respectively. With this new desired transmission torque, equation (1a) becomes

$$\begin{aligned}
\mathbf{D}_x \dot{\mathbf{s}} + \mathbf{C}_x \mathbf{s} + \mathbf{K}_d \mathbf{s} &= -\tilde{\mathbf{D}}_x \dot{\mathbf{v}} - \tilde{\mathbf{C}}_x \mathbf{v} - \tilde{\mathbf{g}}_x + \mathbf{J}_a^{-T}(\boldsymbol{\tau}_t - \boldsymbol{\tau}_{td}) \\
&= -\mathbf{Y}(\mathbf{x}, \dot{\mathbf{x}}, \mathbf{v}, \dot{\mathbf{v}})\tilde{\mathbf{p}}_x + \mathbf{J}_a^{-T}(\boldsymbol{\tau}_t - \boldsymbol{\tau}_{td})
\end{aligned} \tag{4}$$

where $\tilde{\mathbf{D}}_x = \mathbf{D}_x - \hat{\mathbf{D}}_x$, $\tilde{\mathbf{C}}_x = \mathbf{C}_x - \hat{\mathbf{C}}_x$, $\tilde{\mathbf{g}}_x = \mathbf{g}_x - \hat{\mathbf{g}}_x$, and $\tilde{\mathbf{p}}_x = \mathbf{p}_x - \hat{\mathbf{p}}_x$. Hence, if a proper control torque $\boldsymbol{\tau}_a$ can be constructed such that $\boldsymbol{\tau}_t \to \boldsymbol{\tau}_{td}$, and an update law can be designed to have $\hat{\mathbf{p}}_x \to \mathbf{p}_x$, then (4) implies convergence of $\mathbf{s}$, i.e., the closed-loop system behaves like the target impedance. Here, the MRC rule is going to be used again to complete the design. Let us consider the reference model (7.2-7) and the state space representation (7.2-8). The control torque is selected as (7.2-9) to have the torque tracking loop dynamics (7.2-11)

$$\dot{\mathbf{e}}_m = \mathbf{A}_m \mathbf{e}_m \tag{5a}$$

$$\mathbf{e}_\tau = \mathbf{C}_m \mathbf{e}_m \tag{5b}$$

The Lyapunov-like function candidate is selected as

$$V(\mathbf{s}, \mathbf{e}_m, \tilde{\mathbf{p}}_x) = \frac{1}{2}\mathbf{s}^T \mathbf{D}_x \mathbf{s} + \mathbf{e}_m^T \mathbf{P}_t \mathbf{e}_m + \frac{1}{2}\tilde{\mathbf{p}}_x^T \boldsymbol{\Gamma} \tilde{\mathbf{p}}_x \tag{6}$$

where $\mathbf{P}_t = \mathbf{P}_t^T \in \mathfrak{R}^{2n \times 2n}$ is a positive definite matrix satisfying the Lyapunov equation $\mathbf{A}_m^T \mathbf{P}_t + \mathbf{P}_t \mathbf{A}_m = -\mathbf{C}_m^T \mathbf{C}_m$, and $\mathbf{\Gamma} \in \mathfrak{R}^{r \times r}$ is positive definite. Along the trajectories of (4) and (5), the time derivative of (6) can be computed as

$$\dot{V} = -\mathbf{s}^T \mathbf{K}_d \mathbf{s} + \mathbf{s}^T \mathbf{J}_a^{-T} \mathbf{e}_\tau - \mathbf{e}_\tau^T \mathbf{e}_\tau - \tilde{\mathbf{p}}_x^T (\mathbf{\Gamma} \dot{\hat{\mathbf{p}}}_x + \mathbf{Y}^T \mathbf{s}) \tag{7}$$

If we select the update law as

$$\dot{\hat{\mathbf{p}}}_x = -\mathbf{\Gamma}^{-1} \mathbf{Y}^T \mathbf{s} \tag{8}$$

then (7) becomes (7.2-13), and same performance can be concluded.

Remark 2: In this design, we do not need to know the system parameters, but the regressor matrix should be known. Again, in realization of the control torque $\boldsymbol{\tau}_a$, we need to feedback the accelerations and external forces as well as their higher order derivatives. Therefore, this strategy is not practical either.

7.4 Regressor-Free Adaptive Impedance Control of Flexible-Joint Robots

Let us consider the uncertain system equation (7.3-1) again

$$\mathbf{D}_x \dot{\mathbf{s}} + \mathbf{C}_x \mathbf{s} + \mathbf{g}_x + \mathbf{D}_x \dot{\mathbf{v}} + \mathbf{C}_x \mathbf{v} = \mathbf{J}_a^{-T} \boldsymbol{\tau}_t - \mathbf{F}_{ext} \tag{1a}$$

$$\mathbf{J}_t \ddot{\boldsymbol{\tau}}_t + \mathbf{B}_t \dot{\boldsymbol{\tau}}_t + \boldsymbol{\tau}_t = \boldsymbol{\tau}_a - \overline{\mathbf{q}}(\dot{\mathbf{q}}, \ddot{\mathbf{q}}) \tag{1b}$$

The same transmission torque in (7.3-3) is to be used without the regressor representation

$$\boldsymbol{\tau}_{td} = \mathbf{J}_a^T (\mathbf{F}_{ext} + \hat{\mathbf{g}}_x + \hat{\mathbf{D}}_x \dot{\mathbf{v}} + \hat{\mathbf{C}}_x \mathbf{v} - \mathbf{K}_d \mathbf{s}) \tag{2}$$

Again, equation (1a) can be further written as

$$\mathbf{D}_x \dot{\mathbf{s}} + \mathbf{C}_x \mathbf{s} + \mathbf{K}_d \mathbf{s} = -\tilde{\mathbf{D}}_x \dot{\mathbf{v}} - \tilde{\mathbf{C}}_x \mathbf{v} - \tilde{\mathbf{g}}_x + \mathbf{J}_a^{-T} (\boldsymbol{\tau}_t - \boldsymbol{\tau}_{td}) \tag{3}$$

Along the same line as in the previous section, we would like to employ the MRC rule to have $\boldsymbol{\tau}_t \to \boldsymbol{\tau}_{td}$, and the reference model in (7.2-7) is to be used again

$$\mathbf{J}_r \ddot{\boldsymbol{\tau}}_r + \mathbf{B}_r \dot{\boldsymbol{\tau}}_r + \mathbf{K}_r \boldsymbol{\tau}_r = \mathbf{J}_r \ddot{\boldsymbol{\tau}}_{td} + \mathbf{B}_r \dot{\boldsymbol{\tau}}_{td} + \mathbf{K}_r \boldsymbol{\tau}_{td} \tag{4}$$

Similar to (7.2-8), we have the state space representation

$$\dot{\mathbf{x}}_p = \mathbf{A}_p \mathbf{x}_p + \mathbf{B}_p \boldsymbol{\tau}_a - \mathbf{B}_p \bar{\mathbf{q}} \tag{5a}$$

$$\dot{\mathbf{x}}_m = \mathbf{A}_m \mathbf{x}_m + \mathbf{B}_m (\boldsymbol{\tau}_{td} + \bar{\boldsymbol{\tau}}_{td}) \tag{5b}$$

The control torque can thus be designed based on the MRC rule as

$$\boldsymbol{\tau}_a = \boldsymbol{\Theta} \mathbf{x}_p + \boldsymbol{\Phi} \boldsymbol{\tau}_{td} + \hat{\mathbf{h}} \tag{6}$$

where $\boldsymbol{\Theta} \in \mathfrak{R}^{n \times 2n}$ and $\boldsymbol{\Phi} \in \mathfrak{R}^{n \times n}$ satisfy $\mathbf{A}_p + \mathbf{B}_p \boldsymbol{\Theta} = \mathbf{A}_m$ and $\mathbf{B}_p \boldsymbol{\Phi} = \mathbf{B}_m$, respectively, and $\hat{\mathbf{h}}$ is the estimate of $\mathbf{h}(\bar{\boldsymbol{\tau}}_{td}, \bar{\mathbf{q}}) = \boldsymbol{\Phi} \bar{\boldsymbol{\tau}}_{td} + \bar{\mathbf{q}}$. Plugging (6) into (5a) gives

$$\dot{\mathbf{x}}_p = \mathbf{A}_m \mathbf{x}_p + \mathbf{B}_m (\boldsymbol{\tau}_{td} + \bar{\boldsymbol{\tau}}_{td}) + \mathbf{B}_p (\hat{\mathbf{h}} - \mathbf{h}) \tag{7}$$

With the definition of $\mathbf{e}_m = \mathbf{x}_p - \mathbf{x}_m$ and $\mathbf{e}_\tau = \boldsymbol{\tau}_t - \boldsymbol{\tau}_r$, we have the dynamics for the torque tracking loop

$$\dot{\mathbf{e}}_m = \mathbf{A}_m \mathbf{e}_m + \mathbf{B}_p (\hat{\mathbf{h}} - \mathbf{h}) \tag{8a}$$

$$\mathbf{e}_\tau = \mathbf{C}_m \mathbf{e}_m \tag{8b}$$

Therefore, if we may find a proper update law to have $\hat{\mathbf{h}} \to \mathbf{h}$, then the torque tracking can be achieved. To proceed, let us consider the function approximation representations of $\mathbf{D}_x$, $\mathbf{C}_x$, $\mathbf{g}_x$, and $\mathbf{h}$ as

$$\mathbf{D}_x = \mathbf{W}_{\mathbf{D}_x}^T \mathbf{Z}_{\mathbf{D}_x} + \boldsymbol{\varepsilon}_{\mathbf{D}_x} \tag{9a}$$

$$\mathbf{C}_x = \mathbf{W}_{\mathbf{C}_x}^T \mathbf{Z}_{\mathbf{C}_x} + \boldsymbol{\varepsilon}_{\mathbf{C}_x} \tag{9b}$$

$$\mathbf{g}_x = \mathbf{W}_{\mathbf{g}_x}^T \mathbf{z}_{\mathbf{g}_x} + \boldsymbol{\varepsilon}_{\mathbf{g}_x} \tag{9c}$$

$$\mathbf{h} = \mathbf{W}_{\mathbf{h}}^T \mathbf{z}_{\mathbf{h}} + \boldsymbol{\varepsilon}_{\mathbf{h}} \tag{9d}$$

Their estimates are respectively represented as

$$\hat{\mathbf{D}}_x = \hat{\mathbf{W}}_{\mathbf{D}_x}^T \mathbf{Z}_{\mathbf{D}_x} \tag{9e}$$

$$\hat{\mathbf{C}}_x = \hat{\mathbf{W}}_{\mathbf{C}_x}^T \mathbf{Z}_{\mathbf{C}_x} \tag{9f}$$

$$\hat{\mathbf{g}}_x = \hat{\mathbf{W}}_{\mathbf{g}_x}^T \mathbf{z}_{\mathbf{g}_x} \tag{9g}$$

$$\hat{\mathbf{h}} = \hat{\mathbf{W}}_{\mathbf{h}}^T \mathbf{z}_{\mathbf{h}} \tag{9h}$$

Therefore, (3) and (8a) can be rewritten as

$$\mathbf{D}_x \dot{\mathbf{s}} + \mathbf{C}_x \mathbf{s} + \mathbf{K}_d \mathbf{s} = \mathbf{J}_a^{-T}(\boldsymbol{\tau}_t - \boldsymbol{\tau}_{td}) - \tilde{\mathbf{W}}_{\mathbf{D}_x}^T \mathbf{Z}_{\mathbf{D}_x} \dot{\mathbf{v}}$$
$$- \tilde{\mathbf{W}}_{\mathbf{C}_x}^T \mathbf{Z}_{\mathbf{C}_x} \mathbf{v} - \tilde{\mathbf{W}}_{\mathbf{g}_x}^T \mathbf{z}_{\mathbf{g}_x} + \boldsymbol{\varepsilon}_1 \tag{10a}$$

$$\dot{\mathbf{e}}_m = \mathbf{A}_m \mathbf{e}_m - \mathbf{B}_p \tilde{\mathbf{W}}_{\mathbf{h}}^T \mathbf{z}_{\mathbf{h}} + \mathbf{B}_p \boldsymbol{\varepsilon}_2 \tag{10b}$$

where $\boldsymbol{\varepsilon}_1 = \boldsymbol{\varepsilon}_1(\boldsymbol{\varepsilon}_{\mathbf{D}_x}, \boldsymbol{\varepsilon}_{\mathbf{C}_x}, \boldsymbol{\varepsilon}_{\mathbf{g}_x}, \mathbf{s}, \ddot{\mathbf{x}}_i)$ and $\boldsymbol{\varepsilon}_2 = \boldsymbol{\varepsilon}_2(\boldsymbol{\varepsilon}_{\mathbf{h}}, \mathbf{e}_m)$ are lumped approximation errors. Select the Lyapunov-like function candidate as

$$V(\mathbf{s}, \mathbf{e}_m, \tilde{\mathbf{W}}_{\mathbf{D}_x}, \tilde{\mathbf{W}}_{\mathbf{C}_x}, \tilde{\mathbf{W}}_{\mathbf{g}_x}, \tilde{\mathbf{W}}_{\mathbf{h}}) = \frac{1}{2}\mathbf{s}^T \mathbf{D}_x \mathbf{s} + \mathbf{e}_m^T \mathbf{P}_t \mathbf{e}_m + \frac{1}{2}Tr(\tilde{\mathbf{W}}_{\mathbf{D}_x}^T \mathbf{Q}_{\mathbf{D}_x} \tilde{\mathbf{W}}_{\mathbf{D}_x}$$
$$+ \tilde{\mathbf{W}}_{\mathbf{C}_x}^T \mathbf{Q}_{\mathbf{C}_x} \tilde{\mathbf{W}}_{\mathbf{C}_x} + \tilde{\mathbf{W}}_{\mathbf{g}_x}^T \mathbf{Q}_{\mathbf{g}_x} \tilde{\mathbf{W}}_{\mathbf{g}_x} + \tilde{\mathbf{W}}_{\mathbf{h}}^T \mathbf{Q}_{\mathbf{h}} \tilde{\mathbf{W}}_{\mathbf{h}}) \tag{11}$$

where $\mathbf{Q}_{\mathbf{D}_x} \in \Re^{n^2 \beta_{\mathbf{D}} \times n^2 \beta_{\mathbf{D}}}$, $\mathbf{Q}_{\mathbf{C}_x} \in \Re^{n^2 \beta_C \times n^2 \beta_C}$, $\mathbf{Q}_{\mathbf{g}_x} \in \Re^{n\beta_g \times n\beta_g}$, and $\mathbf{Q}_{\mathbf{h}} \in \Re^{n\beta_h \times n\beta_h}$ are positive definite matrices. $\mathbf{P}_t = \mathbf{P}_t^T \in \Re^{2n \times 2n}$ is a positive definite matrix satisfying the Lyapunov equation $\mathbf{A}_m^T \mathbf{P}_t + \mathbf{P}_t \mathbf{A}_m = -\mathbf{C}_m^T \mathbf{C}_m$. Along the trajectory of (10), the time derivative of (11) can be found as

$$\dot{V} = -\mathbf{s}^T \mathbf{K}_d \mathbf{s} + \mathbf{s}^T \mathbf{J}_a^{-T} \mathbf{e}_\tau - \mathbf{e}_\tau^T \mathbf{e}_\tau + \mathbf{s}^T \boldsymbol{\varepsilon}_1 + \mathbf{e}_m^T \mathbf{P}_t \mathbf{B}_p \boldsymbol{\varepsilon}_2$$
$$- Tr[\tilde{\mathbf{W}}_{\mathbf{D}_x}^T (\mathbf{Z}_{\mathbf{D}_x} \dot{\mathbf{v}}\mathbf{s}^T + \mathbf{Q}_{\mathbf{D}_x} \dot{\hat{\mathbf{W}}}_{\mathbf{D}_x}) + \tilde{\mathbf{W}}_{\mathbf{C}_x}^T (\mathbf{Z}_{\mathbf{C}_x} \mathbf{v}\mathbf{s}^T + \mathbf{Q}_{\mathbf{C}_x} \dot{\hat{\mathbf{W}}}_{\mathbf{C}_x})]$$
$$- Tr[\tilde{\mathbf{W}}_{\mathbf{g}_x}^T (\mathbf{z}_{\mathbf{g}_x} \mathbf{s}^T + \mathbf{Q}_{\mathbf{g}_x} \dot{\hat{\mathbf{W}}}_{\mathbf{g}_x}) + \tilde{\mathbf{W}}_{\mathbf{h}}^T (\mathbf{z}_{\mathbf{h}} \mathbf{e}_m^T \mathbf{P}_t \mathbf{B}_p + \mathbf{Q}_{\mathbf{h}} \dot{\hat{\mathbf{W}}}_{\mathbf{h}})] \tag{12}$$

By defining $\mathbf{B}_m = \mathbf{B}_p$ to have $\mathbf{e}_m^T \mathbf{P}_t \mathbf{B}_p = \mathbf{e}_\tau^T$, and selecting the update laws as

$$\dot{\hat{\mathbf{W}}}_{\mathbf{D}_x} = -\mathbf{Q}_{\mathbf{D}_x}^{-1}(\mathbf{Z}_{\mathbf{D}_x} \dot{\mathbf{v}}\mathbf{s}^T + \sigma_{\mathbf{D}_x} \hat{\mathbf{W}}_{\mathbf{D}_x}) \tag{13a}$$

$$\dot{\hat{\mathbf{W}}}_{\mathbf{C}_x} = -\mathbf{Q}_{\mathbf{C}_x}^{-1}(\mathbf{Z}_{\mathbf{C}_x} \mathbf{v}\mathbf{s}^T + \sigma_{\mathbf{C}_x} \hat{\mathbf{W}}_{\mathbf{C}_x}) \tag{13b}$$

$$\dot{\hat{\mathbf{W}}}_{\mathbf{g}_x} = -\mathbf{Q}_{\mathbf{g}_x}^{-1}(\mathbf{z}_{\mathbf{g}_x} \mathbf{s}^T + \sigma_{\mathbf{g}_x} \hat{\mathbf{W}}_{\mathbf{g}_x}) \tag{13c}$$

$$\dot{\hat{\mathbf{W}}}_{\mathbf{h}} = -\mathbf{Q}_{\mathbf{h}}^{-1}(\mathbf{z}_{\mathbf{h}} \mathbf{e}_\tau^T + \sigma_{\mathbf{h}} \hat{\mathbf{W}}_{\mathbf{h}}) \tag{13d}$$

then (12) becomes

$$\dot{V} = -[\mathbf{s}^T \quad \mathbf{e}_\tau^T]\mathbf{Q}\begin{bmatrix} \mathbf{s} \\ \mathbf{e}_\tau \end{bmatrix} + [\mathbf{s}^T \quad \mathbf{e}_\tau^T]\begin{bmatrix} \boldsymbol{\varepsilon}_1 \\ \boldsymbol{\varepsilon}_2 \end{bmatrix} + \sigma_{\mathbf{D}_x} Tr(\tilde{\mathbf{W}}_{\mathbf{D}_x}^T \hat{\mathbf{W}}_{\mathbf{D}_x})$$

$$+ \sigma_{\mathbf{C}_x} Tr(\tilde{\mathbf{W}}_{\mathbf{C}_x}^T \hat{\mathbf{W}}_{\mathbf{C}_x}) + \sigma_{\mathbf{g}_x} Tr(\tilde{\mathbf{W}}_{\mathbf{g}_x}^T \hat{\mathbf{W}}_{\mathbf{g}_x}) + \sigma_{\mathbf{h}} Tr(\tilde{\mathbf{W}}_{\mathbf{h}}^T \hat{\mathbf{W}}_{\mathbf{h}}) \qquad (14)$$

where $\mathbf{Q} = \begin{bmatrix} \mathbf{K}_d & -\dfrac{1}{2}\mathbf{J}_a^{-T} \\ -\dfrac{1}{2}\mathbf{J}_a^{-T} & \mathbf{I}_{n \times n} \end{bmatrix}$ is positive definite by proper selection of $\mathbf{K}_d$.

Remark 3: If a sufficient number of basis functions are employed in the function approximation so that $\boldsymbol{\varepsilon}_1 \approx 0$ and $\boldsymbol{\varepsilon}_2 \approx 0$, the σ-modification terms in (13) can be eliminated and (14) becomes $\dot{V} = -[\mathbf{s}^T \quad \mathbf{e}_\tau^T]\mathbf{Q}\begin{bmatrix} \mathbf{s} \\ \mathbf{e}_\tau \end{bmatrix} \leq 0$ which implies both $\mathbf{s}$ and $\mathbf{e}_\tau$ are uniformly bounded and square integrable. It is straightforward to prove $\dot{\mathbf{s}}$ and $\dot{\mathbf{e}}_\tau$ to be uniformly bounded, and hence convergence of $\mathbf{s}$ and $\mathbf{e}_\tau$ can be concluded by Barbalat's lemma, i.e., the closed-loop system converges to the target impedance.

Remark 4: Suppose $\boldsymbol{\varepsilon}_1$ and $\boldsymbol{\varepsilon}_2$ cannot be ignored and there exist positive numbers δ_1 and δ_2 such that $\|\boldsymbol{\varepsilon}_1\| \leq \delta_1$ and $\|\boldsymbol{\varepsilon}_2\| \leq \delta_2$ for all $t \geq 0$, then, instead of (2) and (6), the modified controllers can be constructed as

$$\boldsymbol{\tau}_{td} = \mathbf{J}_a^T (\mathbf{F}_{ext} + \hat{\mathbf{g}}_x + \hat{\mathbf{D}}_x \dot{\mathbf{v}} + \hat{\mathbf{C}}_x \mathbf{v} - \mathbf{K}_d \mathbf{s} + \boldsymbol{\tau}_{robust1})$$

$$\boldsymbol{\tau}_a = \boldsymbol{\Theta}\mathbf{x}_p + \boldsymbol{\Phi}\boldsymbol{\tau}_{td} + \hat{\mathbf{h}} + \boldsymbol{\tau}_{robust2}$$

where $\boldsymbol{\tau}_{robust1}$ and $\boldsymbol{\tau}_{robust2}$ are robust terms to be designed. Let us consider the Lyapunov function candidate (11) and the update law (13) again but with $\sigma_{(\cdot)} = 0$. The time derivative of V can be computed as

$$\dot{V} = -[\mathbf{s}^T \quad \mathbf{e}_\tau^T]\mathbf{Q}\begin{bmatrix} \mathbf{s} \\ \mathbf{e}_\tau \end{bmatrix} + \delta_1\|\mathbf{s}\| + \delta_2\|\mathbf{e}_\tau\| + \mathbf{s}^T\boldsymbol{\tau}_{robust1} + \mathbf{e}_\tau^T\boldsymbol{\tau}_{robust2}$$

By picking $\boldsymbol{\tau}_{robust1} = -\delta_1[\mathrm{sgn}(s_1) \quad \cdots \quad \mathrm{sgn}(s_n)]^T$, where s_i, $i=1,\ldots,n$ is the i-th element of the vector $\mathbf{s}$, and $\boldsymbol{\tau}_{robust2} = -\delta_2[\mathrm{sgn}(e_{\tau_1}) \quad \cdots \quad \mathrm{sgn}(e_{\tau_{2n}})]^T$, where e_{τ_k}, $k=1,\ldots,2n$ is the k-th entry of $\mathbf{e}_\tau$, we may have $\dot{V} \le 0$, and asymptotic convergence of $\mathbf{s}$ and $\mathbf{e}_\tau$ can be concluded by Barbalat's lemma.

Due to existence of $\boldsymbol{\varepsilon}_1$ and $\boldsymbol{\varepsilon}_2$, we may not determine definiteness of $\dot{V}$. Consider the inequality

$$V \le \frac{1}{2}\lambda_{\max}(\mathbf{A})\left\|\begin{bmatrix} \mathbf{s} \\ \mathbf{e}_\tau \end{bmatrix}\right\|^2 + \frac{1}{2}[\lambda_{\max}(\mathbf{Q}_{\mathbf{D}_x})Tr(\tilde{\mathbf{W}}_{\mathbf{D}_x}^T\tilde{\mathbf{W}}_{\mathbf{D}_x}) + \lambda_{\max}(\mathbf{Q}_{\mathbf{C}_x})Tr(\tilde{\mathbf{W}}_{\mathbf{C}_x}^T\tilde{\mathbf{W}}_{\mathbf{C}_x})$$

$$+ \lambda_{\max}(\mathbf{Q}_{\mathbf{g}_x})Tr(\tilde{\mathbf{W}}_{\mathbf{g}_x}^T\tilde{\mathbf{W}}_{\mathbf{g}_x}) + \lambda_{\max}(\mathbf{Q}_\mathbf{h})Tr(\tilde{\mathbf{W}}_\mathbf{h}^T\tilde{\mathbf{W}}_\mathbf{h})]$$

where $\mathbf{A} = \begin{bmatrix} \mathbf{D} & \mathbf{0} \\ \mathbf{0} & 2\mathbf{C}_m^T\mathbf{P}_t\mathbf{C}_m \end{bmatrix}$, we may rewrite (14) to be

$$\dot{V} \le -\alpha V + \frac{1}{2}[\alpha\lambda_{\max}(\mathbf{A}) - \lambda_{\min}(\mathbf{Q})]\left\|\begin{bmatrix} \mathbf{s} \\ \mathbf{e}_\tau \end{bmatrix}\right\|^2 + \frac{1}{2\lambda_{\min}(\mathbf{Q})}\left\|\begin{bmatrix} \boldsymbol{\varepsilon}_1 \\ \boldsymbol{\varepsilon}_2 \end{bmatrix}\right\|^2$$

$$+ \frac{1}{2}\{[\alpha\lambda_{\max}(\mathbf{Q}_{\mathbf{D}_x}) - \sigma_{\mathbf{D}_x}]Tr(\tilde{\mathbf{W}}_{\mathbf{D}_x}^T\tilde{\mathbf{W}}_{\mathbf{D}_x}) + [\alpha\lambda_{\max}(\mathbf{Q}_{\mathbf{C}_x})$$

$$- \sigma_{\mathbf{C}_x}]Tr(\tilde{\mathbf{W}}_{\mathbf{C}_x}^T\tilde{\mathbf{W}}_{\mathbf{C}_x}) + [\alpha\lambda_{\max}(\mathbf{Q}_{\mathbf{g}_x}) - \sigma_{\mathbf{g}_x}]Tr(\tilde{\mathbf{W}}_{\mathbf{g}_x}^T\tilde{\mathbf{W}}_{\mathbf{g}_x})$$

$$+ [\lambda_{\max}(\mathbf{Q}_\mathbf{h}) - \sigma_\mathbf{h}]Tr(\tilde{\mathbf{W}}_\mathbf{h}^T\tilde{\mathbf{W}}_\mathbf{h})\} + \frac{1}{2}[\sigma_{\mathbf{D}_x}Tr(\mathbf{W}_{\mathbf{D}_x}^T\mathbf{W}_{\mathbf{D}_x})$$

$$+ \sigma_{\mathbf{C}_x}Tr(\mathbf{W}_{\mathbf{C}_x}^T\mathbf{W}_{\mathbf{C}_x}) + \sigma_{\mathbf{g}_x}Tr(\mathbf{W}_{\mathbf{g}_x}^T\mathbf{W}_{\mathbf{g}_x}) + \sigma_\mathbf{h}Tr(\mathbf{W}_\mathbf{h}^T\mathbf{W}_\mathbf{h})]$$

where α is selected to satisfy

$$\alpha \le \min\left\{\frac{\lambda_{\min}(\mathbf{Q})}{\lambda_{\max}(\mathbf{A})}, \frac{\sigma_{\mathbf{D}_x}}{\lambda_{\max}(\mathbf{Q}_{\mathbf{D}_x})}, \frac{\sigma_{\mathbf{C}_x}}{\lambda_{\max}(\mathbf{Q}_{\mathbf{C}_x})}, \frac{\sigma_{\mathbf{g}_x}}{\lambda_{\max}(\mathbf{Q}_{\mathbf{g}_x})}, \frac{\sigma_\mathbf{h}}{\lambda_{\max}(\mathbf{Q}_\mathbf{h})}\right\}.$$

Hence, we may have

$$\dot{V} \le -\alpha V + \frac{1}{2\lambda_{\min}(\mathbf{Q})}\left\|\begin{bmatrix} \boldsymbol{\varepsilon}_1 \\ \boldsymbol{\varepsilon}_2 \end{bmatrix}\right\|^2 + \frac{1}{2}[\sigma_{\mathbf{D}_x}Tr(\mathbf{W}_{\mathbf{D}_x}^T\mathbf{W}_{\mathbf{D}_x})$$

$$+ \sigma_{\mathbf{C}_x}Tr(\mathbf{W}_{\mathbf{C}_x}^T\mathbf{W}_{\mathbf{C}_x}) + \sigma_{\mathbf{g}_x}Tr(\mathbf{W}_{\mathbf{g}_x}^T\mathbf{W}_{\mathbf{g}_x}) + \sigma_\mathbf{h}Tr(\mathbf{W}_\mathbf{h}^T\mathbf{W}_\mathbf{h})] \qquad (15)$$

This implies that $\dot{V} \leq 0$ whenever

$$V > \frac{1}{2\alpha\lambda_{\min}(\mathbf{Q})} \sup_{\tau \geq t_0} \left\| \begin{bmatrix} \boldsymbol{\varepsilon}_1(\tau) \\ \boldsymbol{\varepsilon}_2(\tau) \end{bmatrix} \right\|^2 + \frac{1}{2\alpha}[\sigma_{\mathbf{D}_x} Tr(\mathbf{W}_{\mathbf{D}_x}^T \mathbf{W}_{\mathbf{D}_x})$$
$$+ \sigma_{\mathbf{C}_x} Tr(\mathbf{W}_{\mathbf{C}_x}^T \mathbf{W}_{\mathbf{C}_x}) + \sigma_{\mathbf{g}_x} Tr(\mathbf{W}_{\mathbf{g}_x}^T \mathbf{W}_{\mathbf{g}_x}) + \sigma_{\mathbf{h}} Tr(\mathbf{W}_{\mathbf{h}}^T \mathbf{W}_{\mathbf{h}})]$$

i.e., $\mathbf{s}$, $\mathbf{e}_\tau$, $\tilde{\mathbf{W}}_{\mathbf{D}_x}$, $\tilde{\mathbf{W}}_{\mathbf{C}_x}$, $\tilde{\mathbf{W}}_{\mathbf{g}_x}$, and $\tilde{\mathbf{W}}_{\mathbf{h}}$ are uniformly ultimately bounded. The differential inequality (15) can be solved to have

$$V(t) \leq e^{-\alpha(t-t_0)}V(t_0) + \frac{1}{2\alpha\lambda_{\min}(\mathbf{Q})} \sup_{t_0 < \tau < t} \left\| \begin{bmatrix} \boldsymbol{\varepsilon}_1(\tau) \\ \boldsymbol{\varepsilon}_2(\tau) \end{bmatrix} \right\|^2$$
$$+ \frac{1}{2\alpha}[\sigma_{\mathbf{D}_x} Tr(\mathbf{W}_{\mathbf{D}_x}^T \mathbf{W}_{\mathbf{D}_x}) + \sigma_{\mathbf{C}_x} Tr(\mathbf{W}_{\mathbf{C}_x}^T \mathbf{W}_{\mathbf{C}_x})$$
$$+ \sigma_{\mathbf{g}_x} Tr(\mathbf{W}_{\mathbf{g}_x}^T \mathbf{W}_{\mathbf{g}_x}) + \sigma_{\mathbf{h}} Tr(\mathbf{W}_{\mathbf{h}}^T \mathbf{W}_{\mathbf{h}})]$$

Together with the inequality

$$V \geq \frac{1}{2}\lambda_{\min}(\mathbf{A}) \left\| \begin{bmatrix} \mathbf{s} \\ \mathbf{e}_\tau \end{bmatrix} \right\|^2 + \frac{1}{2}[\lambda_{\min}(\mathbf{Q}_{\mathbf{D}_x})Tr(\tilde{\mathbf{W}}_{\mathbf{D}_x}^T \tilde{\mathbf{W}}_{\mathbf{D}_x}) + \lambda_{\min}(\mathbf{Q}_{\mathbf{C}_x})Tr(\tilde{\mathbf{W}}_{\mathbf{C}_x}^T \tilde{\mathbf{W}}_{\mathbf{C}_x})$$
$$+ \lambda_{\min}(\mathbf{Q}_{\mathbf{g}_x})Tr(\tilde{\mathbf{W}}_{\mathbf{g}_x}^T \tilde{\mathbf{W}}_{\mathbf{g}_x}) + \lambda_{\min}(\mathbf{Q}_{\mathbf{h}})Tr(\tilde{\mathbf{W}}_{\mathbf{h}}^T \tilde{\mathbf{W}}_{\mathbf{h}})]$$

we may find the error bound

$$\left\| \begin{bmatrix} \mathbf{s}(t) \\ \mathbf{e}_\tau(t) \end{bmatrix} \right\| \leq \sqrt{\frac{2V(t_0)}{\lambda_{\min}(\mathbf{A})}} e^{-\frac{\alpha}{2}(t-t_0)} + \frac{1}{\sqrt{\alpha\lambda_{\min}(\mathbf{A})\lambda_{\min}(\mathbf{Q})}} \sup_{t_0 < \tau < t} \left\| \begin{bmatrix} \boldsymbol{\varepsilon}_1(\tau) \\ \boldsymbol{\varepsilon}_2(\tau) \end{bmatrix} \right\|$$
$$+ \frac{1}{\sqrt{\alpha\lambda_{\min}(\mathbf{A})}}[\sigma_{\mathbf{D}_x} Tr(\mathbf{W}_{\mathbf{D}_x}^T \mathbf{W}_{\mathbf{D}_x}) + \sigma_{\mathbf{C}_x} Tr(\mathbf{W}_{\mathbf{C}_x}^T \mathbf{W}_{\mathbf{C}_x})$$
$$+ \sigma_{\mathbf{g}_x} Tr(\mathbf{W}_{\mathbf{g}_x}^T \mathbf{W}_{\mathbf{g}_x}) + \sigma_{\mathbf{h}} Tr(\mathbf{W}_{\mathbf{h}}^T \mathbf{W}_{\mathbf{h}})]^{\frac{1}{2}}$$

Remark 5: Realization of the strategy does not need the information of the regressor matrix, accelerations, or time derivatives of the external force, which largely simplifies the implementation.

Table 7.1 summarizes the adaptive impedance control designs for FJR in this chapter in terms of the controller forms, adaptive laws and implementation issues.

Table 7.1 Summary of the adaptive impedance control for FJR

	Flexible-joint robots interacting with environment $$\mathbf{D}_x \dot{\mathbf{s}} + \mathbf{C}_x \mathbf{s} + \mathbf{g}_x + \mathbf{D}_x \dot{\mathbf{v}} + \mathbf{C}_x \mathbf{v} = \mathbf{J}_a^{-T}\boldsymbol{\tau}_t - \mathbf{F}_{ext} \quad (7.3\text{-}1)$$ $$\mathbf{J}_t \ddot{\boldsymbol{\tau}}_t + \mathbf{B}_t \dot{\boldsymbol{\tau}}_t + \boldsymbol{\tau}_t = \boldsymbol{\tau}_a - \bar{\mathbf{q}}(\dot{\mathbf{q}}, \ddot{\mathbf{q}})$$	
	Regressor-based	Regressor-free
Controller	$\boldsymbol{\tau}_{td} = \mathbf{J}_a^T(\mathbf{F}_{ext} + \hat{\mathbf{g}}_x + \hat{\mathbf{D}}_x \dot{\mathbf{v}}$ $+ \hat{\mathbf{C}}_x \mathbf{v} - \mathbf{K}_d \mathbf{s})$ $= \mathbf{J}_a^T[\mathbf{F}_{ext} + \mathbf{Y}(\mathbf{x}, \dot{\mathbf{x}}, \mathbf{v}, \dot{\mathbf{v}})\hat{\mathbf{p}}_x$ $- \mathbf{K}_d \mathbf{s}]$ $\boldsymbol{\tau}_a = \boldsymbol{\Theta}\mathbf{x}_p + \boldsymbol{\Phi}\boldsymbol{\tau}_{td} + \mathbf{h}(\bar{\boldsymbol{\tau}}_{td}, \bar{\mathbf{q}})$ (7.3-3), (7.2-9)	$\boldsymbol{\tau}_{td} = \mathbf{J}_a^T(\mathbf{F}_{ext} + \hat{\mathbf{g}}_x + \hat{\mathbf{D}}_x \dot{\mathbf{v}}$ $+ \hat{\mathbf{C}}_x \mathbf{v} - \mathbf{K}_d \mathbf{s})$ $\boldsymbol{\tau}_a = \boldsymbol{\Theta}\mathbf{x}_p + \boldsymbol{\Phi}\boldsymbol{\tau}_{td} + \hat{\mathbf{h}}$ (7.4-2), (7.4-6)
Adaptive Law	$\dot{\hat{\mathbf{p}}}_x = -\boldsymbol{\Gamma}^{-1}\mathbf{Y}^T\mathbf{s}$ (7.3-8)	$\dot{\hat{\mathbf{W}}}_{\mathbf{D}_x} = -\mathbf{Q}_{\mathbf{D}_x}^{-1}(\mathbf{Z}_{\mathbf{D}_x}\dot{\mathbf{v}}\mathbf{s}^T + \sigma_{\mathbf{D}_x}\hat{\mathbf{W}}_{\mathbf{D}_x})$ $\dot{\hat{\mathbf{W}}}_{\mathbf{C}_x} = -\mathbf{Q}_{\mathbf{C}_x}^{-1}(\mathbf{Z}_{\mathbf{C}_x}\mathbf{v}\mathbf{s}^T + \sigma_{\mathbf{C}_x}\hat{\mathbf{W}}_{\mathbf{C}_x})$ $\dot{\hat{\mathbf{W}}}_{\mathbf{g}_x} = -\mathbf{Q}_{\mathbf{g}_x}^{-1}(\mathbf{z}_{\mathbf{g}_x}\mathbf{s}^T + \sigma_{\mathbf{g}_x}\hat{\mathbf{W}}_{\mathbf{g}_x})$ $\dot{\hat{\mathbf{W}}}_{\mathbf{h}} = -\mathbf{Q}_{\mathbf{h}}^{-1}(\mathbf{z}_{\mathbf{h}}\mathbf{e}_\tau^T + \sigma_{\mathbf{h}}\hat{\mathbf{W}}_{\mathbf{h}})$ (7.4-13)
Realization Issue	Need to feedback joint accelerations, external force, and their higher derivatives.	Does not need the information for the joint accelerations. Does not need to know the higher derivatives of the joint accelerations or external force.

Example 7.1:

Consider the flexible-joint robot (3.6-3), and we would like to verify the efficacy of the strategy developed in this section by computer simulation. Actual values of the system parameters are $m_1 = m_2 = 0.5(kg)$, $l_1 = l_2 = 0.75(m)$, $l_{c1} = l_{c2} = 0.375(m)$, $I_1 = I_2 = 0.0234(kg\text{-}m^2)$, and $k_1 = k_2 = 100(\text{N-}m/\text{rad})$. Parameters at the motor side are $j_1 = 0.02(kg\text{-}m^2)$, $j_2 = 0.01(kg\text{-}m^2)$, $b_1 = 5(\text{N-}m\text{-}sec/rad)$, and $b_2 = 4(\text{N-}m\text{-}sec/rad)$. The endpoint and the motor angle start from the initial value $\mathbf{x}(0) = [0.8m \quad 0.75m \quad 0 \quad 0]^T$ and $\boldsymbol{\theta}(0) = [0.0022 \quad 1.5019 \quad 0 \quad 0]^T$ respectively to track a 0.2m radius circle centered at (0.8m, 1.0m) in 10 seconds without knowing its precise model. The initial state for the reference model is $\boldsymbol{\tau}_r(0) = [9.8 \quad 3.2 \quad 0 \quad 0]^T$, which is the same as the initial value for the desired transmission torque. The constraint surface is smooth and can be modeled as a linear spring $f_{ext}=k_w(x-x_w)$ where f_{ext} is the force acting on the surface, $k_w=5000\text{N/m}$ is the environmental stiffness, x is the coordinate of the end-point in the X direction, and $x_w=0.95m$ is the position of the surface. Since the surface is away from the desired initial endpoint position $(0.8m,\ 0,8m)$, different phases of operations can be observed. The controller is applied with the gain matrices

$$\mathbf{K}_d = \begin{bmatrix} 20 & 0 \\ 0 & 20 \end{bmatrix}, \text{ and } \mathbf{\Lambda} = \begin{bmatrix} 10 & 0 \\ 0 & 10 \end{bmatrix}.$$

Parameter matrices in the target impedance are selected as

$$\mathbf{M}_i = \begin{bmatrix} 0.5 & 0 \\ 0 & 0.5 \end{bmatrix}, \mathbf{B}_i = \begin{bmatrix} 100 & 0 \\ 0 & 100 \end{bmatrix}, \text{ and } \mathbf{K}_i = \begin{bmatrix} 1000 & 0 \\ 0 & 1000 \end{bmatrix}.$$

The 11-term Fourier series is selected as the basis function for the approximation of entries in $\mathbf{D}_x$, $\mathbf{C}_x$, $\mathbf{g}_x$, and $\mathbf{h}$. Therefore, $\hat{\mathbf{W}}_\mathbf{D}$ and $\hat{\mathbf{W}}_\mathbf{C}$ are in $\mathfrak{R}^{44\times2}$, and $\hat{\mathbf{W}}_\mathbf{g}$ and $\hat{\mathbf{W}}_\mathbf{h}$ are in $\mathfrak{R}^{22\times2}$. The initial weighting vectors for the entries are assigned to be

$$\hat{\mathbf{w}}_{D_{x11}}(0) = [0.05 \quad 0 \quad \cdots \quad 0]^T \in \mathfrak{R}^{11\times1}$$

$$\hat{\mathbf{w}}_{D_{x12}}(0) = \hat{\mathbf{w}}_{D_{x21}}(0) = [-0.05 \quad 0 \quad \cdots \quad 0]^T \in \mathfrak{R}^{11\times1}$$

$$\hat{\mathbf{w}}_{D_{x22}}(0) = [0.1 \quad 0 \quad \cdots \quad 0]^T \in \mathfrak{R}^{11\times1}$$

$$\hat{\mathbf{w}}_{C_{x11}}(0) = [0.05 \quad 0 \quad \cdots \quad 0]^T \in \mathfrak{R}^{11\times1}$$

$$\hat{\mathbf{w}}_{C_{x12}}(0) = \hat{\mathbf{w}}_{C_{x21}}(0) = [-0.05 \quad 0 \quad \cdots \quad 0]^T \in \mathfrak{R}^{11\times1}$$

$$\hat{\mathbf{w}}_{C_{x22}}(0) = [0.1 \quad 0 \quad \cdots \quad 0]^T \in \mathfrak{R}^{11\times1}$$

$$\hat{\mathbf{w}}_{g_{x1}}(0) = \hat{\mathbf{w}}_{g_{x2}}(0) = [0 \quad 0 \quad \cdots \quad 0]^T \in \mathfrak{R}^{11\times1}$$

$$\hat{\mathbf{w}}_{h_1}(0) = \hat{\mathbf{w}}_{h_2}(0) = [0 \quad 0 \quad \cdots \quad 0]^T \in \mathfrak{R}^{11\times1}$$

The gain matrices in the update laws (9) are designed as

$$\mathbf{Q_{D_x}^{-1}} = 0.01\mathbf{I}_{44}, \ \mathbf{Q_{C_x}^{-1}} = 0.01\mathbf{I}_{44}, \ \mathbf{Q_{g_x}^{-1}} = 50\mathbf{I}_{22}, \text{ and } \mathbf{Q_h^{-1}} = 10\mathbf{I}_{22}.$$

We assume in the simulation that the approximation error can be neglected, and hence the σ-modification parameters are chosen as $\sigma_{(.)} = 0$. The simulation results are shown in Figure 7.1 to 7.9. Figure 7.1 shows the robot endpoint tracking performance in the Cartesian space. It can be seen that after some transient response the endpoint converges to the desired trajectory in the free space nicely. Afterwards, the endpoint contacts with the constraint surface at $x_w = 0.95(m)$ compliantly. When entering the free space again, the endpoint follows the desired trajectory with very small tracking error regardless of the system uncertainties. Figure 7.2 presents the time history of the joint space tracking performance. The transient states converge very fast without unwanted oscillations. The joint space trajectory in the constraint motion phase is smooth. Figure 7.3 gives the torque tracking performance. The control efforts to the two joints are reasonable that can be verified in Figure 7.4. The external forces exerted on the endpoint during the constraint motion phase are shown in Figure 7.5. Figure 7.6 to 7.9 are the performance of function approximation. Although most parameters do not converge to their actual values, they still remain bounded as desired.

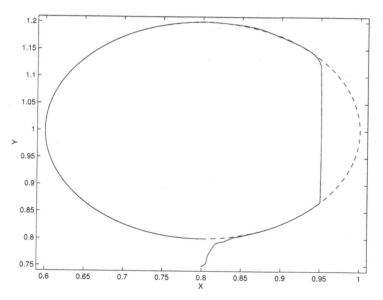

Figure 7.1 Robot endpoint tracking performance in the Cartesian space. After some transient the endpoint converges to the desired trajectory in the free space nicely. Afterwards, the endpoint contacts with the constraint surface compliantly. When entering the free space again, the endpoint follows the desired trajectory with very small tracking error regardless of the system uncertainties

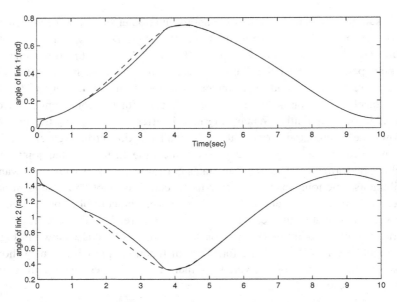

Figure 7.2 The joint space tracking performance. The transient is very fast and the constraint motion phase is smooth

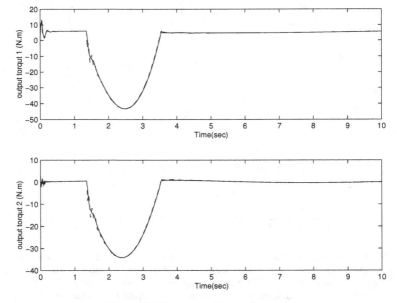

Figure 7.3 Torque tracking performance

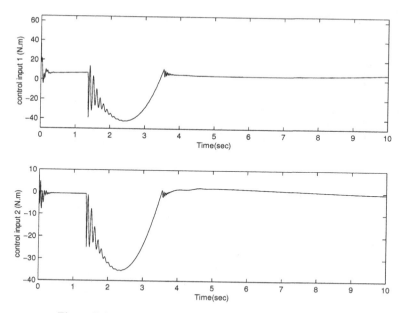

Figure 7.4 The control efforts for both joints are all reasonable

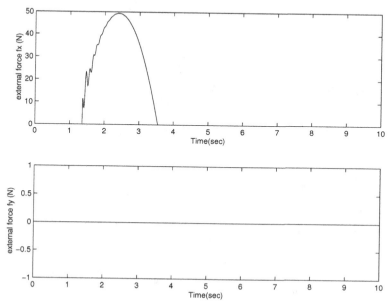

Figure 7.5 Time histories of the external forces in the Cartesian space

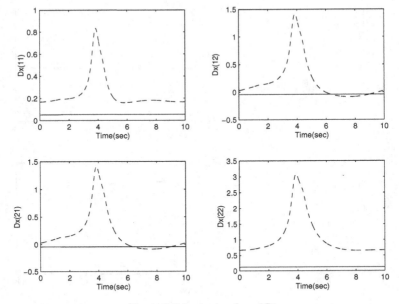

Figure 7.6 Approximation of $\mathbf{D}_x$

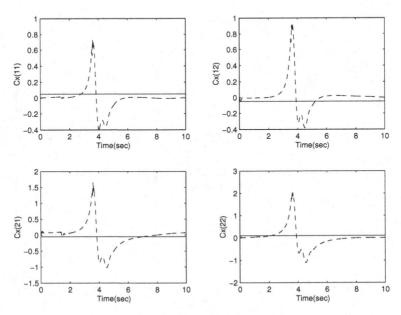

Figure 7.7 Approximation of $\mathbf{C}_x$

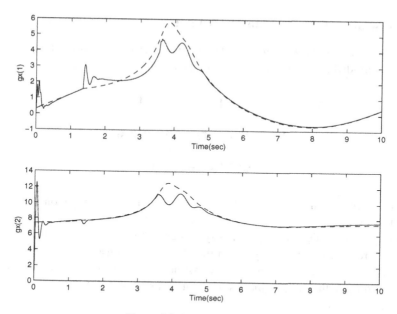

Figure 7.8 Approximation of $\mathbf{g}_x$

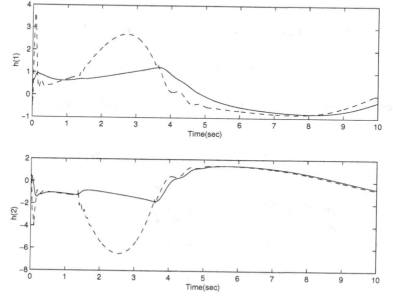

Figure 7.9 Approximation of $\mathbf{h}$

7.5 Consideration of Actuator Dynamics

According to (3.9-1) and (7.2-4), the dynamics of a rigid-link flexible-joint electrically-driven robot interacting with the environment can be described by

$$\mathbf{D}_x \dot{\mathbf{s}} + \mathbf{C}_x \mathbf{s} + \mathbf{g}_x + \mathbf{D}_x \dot{\mathbf{v}} + \mathbf{C}_x \mathbf{v} = \mathbf{J}_a^{-T} \boldsymbol{\tau}_t - \mathbf{F}_{ext} \tag{1a}$$

$$\mathbf{J}_t \ddot{\boldsymbol{\tau}}_t + \mathbf{B}_t \dot{\boldsymbol{\tau}}_t + \boldsymbol{\tau}_t = \mathbf{H}\mathbf{i} - \overline{\mathbf{q}}(\dot{\mathbf{q}}, \ddot{\mathbf{q}}) \tag{1b}$$

$$\mathbf{L}\dot{\mathbf{i}} + \mathbf{R}\mathbf{i} + \mathbf{K}_b \dot{\mathbf{q}} = \mathbf{u} \tag{1c}$$

This system is in a cascade form similar to the configuration shown in Figure 6.9, and hence the backstepping-like procedure can be applied here. A desired torque trajectory $\boldsymbol{\tau}_{td}$ is firstly designed for convergence of $\mathbf{s}$ in (1a). The desired current trajectory $\mathbf{i}_d$ can then be found to ensure $\boldsymbol{\tau}_t \rightarrow \boldsymbol{\tau}_{td}$ in (1b). Finally, the control effort $\mathbf{u}$ in (1c) is constructed to have convergence of $\mathbf{i}$ to $\mathbf{i}_d$.

Assuming that all parameters in (1) are known, and then the desired torque can be designed as

$$\boldsymbol{\tau}_{td} = \mathbf{J}_a^T (\mathbf{F}_{ext} + \mathbf{g}_x + \mathbf{D}_x \dot{\mathbf{v}} + \mathbf{C}_x \mathbf{v} - \mathbf{K}_d \mathbf{s}) \tag{2}$$

Therefore, the dynamics for output error tracking is found to be

$$\mathbf{D}_x \dot{\mathbf{s}} + \mathbf{C}_x \mathbf{s} + \mathbf{K}_d \mathbf{s} = \mathbf{J}_a^{-T} (\boldsymbol{\tau}_t - \boldsymbol{\tau}_{td}) \tag{3}$$

To ensure torque tracking in (1b), the MRC rule is applied with the reference model

$$\mathbf{J}_r \ddot{\boldsymbol{\tau}}_r + \mathbf{B}_r \dot{\boldsymbol{\tau}}_r + \mathbf{K}_r \boldsymbol{\tau}_r = \mathbf{J}_r \ddot{\boldsymbol{\tau}}_{td} + \mathbf{B}_r \dot{\boldsymbol{\tau}}_{td} + \mathbf{K}_r \boldsymbol{\tau}_{td} \tag{4}$$

where $\boldsymbol{\tau}_r \in \mathfrak{R}^n$ is the state vector of the reference model, and $\mathbf{J}_r \in \mathfrak{R}^{n \times n}$, $\mathbf{B}_r \in \mathfrak{R}^{n \times n}$, and $\mathbf{K}_r \in \mathfrak{R}^{n \times n}$ are selected to give convergence of $\boldsymbol{\tau}_r$ to $\boldsymbol{\tau}_{td}$. With the definition of $\overline{\boldsymbol{\tau}}_{td}(\dot{\boldsymbol{\tau}}_{td}, \ddot{\boldsymbol{\tau}}_{td}) = \mathbf{K}_r^{-1}(\mathbf{B}_r \dot{\boldsymbol{\tau}}_{td} + \mathbf{J}_r \ddot{\boldsymbol{\tau}}_{td})$, we may rewrite (1b) and (4) into the state space representation

$$\dot{\mathbf{x}}_p = \mathbf{A}_p \mathbf{x}_p + \mathbf{B}_p \mathbf{H}\mathbf{i} - \mathbf{B}_p \overline{\mathbf{q}} \tag{5a}$$

$$\dot{\mathbf{x}}_m = \mathbf{A}_m \mathbf{x}_m + \mathbf{B}_m (\boldsymbol{\tau}_{td} + \overline{\boldsymbol{\tau}}_{td}) \tag{5b}$$

where $\mathbf{x}_p = [\boldsymbol{\tau}_t^T \quad \dot{\boldsymbol{\tau}}_t^T]^T \in \mathfrak{R}^{2n}$ and $\mathbf{x}_m = [\boldsymbol{\tau}_r^T \quad \dot{\boldsymbol{\tau}}_r^T]^T \in \mathfrak{R}^{2n}$ are augmented state

vectors. $\mathbf{A}_p = \begin{bmatrix} \mathbf{0} & \mathbf{I}_{n \times n} \\ -\mathbf{J}_t^{-1} & -\mathbf{J}_t^{-1}\mathbf{B}_t \end{bmatrix} \in \mathfrak{R}^{2n \times 2n}$ and $\mathbf{A}_m = \begin{bmatrix} \mathbf{0} & \mathbf{I}_{n \times n} \\ -\mathbf{J}_r^{-1}\mathbf{K}_r & -\mathbf{J}_r^{-1}\mathbf{B}_r \end{bmatrix} \in \mathfrak{R}^{2n \times 2n}$

are augmented system matrices, and $\mathbf{B}_p = \begin{bmatrix} \mathbf{0} \\ \mathbf{J}_t^{-1} \end{bmatrix} \in \mathfrak{R}^{2n \times n}$ and

$\mathbf{B}_m = \begin{bmatrix} \mathbf{0} \\ \mathbf{J}_r^{-1}\mathbf{K}_r \end{bmatrix} \in \mathfrak{R}^{2n \times n}$ are augmented input gain matrices. Let $\boldsymbol{\tau}_t = \mathbf{C}_p\mathbf{x}_p$

and $\boldsymbol{\tau}_r = \mathbf{C}_m\mathbf{x}_m$ be respectively the output signal vector for (5a) and (5b), where $\mathbf{C}_p = \mathbf{C}_m = [\mathbf{I}_{n \times n} \quad \mathbf{0}] \in \mathfrak{R}^{n \times 2n}$ are augmented output signal matrices. The pair $(\mathbf{A}_m, \mathbf{B}_m)$ is controllable, $(\mathbf{A}_m, \mathbf{C}_m)$ is observable, and the transfer function $\mathbf{C}_m(s\mathbf{I} - \mathbf{A}_m)^{-1}\mathbf{B}_m$ is SPR. According to the MRC design, the desired current $\mathbf{i}_d$ is selected as

$$\mathbf{i}_d = \mathbf{H}^{-1}[\boldsymbol{\Theta}\mathbf{x}_p + \boldsymbol{\Phi}\boldsymbol{\tau}_{td} + \mathbf{h}(\bar{\boldsymbol{\tau}}_{td}, \bar{\mathbf{q}})] \tag{6}$$

where $\boldsymbol{\Theta}$ and $\boldsymbol{\Phi}$ are matrices satisfying $\mathbf{A}_p + \mathbf{B}_p\boldsymbol{\Theta} = \mathbf{A}_m$ and $\mathbf{B}_p\boldsymbol{\Phi} = \mathbf{B}_m$, respectively, and $\mathbf{h}$ is defined as $\mathbf{h}(\bar{\boldsymbol{\tau}}_{td}, \bar{\mathbf{q}}) = \boldsymbol{\Phi}\bar{\boldsymbol{\tau}}_{td} + \bar{\mathbf{q}}$. Substituting (6) into (5a), we may obtain

$$\dot{\mathbf{x}}_p = \mathbf{A}_m\dot{\mathbf{x}}_p + \mathbf{B}_m(\boldsymbol{\tau}_d + \bar{\boldsymbol{\tau}}_{td}) + \mathbf{B}_p\mathbf{H}(\mathbf{i} - \mathbf{i}_d) \tag{7}$$

With the definition $\mathbf{e}_m = \mathbf{x}_p - \mathbf{x}_m$ and $\mathbf{e}_\tau = \boldsymbol{\tau}_t - \boldsymbol{\tau}_r$, the dynamics for the torque tracking loop becomes

$$\dot{\mathbf{e}}_m = \mathbf{A}_m\mathbf{e}_m + \mathbf{B}_p\mathbf{H}(\mathbf{i} - \mathbf{i}_d) \tag{8a}$$

$$\mathbf{e}_\tau = \mathbf{C}_m\mathbf{e}_m \tag{8b}$$

In order to ensure $\boldsymbol{\tau}_t \to \boldsymbol{\tau}_{td}$ and $\mathbf{i} \to \mathbf{i}_d$, the control law in (1c) is designed as

$$\mathbf{u} = \mathbf{L}\dot{\mathbf{i}}_d + \mathbf{R}\mathbf{i} + \mathbf{K}_b\dot{\mathbf{q}} - \mathbf{K}_c\mathbf{e}_i \tag{9}$$

where $\mathbf{e}_i = \mathbf{i} - \mathbf{i}_d$ is the current error vector, and $\mathbf{K}_c$ is a positive definite matrix. Plugging, (9) into (1c), we may have the dynamics for the current tracking loop

$$\mathbf{L}\dot{\mathbf{e}}_i + \mathbf{K}_c\mathbf{e}_i = \mathbf{0} \tag{10}$$

To prove the closed loop stability of the whole system, let us consider a Lyapunov-like function candidate

$$V(\mathbf{s},\mathbf{e}_m,\mathbf{e}_i) = \frac{1}{2}\mathbf{s}^T\mathbf{D}\mathbf{s} + \mathbf{e}_m^T\mathbf{P}_t\mathbf{e}_m + \frac{1}{2}\mathbf{e}_i^T\mathbf{L}\mathbf{e}_i \tag{11}$$

where $\mathbf{P}_t = \mathbf{P}_t^T \in \mathfrak{R}^{2n\times 2n}$ is a positive definite matrix satisfying the Lyapunov equation $\mathbf{A}_m^T\mathbf{P}_t + \mathbf{P}_t\mathbf{A}_m = -\mathbf{C}_m^T\mathbf{C}_m$. Along the trajectories of (3), (8) and (10), the time derivative of V can be computed as

$$\dot{V} = -\mathbf{s}^T\mathbf{K}_d\mathbf{s} + \frac{1}{2}\mathbf{s}^T(\dot{\mathbf{D}}_x - 2\mathbf{C}_x)\mathbf{s} + \mathbf{s}^T\mathbf{J}_a^{-T}\mathbf{e}_\tau$$

$$-\mathbf{e}_\tau^T\mathbf{e}_\tau + \mathbf{e}_m^T\mathbf{P}_t\mathbf{B}_p\mathbf{H}\mathbf{e}_i - \mathbf{e}_i^T\mathbf{K}_c\mathbf{e}_i \tag{12}$$

Selecting $\mathbf{B}_m = \mathbf{B}_p$ such that $\mathbf{e}_m^T\mathbf{P}_t\mathbf{B}_p = \mathbf{e}_\tau^T$, equation (12) becomes

$$\dot{V} = -[\mathbf{s}^T \quad \mathbf{e}_\tau^T \quad \mathbf{e}_i^T]\mathbf{Q}\begin{bmatrix}\mathbf{s}\\ \mathbf{e}_\tau \\ \mathbf{e}_i\end{bmatrix} \le 0 \tag{13}$$

where $\mathbf{Q} = \begin{bmatrix} \mathbf{K}_d & -\dfrac{1}{2}\mathbf{J}_a^{-T} & \mathbf{0} \\[2mm] -\dfrac{1}{2}\mathbf{J}_a^{-T} & \mathbf{I}_{n\times n} & -\dfrac{1}{2}\mathbf{H} \\[2mm] \mathbf{0} & -\dfrac{1}{2}\mathbf{H} & \mathbf{K}_c \end{bmatrix}$ is positive definite due to proper

selection of $\mathbf{K}_d$ and $\mathbf{K}_c$. Therefore, we have proved that $\mathbf{s}$, $\mathbf{e}_\tau$ and $\mathbf{e}_i$ are uniformly bounded, and their square integrability can also be proved from (13). Furthermore, uniformly boundedness of $\dot{\mathbf{s}}$, $\dot{\mathbf{e}}_\tau$, and $\dot{\mathbf{e}}_i$ are also easy to be proved, and $\mathbf{q} \to \mathbf{q}_d$, $\boldsymbol{\tau}_t \to \boldsymbol{\tau}_{td}$, and $\mathbf{i} \to \mathbf{i}_d$ follow by Barbalat's lemma. Hence, the closed-loop system behaves like the target impedance.

Remark 6: To implement the control strategy, all system parameters are required to be available, and we need to feedback $\ddot{\mathbf{q}}$ and its higher order derivatives. Therefore, the design introduced in this section is not feasible for practical applications.

7.5.1 Regressor-based adaptive controller design

Consider the system in (1) again, but $\mathbf{D}_x$, $\mathbf{C}_x$, $\mathbf{g}_x$, $\mathbf{L}$, $\mathbf{R}$ and $\mathbf{K}_b$ are unavailable. The desired torque trajectory in (2) is not feasible, and we modify it as

$$\boldsymbol{\tau}_{td} = \mathbf{J}_a^T (\mathbf{F}_{ext} + \hat{\mathbf{g}}_x + \hat{\mathbf{D}}_x \dot{\mathbf{v}} + \hat{\mathbf{C}}_x \mathbf{v} - \mathbf{K}_d \mathbf{s})$$
$$= \mathbf{J}_a^T [\mathbf{F}_{ext} + \mathbf{Y}(\mathbf{x}, \dot{\mathbf{x}}, \mathbf{v}, \dot{\mathbf{v}}) \hat{\mathbf{p}}_x - \mathbf{K}_d \mathbf{s}] \tag{14}$$

where $\hat{\mathbf{D}}_x$, $\hat{\mathbf{C}}_x$, $\hat{\mathbf{g}}_x$, and $\hat{\mathbf{p}}_x$ are estimates of $\mathbf{D}_x$, $\mathbf{C}_x$, $\mathbf{g}_x$, and $\mathbf{p}_x$, respectively. Plugging (14) into (1a), and we may obtain the error dynamics for the output tracking loop

$$\mathbf{D}_x \dot{\mathbf{s}} + \mathbf{C}_x \mathbf{s} + \mathbf{K}_d \mathbf{s} = -\tilde{\mathbf{D}}_x \dot{\mathbf{v}} - \tilde{\mathbf{C}}_x \mathbf{v} - \tilde{\mathbf{g}}_x + \mathbf{J}_a^{-T}(\boldsymbol{\tau}_t - \boldsymbol{\tau}_{td})$$
$$= -\mathbf{Y}(\mathbf{x}, \dot{\mathbf{x}}, \mathbf{v}, \dot{\mathbf{v}}) \tilde{\mathbf{p}}_x + \mathbf{J}_a^{-T}(\boldsymbol{\tau}_t - \boldsymbol{\tau}_{td}) \tag{15}$$

where $\tilde{\mathbf{D}}_x = \mathbf{D}_x - \hat{\mathbf{D}}_x$, $\tilde{\mathbf{C}}_x = \mathbf{C}_x - \hat{\mathbf{C}}_x$, $\tilde{\mathbf{g}}_x = \mathbf{g}_x - \hat{\mathbf{g}}_x$, and $\tilde{\mathbf{p}}_x = \mathbf{p}_x - \hat{\mathbf{p}}_x$. Therefore, if we may find a controller so that $\boldsymbol{\tau}_t \to \boldsymbol{\tau}_{td}$ and an update law to have $\hat{\mathbf{p}} \to \mathbf{p}$, then (15) implies asymptotic convergence of the output error. To this end, we would like to use the MRC rule with the reference model in (4) and the state space representation in (5). The desired current trajectory is designed as the one in (6) to have the dynamics for the torque tracking loop as in (8). Instead of (9), a new controller is designed as

$$\mathbf{u} = \hat{\mathbf{L}} \mathbf{i}_d + \hat{\mathbf{R}} \mathbf{i} + \hat{\mathbf{K}}_b \dot{\mathbf{q}} - \mathbf{K}_c \mathbf{e}_i$$
$$= \hat{\mathbf{p}}_i^T \boldsymbol{\varphi} - \mathbf{K}_c \mathbf{e}_i \tag{16}$$

where $\boldsymbol{\varphi} = [\mathbf{i}_d^T \quad \mathbf{i}^T \quad \dot{\mathbf{q}}^T]^T \in \Re^{3n}$, $\hat{\mathbf{p}}_i = [\hat{\mathbf{L}}^T \quad \hat{\mathbf{R}}^T \quad \hat{\mathbf{K}}_b^T]^T \in \Re^{3n \times n}$, and we may have the dynamics for the current tracking loop as

$$\mathbf{L} \dot{\mathbf{e}}_i + \mathbf{K}_c \mathbf{e}_i = -\tilde{\mathbf{p}}_i^T \boldsymbol{\varphi} \tag{17}$$

where $\tilde{\mathbf{p}}_i = \mathbf{p}_i - \hat{\mathbf{p}}_i$. To prove stability, we select the Lyapunov-like function candidate

$$V(\mathbf{s}, \mathbf{e}_m, \mathbf{e}_i, \tilde{\mathbf{p}}_x, \tilde{\mathbf{p}}_i) = \frac{1}{2} \mathbf{s}^T \mathbf{D}_x \mathbf{s} + \mathbf{e}_m^T \mathbf{P}_t \mathbf{e}_m$$
$$+ \frac{1}{2} \mathbf{e}_i^T \mathbf{L} \mathbf{e}_i + \frac{1}{2} \tilde{\mathbf{p}}_x^T \boldsymbol{\Gamma} \tilde{\mathbf{p}}_x + Tr(\tilde{\mathbf{p}}_i^T \boldsymbol{\Gamma}_i \tilde{\mathbf{p}}_i) \tag{18}$$

where $\mathbf{P}_t = \mathbf{P}_t^T \in \mathfrak{R}^{2n \times 2n}$ is positive definite satisfying the Lyapunov equation $\mathbf{A}_m^T \mathbf{P}_t + \mathbf{P}_t \mathbf{A}_m = -\mathbf{C}_m^T \mathbf{C}_m$. Taking time derivative of (18) along the trajectories of (8), (15) and (17), we have

$$\dot{V} = -\mathbf{s}^T \mathbf{K}_d \mathbf{s} + \mathbf{s}^T \mathbf{J}_a^{-T} \mathbf{e}_\tau - \mathbf{e}_\tau^T \mathbf{e}_\tau + \mathbf{e}_m^T \mathbf{P}_t \mathbf{B}_p \mathbf{H} \mathbf{e}_i - \mathbf{e}_i^T \mathbf{K}_c \mathbf{e}_i$$
$$- \tilde{\mathbf{p}}_x^T (\mathbf{\Gamma} \dot{\hat{\mathbf{p}}}_x + \mathbf{Y}^T \mathbf{s}) - Tr[\tilde{\mathbf{p}}_i^T (\mathbf{\Gamma}_i \dot{\hat{\mathbf{p}}}_i + \boldsymbol{\varphi} \mathbf{e}_i^T)] \tag{19}$$

Pick $\mathbf{B}_m = \mathbf{B}_p$ to have $\mathbf{e}_m^T \mathbf{P}_t \mathbf{B}_p = \mathbf{e}_\tau^T$, then the update law can be selected to be

$$\dot{\hat{\mathbf{p}}}_x = -\mathbf{\Gamma}^{-1} \mathbf{Y}^T \mathbf{s} \tag{20a}$$

$$\dot{\hat{\mathbf{p}}}_i = -\mathbf{\Gamma}_i^{-1} \boldsymbol{\varphi} \mathbf{e}_i^T \tag{20b}$$

Thus, (19) becomes (13); therefore, we have proved that $\mathbf{s}$, $\mathbf{e}_\tau$ and $\mathbf{e}_i$ are uniformly bounded, and their square integrability can also be proved from (13). Furthermore, uniformly boundedness of $\dot{\mathbf{s}}$, $\dot{\mathbf{e}}_\tau$, and $\dot{\mathbf{e}}_i$ are also easy to be proved, and hence, $\mathbf{q} \to \mathbf{q}_d$, $\boldsymbol{\tau}_t \to \boldsymbol{\tau}_{td}$, and $\mathbf{i} \to \mathbf{i}_d$ follow by Barbalat's lemma. Consequently, we may conclude that the closed-loop system will behave like the target impedance regardless of the system uncertainties.

Remark 7: To implement the controller strategy, we do not need to have the knowledge of most system parameters, but we have to feedback $\ddot{\mathbf{q}}$ and calculate the regressor matrix and their higher order derivatives. Therefore, the design introduced in this section is not feasible for practical applications, either.

7.5.2 Regressor-free adaptive controller design

Consider the system in (1) again, but $\mathbf{D}_x$, $\mathbf{C}_x$, $\mathbf{g}_x$, $\mathbf{L}$, $\mathbf{R}$ and $\mathbf{K}_b$ are unavailable. The desired torque trajectory in (2) is modified as

$$\boldsymbol{\tau}_{td} = \mathbf{J}_a^T (\mathbf{F}_{ext} + \hat{\mathbf{g}}_x + \hat{\mathbf{D}}_x \dot{\mathbf{v}} + \hat{\mathbf{C}}_x \mathbf{v} - \mathbf{K}_d \mathbf{s}) \tag{21}$$

The dynamics for the output tracking loop can thus be written as

$$\mathbf{D}_x \dot{\mathbf{s}} + \mathbf{C}_x \mathbf{s} + \mathbf{K}_d \mathbf{s} = -\tilde{\mathbf{D}}_x \dot{\mathbf{v}} - \tilde{\mathbf{C}}_x \mathbf{v} - \tilde{\mathbf{g}}_x + \mathbf{J}_a^{-T} (\boldsymbol{\tau}_t - \boldsymbol{\tau}_{td}) \tag{22}$$

Therefore, if we may find a control law to drive $\boldsymbol{\tau}_t \to \boldsymbol{\tau}_{td}$ and update laws to have $\hat{\mathbf{D}}_x \to \mathbf{D}_x$, $\hat{\mathbf{C}}_x \to \mathbf{C}_x$, and $\hat{\mathbf{g}}_x \to \mathbf{g}_x$, then (22) implies convergence of the output error. To this end, we would like to use the MRC rule again with the reference model in (4) and the state space representation in (5). The desired current trajectory is designed according to (6) to be

$$\mathbf{i}_d = \mathbf{H}^{-1}[\boldsymbol{\Theta}\mathbf{x}_p + \boldsymbol{\Phi}\boldsymbol{\tau}_{td} + \hat{\mathbf{h}}] \tag{23}$$

where $\hat{\mathbf{h}}$ is an estimate of $\mathbf{h}(\overline{\boldsymbol{\tau}}_{td}, \overline{\mathbf{q}}) = \boldsymbol{\Phi}\overline{\boldsymbol{\tau}}_{td} + \overline{\mathbf{q}}$. Consequently, the dynamics for the torque tracking loop becomes

$$\dot{\mathbf{e}}_m = \mathbf{A}_m\mathbf{e}_m + \mathbf{B}_p\mathbf{H}(\mathbf{i} - \mathbf{i}_d) + \mathbf{B}_p(\hat{\mathbf{h}} - \mathbf{h}) \tag{24a}$$

$$\mathbf{e}_\tau = \mathbf{C}_m\mathbf{e}_m \tag{24b}$$

Hence, if we may design a control law to ensure $\mathbf{i} \to \mathbf{i}_d$ and an update law to have $\hat{\mathbf{h}} \to \mathbf{h}$, then we may have convergence of the torque tracking loop. The control strategy can be constructed as

$$\mathbf{u} = \hat{\mathbf{f}} - \mathbf{K}_c\mathbf{e}_i \tag{25}$$

where $\mathbf{e}_i = \mathbf{i} - \mathbf{i}_d$ is the current error, $\mathbf{K}_c$ is a positive definite matrix and $\hat{\mathbf{f}}$ is an estimate of $\mathbf{f}(\dot{\mathbf{i}}_d, \mathbf{i}, \dot{\mathbf{q}}) = \mathbf{L}\dot{\mathbf{i}}_d + \mathbf{R}\mathbf{i} + \mathbf{K}_b\dot{\mathbf{q}}$. With this control law, the dynamics for the current tracking loop can be found as

$$\mathbf{L}\dot{\mathbf{e}}_i + \mathbf{K}_c\mathbf{e}_i = \hat{\mathbf{f}} - \mathbf{f} \tag{26}$$

If we may select a proper update law to have $\hat{\mathbf{f}} \to \mathbf{f}$, (26) ensures convergence in the current tracking loop. Since $\mathbf{D}_x$, $\mathbf{C}_x$, $\mathbf{g}_x$, $\mathbf{h}(\overline{\boldsymbol{\tau}}_{td}, \overline{\mathbf{q}})$ and $\mathbf{f}(\dot{\mathbf{i}}_d, \mathbf{i}, \dot{\mathbf{q}})$ are time-varying functions and their variation bounds are not given, their function approximation representations are employed as

$$\mathbf{D}_x = \mathbf{W}_{\mathbf{D}_x}^T \mathbf{Z}_{\mathbf{D}_x} + \boldsymbol{\varepsilon}_{\mathbf{D}_x} \tag{27a}$$

$$\mathbf{C}_x = \mathbf{W}_{\mathbf{C}_x}^T \mathbf{Z}_{\mathbf{C}_x} + \boldsymbol{\varepsilon}_{\mathbf{C}_x} \tag{27b}$$

$$\mathbf{g}_x = \mathbf{W}_{\mathbf{g}_x}^T \mathbf{z}_{\mathbf{g}_x} + \boldsymbol{\varepsilon}_{\mathbf{g}_x} \tag{27c}$$

$$\mathbf{h} = \mathbf{W}_{\mathbf{h}}^T \mathbf{z}_{\mathbf{h}} + \boldsymbol{\varepsilon}_{\mathbf{h}} \tag{27d}$$

$$\mathbf{f} = \mathbf{W}_{\mathbf{f}}^T \mathbf{z}_{\mathbf{f}} + \boldsymbol{\varepsilon}_{\mathbf{f}} \tag{27e}$$

where $\mathbf{W}_{\mathbf{D}_x} \in \Re^{n^2\beta_D \times n}$, $\mathbf{W}_{\mathbf{C}_x} \in \Re^{n^2\beta_C \times n}$, $\mathbf{W}_{\mathbf{g}_x} \in \Re^{n\beta_g \times n}$, $\mathbf{W}_{\mathbf{h}} \in \Re^{n\beta_h \times n}$, and $\mathbf{W}_{\mathbf{f}} \in \Re^{n\beta_f \times n}$ are weighting matrices for $\mathbf{D}_x$, $\mathbf{C}_x$, $\mathbf{g}_x$, $\mathbf{h}$, and $\mathbf{f}$, respectively, while $\mathbf{Z}_{\mathbf{D}_x} \in \Re^{n^2\beta_D \times n}$, $\mathbf{Z}_{\mathbf{C}_x} \in \Re^{n^2\beta_C \times n}$, $\mathbf{z}_{\mathbf{g}_x} \in \Re^{n\beta_g \times 1}$, $\mathbf{z}_{\mathbf{h}} \in \Re^{n\beta_h \times 1}$, and $\mathbf{z}_{\mathbf{f}} \in \Re^{n\beta_f \times 1}$ are basis function matrices. Likewise, we have the representations for the estimates as

$$\hat{\mathbf{D}}_x = \hat{\mathbf{W}}_{\mathbf{D}_x}^T \mathbf{Z}_{\mathbf{D}_x} \tag{27f}$$

$$\hat{\mathbf{C}}_x = \hat{\mathbf{W}}_{\mathbf{C}_x}^T \mathbf{Z}_{\mathbf{C}_x} \tag{27g}$$

$$\hat{\mathbf{g}}_x = \hat{\mathbf{W}}_{\mathbf{g}_x}^T \mathbf{z}_{\mathbf{g}_x} \tag{27h}$$

$$\hat{\mathbf{h}} = \hat{\mathbf{W}}_{\mathbf{h}}^T \mathbf{z}_{\mathbf{h}} \tag{27i}$$

$$\hat{\mathbf{f}} = \hat{\mathbf{W}}_{\mathbf{f}}^T \mathbf{z}_{\mathbf{f}} \tag{27j}$$

Thus, the output error dynamics (22), torque tracking error dynamics (24a), and current tracking error dynamics (26) can be rewritten as

$$\mathbf{D}_x\dot{\mathbf{s}} + \mathbf{C}_x\mathbf{s} + \mathbf{K}_d\mathbf{s} = \mathbf{J}_a^{-T}(\boldsymbol{\tau}_t - \boldsymbol{\tau}_{td}) - \tilde{\mathbf{W}}_{\mathbf{D}_x}^T \mathbf{Z}_{\mathbf{D}_x}\dot{\mathbf{v}}$$
$$- \tilde{\mathbf{W}}_{\mathbf{C}_x}^T \mathbf{Z}_{\mathbf{C}_x}\mathbf{v} - \tilde{\mathbf{W}}_{\mathbf{g}_x}^T \mathbf{z}_{\mathbf{g}_x} + \boldsymbol{\varepsilon}_1 \tag{28a}$$

$$\dot{\mathbf{e}}_m = \mathbf{A}_m\mathbf{e}_m - \mathbf{B}_p\tilde{\mathbf{W}}_{\mathbf{h}}^T\mathbf{z}_{\mathbf{h}} + \mathbf{B}_p\mathbf{H}\mathbf{e}_i + \mathbf{B}_p\boldsymbol{\varepsilon}_2 \tag{28b}$$

$$\mathbf{L}\dot{\mathbf{e}}_i + \mathbf{K}_c\mathbf{e}_i = -\tilde{\mathbf{W}}_{\mathbf{f}}^T\mathbf{z}_{\mathbf{f}} + \boldsymbol{\varepsilon}_3 \tag{28c}$$

where $\boldsymbol{\varepsilon}_1 = \boldsymbol{\varepsilon}_1(\boldsymbol{\varepsilon}_{\mathbf{D}_x}, \boldsymbol{\varepsilon}_{\mathbf{C}_x}, \boldsymbol{\varepsilon}_{\mathbf{g}_x}, \mathbf{s}, \ddot{\mathbf{x}}_i)$, $\boldsymbol{\varepsilon}_2 = \boldsymbol{\varepsilon}_2(\boldsymbol{\varepsilon}_{\mathbf{h}}, \mathbf{e}_m)$, and $\boldsymbol{\varepsilon}_3 = \boldsymbol{\varepsilon}_3(\boldsymbol{\varepsilon}_{\mathbf{f}}, \mathbf{e}_i)$ are lumped approximation error vectors. Define the Lyapunov-like function candidate as

$$V(\mathbf{s}, \mathbf{e}_m, \mathbf{e}_i, \tilde{\mathbf{W}}_{\mathbf{D}_x}, \tilde{\mathbf{W}}_{\mathbf{C}_x}, \tilde{\mathbf{W}}_{\mathbf{g}_x}, \tilde{\mathbf{W}}_{\mathbf{h}}, \tilde{\mathbf{W}}_{\mathbf{f}}) = \frac{1}{2}[\mathbf{s}^T\mathbf{D}_x\mathbf{s} + 2\mathbf{e}_m^T\mathbf{P}_t\mathbf{e}_m$$
$$+ \mathbf{e}_i^T\mathbf{L}\mathbf{e}_i + Tr(\tilde{\mathbf{W}}_{\mathbf{D}_x}^T\mathbf{Q}_{\mathbf{D}_x}\tilde{\mathbf{W}}_{\mathbf{D}_x} + \tilde{\mathbf{W}}_{\mathbf{C}_x}^T\mathbf{Q}_{\mathbf{C}_x}\tilde{\mathbf{W}}_{\mathbf{C}_x} + \tilde{\mathbf{W}}_{\mathbf{g}_x}^T\mathbf{Q}_{\mathbf{g}_x}\tilde{\mathbf{W}}_{\mathbf{g}_x})$$
$$+ Tr(\tilde{\mathbf{W}}_{\mathbf{h}}^T\mathbf{Q}_{\mathbf{h}}\tilde{\mathbf{W}}_{\mathbf{h}} + \tilde{\mathbf{W}}_{\mathbf{f}}^T\mathbf{Q}_{\mathbf{f}}\tilde{\mathbf{W}}_{\mathbf{f}})] \tag{29}$$

where matrices $\mathbf{Q}_{\mathbf{D}_x} \in \Re^{n^2\beta_D \times n^2\beta_D}$, $\mathbf{Q}_{\mathbf{C}_x} \in \Re^{n^2\beta_C \times n^2\beta_C}$, $\mathbf{Q}_{\mathbf{g}_x} \in \Re^{n\beta_g \times n\beta_g}$, $\mathbf{Q}_{\mathbf{h}} \in \Re^{n\beta_h \times n\beta_h}$, and $\mathbf{Q}_{\mathbf{f}} \in \Re^{n\beta_f \times n\beta_f}$ are positive definite. Along the trajectory of (28), we may compute the time derivative of V as

$$\dot{V} = -\mathbf{s}^T \mathbf{K}_d \mathbf{s} + \mathbf{s}^T \mathbf{J}_a^{-T} \mathbf{e}_\tau - \mathbf{e}_\tau^T \mathbf{e}_\tau + \mathbf{e}_m^T \mathbf{P}_t \mathbf{B}_p \mathbf{H} \mathbf{e}_i - \mathbf{e}_i^T \mathbf{K}_c \mathbf{e}_i$$

$$+ \mathbf{s}^T \boldsymbol{\varepsilon}_1 + \mathbf{e}_m^T \mathbf{P}_t \mathbf{B}_p \boldsymbol{\varepsilon}_2 + \mathbf{e}_i^T \boldsymbol{\varepsilon}_3 - Tr[\tilde{\mathbf{W}}_{\mathbf{D}_x}^T (\mathbf{Z}_{\mathbf{D}_x} \dot{\mathbf{v}} \mathbf{s}^T + \mathbf{Q}_{\mathbf{D}_x} \dot{\hat{\mathbf{W}}}_{\mathbf{D}_x})]$$

$$- Tr[\tilde{\mathbf{W}}_{\mathbf{C}_x}^T (\mathbf{Z}_{\mathbf{C}_x} \mathbf{v} \mathbf{s}^T + \mathbf{Q}_{\mathbf{C}_x} \dot{\hat{\mathbf{W}}}_{\mathbf{C}_x}) + \tilde{\mathbf{W}}_{\mathbf{g}_x}^T (\mathbf{z}_{\mathbf{g}_x} \mathbf{s}^T + \mathbf{Q}_{\mathbf{g}_x} \dot{\hat{\mathbf{W}}}_{\mathbf{g}_x})]$$

$$- Tr[\tilde{\mathbf{W}}_{\mathbf{h}}^T (\mathbf{z}_\mathbf{h} \mathbf{e}_m^T \mathbf{P}_t \mathbf{B}_p + \mathbf{Q}_\mathbf{h} \dot{\hat{\mathbf{W}}}_\mathbf{h}) + \tilde{\mathbf{W}}_\mathbf{f}^T (\mathbf{z}_\mathbf{f} \mathbf{e}_i^T + \mathbf{Q}_\mathbf{f} \dot{\hat{\mathbf{W}}}_\mathbf{f})] \qquad (30)$$

If we select $\mathbf{B}_m = \mathbf{B}_p$ so that $\mathbf{e}_m^T \mathbf{P}_t \mathbf{B}_p = \mathbf{e}_\tau^T$, and if we pick the update laws as

$$\dot{\hat{\mathbf{W}}}_{\mathbf{D}_x} = -\mathbf{Q}_{\mathbf{D}_x}^{-1} (\mathbf{Z}_{\mathbf{D}_x} \dot{\mathbf{v}} \mathbf{s}^T + \sigma_{\mathbf{D}_x} \hat{\mathbf{W}}_{\mathbf{D}_x}) \qquad (31a)$$

$$\dot{\hat{\mathbf{W}}}_{\mathbf{C}_x} = -\mathbf{Q}_{\mathbf{C}_x}^{-1} (\mathbf{Z}_{\mathbf{C}_x} \mathbf{v} \mathbf{s}^T + \sigma_{\mathbf{C}_x} \hat{\mathbf{W}}_{\mathbf{C}_x}) \qquad (31b)$$

$$\dot{\hat{\mathbf{W}}}_{\mathbf{g}_x} = -\mathbf{Q}_{\mathbf{g}_x}^{-1} (\mathbf{z}_{\mathbf{g}_x} \mathbf{s}^T + \sigma_{\mathbf{g}_x} \hat{\mathbf{W}}_{\mathbf{g}_x}) \qquad (31c)$$

$$\dot{\hat{\mathbf{W}}}_\mathbf{h} = -\mathbf{Q}_\mathbf{h}^{-1} (\mathbf{z}_\mathbf{h} \mathbf{e}_\tau^T + \sigma_\mathbf{h} \hat{\mathbf{W}}_\mathbf{h}) \qquad (31d)$$

$$\dot{\hat{\mathbf{W}}}_\mathbf{f} = -\mathbf{Q}_\mathbf{f}^{-1} (\mathbf{z}_\mathbf{f} \mathbf{e}_i^T + \sigma_\mathbf{f} \hat{\mathbf{W}}_\mathbf{f}) \qquad (31e)$$

then (30) becomes

$$\dot{V} = -[\mathbf{s}^T \quad \mathbf{e}_\tau^T \quad \mathbf{e}_i^T] \mathbf{Q} \begin{bmatrix} \mathbf{s} \\ \mathbf{e}_\tau \\ \mathbf{e}_i \end{bmatrix} + [\mathbf{s}^T \quad \mathbf{e}_\tau^T \quad \mathbf{e}_i^T] \begin{bmatrix} \boldsymbol{\varepsilon}_1 \\ \boldsymbol{\varepsilon}_2 \\ \boldsymbol{\varepsilon}_3 \end{bmatrix}$$

$$+ \sigma_{\mathbf{D}_x} Tr(\tilde{\mathbf{W}}_{\mathbf{D}_x}^T \hat{\mathbf{W}}_{\mathbf{D}_x}) + \sigma_{\mathbf{C}_x} Tr(\tilde{\mathbf{W}}_{\mathbf{C}_x}^T \hat{\mathbf{W}}_{\mathbf{C}_x}) + \sigma_{\mathbf{g}_x} Tr(\tilde{\mathbf{W}}_{\mathbf{g}_x}^T \hat{\mathbf{W}}_{\mathbf{g}_x})$$

$$+ \sigma_\mathbf{h} Tr(\tilde{\mathbf{W}}_\mathbf{h}^T \hat{\mathbf{W}}_\mathbf{h}) + \sigma_\mathbf{f} Tr(\tilde{\mathbf{W}}_\mathbf{f}^T \hat{\mathbf{W}}_\mathbf{f}) \qquad (32)$$

where $\mathbf{Q} = \begin{bmatrix} \mathbf{K}_d & -\dfrac{1}{2}\mathbf{J}_a^{-T} & \mathbf{0} \\ -\dfrac{1}{2}\mathbf{J}_a^{-T} & \mathbf{I}_{n \times n} & -\dfrac{1}{2}\mathbf{H} \\ \mathbf{0} & -\dfrac{1}{2}\mathbf{H} & \mathbf{K}_c \end{bmatrix}$ is positive definite due to proper

selection of $\mathbf{K}_d$ and $\mathbf{K}_c$.

Remark 8: Realization of the control law (25) and update laws (31) does not need the information of joint accelerations, regressor matrix, or their higher order derivatives, which largely simplified their implementation.

Remark 9: Suppose a sufficient number of basis functions are used and the approximation error can be ignored, then it is not necessary to include the σ-modification terms in (31). Hence, (32) can be reduced to (13), and convergence of $\mathbf{s}$, $\mathbf{e}_\tau$ and $\mathbf{e}_i$ can be further proved by Barbalat's lemma. Therefore, the closed loop system behaves like the target impedance.

Remark 10: If the approximation error cannot be ignored, but we can find positive numbers δ_1, δ_2 and δ_3 such that $\|\boldsymbol{\varepsilon}_i\| \le \delta_i$, $i=1,2,3$, then robust terms $\boldsymbol{\tau}_{robust\,1}$, $\boldsymbol{\tau}_{robust\,2}$ and $\boldsymbol{\tau}_{robust\,3}$ can be included into (21), (23) and (25) to have

$$\boldsymbol{\tau}_{td} = \mathbf{J}_a^T (\mathbf{F}_{ext} + \hat{\mathbf{g}}_x + \hat{\mathbf{D}}_x \dot{\mathbf{v}} + \hat{\mathbf{C}}_x \mathbf{v} - \mathbf{K}_d \mathbf{s} + \boldsymbol{\tau}_{robust})$$

$$\mathbf{i}_d = \mathbf{H}^{-1}[\boldsymbol{\Theta}\mathbf{x}_p + \boldsymbol{\Phi}\boldsymbol{\tau}_{td} + \hat{\mathbf{h}} + \boldsymbol{\tau}_{robust\,2}]$$

$$\mathbf{u} = \hat{\mathbf{f}} - \mathbf{K}_c \mathbf{e}_i + \boldsymbol{\tau}_{robust\,3}$$

Consider the Lyapunov-like function candidate (29) again, and the update law (31) without σ-modification; then the time derivative of V becomes

$$\dot{V} = -[\mathbf{s}^T \quad \mathbf{e}_\tau^T \quad \mathbf{e}_i^T] \mathbf{Q} \begin{bmatrix} \mathbf{s} \\ \mathbf{e}_\tau \\ \mathbf{e}_i \end{bmatrix} + \delta_1 \|\mathbf{s}\| + \delta_2 \|\mathbf{e}_\tau\| + \delta_3 \|\mathbf{e}_i\|$$
$$+ \mathbf{s}^T \boldsymbol{\tau}_{robust\,1} + \mathbf{e}_\tau^T \boldsymbol{\tau}_{robust\,2} + \mathbf{e}_i^T \boldsymbol{\tau}_{robust\,3}$$

If we select $\boldsymbol{\tau}_{robust\,1} = -\delta_1[\text{sgn}(s_1) \quad \text{sgn}(s_2) \quad \cdots \quad \text{sgn}(s_n)]^T$ where s_i, $i=1,\dots,n$ is the i-th entry in $\mathbf{s}$, $\boldsymbol{\tau}_{robust\,2} = -\delta_2[\text{sgn}(e_{\tau_1}) \quad \cdots \quad \text{sgn}(e_{\tau_{2n}})]^T$ where e_{τ_j}, $j=1,\dots,n$ is the j-th entry in $\mathbf{e}_\tau$, and $\boldsymbol{\tau}_{robust\,3} = -\delta_3[\text{sgn}(e_{i_1}) \quad \cdots \quad \text{sgn}(e_{i_n})]^T$ where e_{i_k}, $k=1,\dots,n$ is the k-th element in $\mathbf{e}_i$, then we may have (13) again. This will further give convergence of the output error by Barbalat's lemma.

Owing to the existence of the approximation errors, the definiteness of $\dot{V}$ cannot be determined. By considering the inequality

$$V \le \frac{1}{2}\lambda_{\max}(\mathbf{A})\left\|\begin{bmatrix}\mathbf{s}\\ \mathbf{e}_\tau\\ \mathbf{e}_i\end{bmatrix}\right\|^2 + \frac{1}{2}[\lambda_{\max}(\mathbf{Q_{D_x}})Tr(\tilde{\mathbf{W}}_{\mathbf{D_x}}^T\tilde{\mathbf{W}}_{\mathbf{D_x}})$$

$$+\lambda_{\max}(\mathbf{Q_{C_x}})Tr(\tilde{\mathbf{W}}_{\mathbf{C_x}}^T\tilde{\mathbf{W}}_{\mathbf{C_x}}) + \lambda_{\max}(\mathbf{Q_{g_x}})Tr(\tilde{\mathbf{W}}_{\mathbf{g_x}}^T\tilde{\mathbf{W}}_{\mathbf{g_x}})$$

$$+\lambda_{\max}(\mathbf{Q_h})Tr(\tilde{\mathbf{W}}_{\mathbf{h}}^T\tilde{\mathbf{W}}_{\mathbf{h}}) + \lambda_{\max}(\mathbf{Q_f})Tr(\tilde{\mathbf{W}}_{\mathbf{f}}^T\tilde{\mathbf{W}}_{\mathbf{f}})]$$

where $\mathbf{A} = \begin{bmatrix}\mathbf{D} & \mathbf{0} & \mathbf{0}\\ \mathbf{0} & 2\mathbf{C}_m^T\mathbf{P}_t\mathbf{C}_m & \mathbf{0}\\ \mathbf{0} & \mathbf{0} & \mathbf{L}\end{bmatrix}$, we may rewrite (32) as

$$\dot{V} \le -\alpha V + \frac{1}{2}[\alpha\lambda_{\max}(\mathbf{A}) - \lambda_{\min}(\mathbf{Q})]\left\|\begin{bmatrix}\mathbf{s}\\ \mathbf{e}_\tau\\ \mathbf{e}_i\end{bmatrix}\right\|^2 + \frac{1}{2\lambda_{\min}(\mathbf{Q})}\left\|\begin{bmatrix}\varepsilon_1\\ \varepsilon_2\\ \varepsilon_3\end{bmatrix}\right\|^2$$

$$+\frac{1}{2}\{[\alpha\lambda_{\max}(\mathbf{Q_{D_x}}) - \sigma_{\mathbf{D_x}}]Tr(\tilde{\mathbf{W}}_{\mathbf{D_x}}^T\tilde{\mathbf{W}}_{\mathbf{D_x}}) + [\alpha\lambda_{\max}(\mathbf{Q_{C_x}})$$

$$-\sigma_{\mathbf{C_x}}]Tr(\tilde{\mathbf{W}}_{\mathbf{C_x}}^T\tilde{\mathbf{W}}_{\mathbf{C_x}}) + [\alpha\lambda_{\max}(\mathbf{Q_{g_x}}) - \sigma_{\mathbf{g_x}}]Tr(\tilde{\mathbf{W}}_{\mathbf{g_x}}^T\tilde{\mathbf{W}}_{\mathbf{g_x}})$$

$$+[\alpha\lambda_{\max}(\mathbf{Q_h}) - \sigma_{\mathbf{h}}]Tr(\tilde{\mathbf{W}}_{\mathbf{h}}^T\tilde{\mathbf{W}}_{\mathbf{h}}) + [\alpha\lambda_{\max}(\mathbf{Q_f})$$

$$-\sigma_{\mathbf{f}}]Tr(\tilde{\mathbf{W}}_{\mathbf{f}}^T\tilde{\mathbf{W}}_{\mathbf{f}})\} + \frac{1}{2}[\sigma_{\mathbf{D_x}}Tr(\mathbf{W}_{\mathbf{D_x}}^T\mathbf{W}_{\mathbf{D_x}}) + \sigma_{\mathbf{C_x}}Tr(\mathbf{W}_{\mathbf{C_x}}^T\mathbf{W}_{\mathbf{C_x}})$$

$$+\sigma_{\mathbf{g_x}}Tr(\mathbf{W}_{\mathbf{g_x}}^T\mathbf{W}_{\mathbf{g_x}}) + \sigma_{\mathbf{h}}Tr(\mathbf{W}_{\mathbf{h}}^T\mathbf{W}_{\mathbf{h}}) + \sigma_{\mathbf{f}}Tr(\mathbf{W}_{\mathbf{f}}^T\mathbf{W}_{\mathbf{f}})]$$

where α is selected to satisfy

$$\alpha \le \min\left\{\frac{\lambda_{\min}(\mathbf{Q})}{\lambda_{\max}(\mathbf{A})}, \frac{\sigma_{\mathbf{D_x}}}{\lambda_{\max}(\mathbf{Q_{D_x}})}, \frac{\sigma_{\mathbf{C_x}}}{\lambda_{\max}(\mathbf{Q_{C_x}})}, \frac{\sigma_{\mathbf{g_x}}}{\lambda_{\max}(\mathbf{Q_{g_x}})}, \frac{\sigma_{\mathbf{h}}}{\lambda_{\max}(\mathbf{Q_h})}, \frac{\sigma_{\mathbf{f}}}{\lambda_{\max}(\mathbf{Q_f})}\right\}$$

such that we may further have

$$\dot{V} \le -\alpha V + \frac{1}{2\lambda_{\min}(\mathbf{Q})}\left\|\begin{bmatrix}\varepsilon_1\\ \varepsilon_2\\ \varepsilon_3\end{bmatrix}\right\|^2 + \frac{1}{2}[\sigma_{\mathbf{D_x}}Tr(\mathbf{W}_{\mathbf{D_x}}^T\mathbf{W}_{\mathbf{D_x}})$$

$$+\sigma_{\mathbf{C_x}}Tr(\mathbf{W}_{\mathbf{C_x}}^T\mathbf{W}_{\mathbf{C_x}}) + \sigma_{\mathbf{g_x}}Tr(\mathbf{W}_{\mathbf{g_x}}^T\mathbf{W}_{\mathbf{g_x}})$$

$$+\sigma_{\mathbf{h}}Tr(\mathbf{W}_{\mathbf{h}}^T\mathbf{W}_{\mathbf{h}}) + \sigma_{\mathbf{f}}Tr(\mathbf{W}_{\mathbf{f}}^T\mathbf{W}_{\mathbf{f}})] \tag{33}$$

Therefore, we have proved that $\dot{V} < 0$ whenever

$$
\begin{aligned}
V > \ & \frac{1}{2\alpha\lambda_{\min}(\mathbf{Q})} \sup_{\tau \geq t_0} \left\| \begin{bmatrix} \boldsymbol{\varepsilon}_1(\tau) \\ \boldsymbol{\varepsilon}_2(\tau) \\ \boldsymbol{\varepsilon}_3(\tau) \end{bmatrix} \right\|^2 + \frac{1}{2\alpha} [\sigma_{\mathbf{D}_x} Tr(\mathbf{W}_{\mathbf{D}_x}^T \mathbf{W}_{\mathbf{D}_x}) \\
& + \sigma_{\mathbf{C}_x} Tr(\mathbf{W}_{\mathbf{C}_x}^T \mathbf{W}_{\mathbf{C}_x}) + \sigma_{\mathbf{g}_x} Tr(\mathbf{W}_{\mathbf{g}_x}^T \mathbf{W}_{\mathbf{g}_x}) \\
& + \sigma_{\mathbf{h}} Tr(\mathbf{W}_{\mathbf{h}}^T \mathbf{W}_{\mathbf{h}}) + \sigma_{\mathbf{f}} Tr(\mathbf{W}_{\mathbf{f}}^T \mathbf{W}_{\mathbf{f}})]
\end{aligned}
$$

i.e., $\mathbf{s}$, $\mathbf{e}_\tau$, $\mathbf{e}_i$, $\tilde{\mathbf{W}}_{\mathbf{D}_x}$, $\tilde{\mathbf{W}}_{\mathbf{C}_x}$, $\tilde{\mathbf{W}}_{\mathbf{g}_x}$, $\tilde{\mathbf{W}}_{\mathbf{h}}$, and $\tilde{\mathbf{W}}_{\mathbf{f}}$ are uniformly ultimately bounded. In addition, (33) implies

$$
\begin{aligned}
V(t) \leq \ & e^{-\alpha(t-t_0)} V(t_0) + \frac{1}{2\alpha\lambda_{\min}(\mathbf{Q})} \sup_{t_0 < \tau < t} \left\| \begin{bmatrix} \boldsymbol{\varepsilon}_1(\tau) \\ \boldsymbol{\varepsilon}_2(\tau) \\ \boldsymbol{\varepsilon}_3(\tau) \end{bmatrix} \right\|^2 \\
& + \frac{1}{2\alpha} [\sigma_{\mathbf{D}_x} Tr(\mathbf{W}_{\mathbf{D}_x}^T \mathbf{W}_{\mathbf{D}_x}) + \sigma_{\mathbf{C}_x} Tr(\mathbf{W}_{\mathbf{C}_x}^T \mathbf{W}_{\mathbf{C}_x}) \\
& + \sigma_{\mathbf{g}_x} Tr(\mathbf{W}_{\mathbf{g}_x}^T \mathbf{W}_{\mathbf{g}_x}) + \sigma_{\mathbf{h}} Tr(\mathbf{W}_{\mathbf{h}}^T \mathbf{W}_{\mathbf{h}}) + \sigma_{\mathbf{f}} Tr(\mathbf{W}_{\mathbf{f}}^T \mathbf{W}_{\mathbf{f}})]
\end{aligned}
$$

Together with the inequality

$$
\begin{aligned}
V \geq \ & \frac{1}{2} \lambda_{\min}(\mathbf{A}) \left\| \begin{bmatrix} \mathbf{s} \\ \mathbf{e}_\tau \\ \mathbf{e}_i \end{bmatrix} \right\|^2 + \frac{1}{2} [\lambda_{\min}(\mathbf{Q}_{\mathbf{D}}) Tr(\tilde{\mathbf{W}}_{\mathbf{D}_x}^T \tilde{\mathbf{W}}_{\mathbf{D}_x}) \\
& + \lambda_{\min}(\mathbf{Q}_{\mathbf{C}_x}) Tr(\tilde{\mathbf{W}}_{\mathbf{C}_x}^T \tilde{\mathbf{W}}_{\mathbf{C}_x}) + \lambda_{\min}(\mathbf{Q}_{\mathbf{g}_x}) Tr(\tilde{\mathbf{W}}_{\mathbf{g}_x}^T \tilde{\mathbf{W}}_{\mathbf{g}_x}) \\
& + \lambda_{\min}(\mathbf{Q}_{\mathbf{h}}) Tr(\tilde{\mathbf{W}}_{\mathbf{h}}^T \tilde{\mathbf{W}}_{\mathbf{h}}) + \lambda_{\min}(\mathbf{Q}_{\mathbf{f}}) Tr(\tilde{\mathbf{W}}_{\mathbf{f}}^T \tilde{\mathbf{W}}_{\mathbf{f}})]
\end{aligned}
$$

we may find the error bound

$$
\begin{aligned}
\left\| \begin{bmatrix} \mathbf{s} \\ \mathbf{e}_\tau \\ \mathbf{e}_i \end{bmatrix} \right\| \leq \ & \sqrt{\frac{2V(t_0)}{\lambda_{\min}(\mathbf{A})}} e^{-\frac{\alpha}{2}(t-t_0)} + \frac{1}{\sqrt{\alpha\lambda_{\min}(\mathbf{A})\lambda_{\min}(\mathbf{Q})}} \sup_{t_0 < \tau < t} \left\| \begin{bmatrix} \boldsymbol{\varepsilon}_1(\tau) \\ \boldsymbol{\varepsilon}_2(\tau) \\ \boldsymbol{\varepsilon}_3(\tau) \end{bmatrix} \right\| \\
& + \frac{1}{\sqrt{\alpha\lambda_{\min}(\mathbf{A})}} [\sigma_{\mathbf{D}_x} Tr(\mathbf{W}_{\mathbf{D}_x}^T \mathbf{W}_{\mathbf{D}_x}) + \sigma_{\mathbf{C}_x} Tr(\mathbf{W}_{\mathbf{C}_x}^T \mathbf{W}_{\mathbf{C}_x}) \\
& + \sigma_{\mathbf{g}_x} Tr(\mathbf{W}_{\mathbf{g}_x}^T \mathbf{W}_{\mathbf{g}_x}) + \sigma_{\mathbf{h}} Tr(\mathbf{W}_{\mathbf{h}}^T \mathbf{W}_{\mathbf{h}}) + \sigma_{\mathbf{f}} Tr(\mathbf{W}_{\mathbf{f}}^T \mathbf{W}_{\mathbf{f}})]^{\frac{1}{2}}
\end{aligned}
$$

Therefore, the error signal is bounded by an exponential function. Table 7.2 summarizes the adaptive impedance control of EDFJR in terms of their controller forms, adaptive laws and implementation issues.

Table 7.2 Summary of the adaptive impedance control for EDFJR

	Electrically driven flexible-joint robots interacting with environment $$\mathbf{D}_x\dot{\mathbf{s}} + \mathbf{C}_x\mathbf{s} + \mathbf{g}_x + \mathbf{D}_x\dot{\mathbf{v}} + \mathbf{C}_x\mathbf{v} = \mathbf{J}_a^{-T}\boldsymbol{\tau}_t - \mathbf{F}_{ext}$$ $$\mathbf{J}_t\ddot{\boldsymbol{\tau}}_t + \mathbf{B}_t\dot{\boldsymbol{\tau}}_t + \boldsymbol{\tau}_t = \mathbf{Hi} - \overline{\mathbf{q}}(\dot{\mathbf{q}}, \ddot{\mathbf{q}}) \qquad (7.5\text{-}1)$$ $$\mathbf{Li} + \mathbf{Ri} + \mathbf{K}_b\dot{\mathbf{q}} = \mathbf{u}$$	
	Regressor-based	Regressor-free
Controller	$\boldsymbol{\tau}_{td} = \mathbf{J}_a^T(\mathbf{F}_{ext} + \hat{\mathbf{g}}_x + \hat{\mathbf{D}}_x\dot{\mathbf{v}}$ $\qquad + \hat{\mathbf{C}}_x\mathbf{v} - \mathbf{K}_d\mathbf{s})$ $\quad = \mathbf{J}_a^T[\mathbf{F}_{ext} + \mathbf{Y}(\mathbf{x}, \dot{\mathbf{x}}, \mathbf{v}, \dot{\mathbf{v}})\hat{\mathbf{p}}_x$ $\qquad - \mathbf{K}_d\mathbf{s}]$ $\mathbf{i}_d = \mathbf{H}^{-1}[\boldsymbol{\Theta}\mathbf{x}_p + \boldsymbol{\Phi}\boldsymbol{\tau}_{td} + \mathbf{h}(\overline{\boldsymbol{\tau}}_{td}, \overline{\mathbf{q}})]$ $\mathbf{u} = \hat{\mathbf{L}}\mathbf{i}_d + \hat{\mathbf{R}}\mathbf{i} + \hat{\mathbf{K}}_b\dot{\mathbf{q}} - \mathbf{K}_c\mathbf{e}_i$ $\quad = \hat{\mathbf{p}}_i^T\boldsymbol{\varphi} - \mathbf{K}_c\mathbf{e}_i$ (7.5-14), (7.5-6), (7.5-16)	$\boldsymbol{\tau}_{td} = \mathbf{J}_a^T(\mathbf{F}_{ext} + \hat{\mathbf{g}}_x + \hat{\mathbf{D}}_x\dot{\mathbf{v}}$ $\qquad + \hat{\mathbf{C}}_x\mathbf{v} - \mathbf{K}_d\mathbf{s})$ $\mathbf{i}_d = \mathbf{H}^{-1}[\boldsymbol{\Theta}\mathbf{x}_p + \boldsymbol{\Phi}\boldsymbol{\tau}_{td} + \hat{\mathbf{h}}]$ $\mathbf{u} = \hat{\mathbf{f}} - \mathbf{K}_c\mathbf{e}_i$ (7.5-21), (7.5-23), (7.5-25)
Adaptive Law	$\dot{\hat{\mathbf{p}}}_x = -\boldsymbol{\Gamma}^{-1}\mathbf{Y}^T\mathbf{s}$ $\dot{\hat{\mathbf{p}}}_i = -\boldsymbol{\Gamma}_i^{-1}\boldsymbol{\varphi}\mathbf{e}_i^T$ (7.5-20)	$\dot{\hat{\mathbf{W}}}_{\mathbf{D}_x} = -\mathbf{Q}_{\mathbf{D}_x}^{-1}(\mathbf{Z}_{\mathbf{D}_x}\dot{\mathbf{v}}\mathbf{s}^T + \sigma_{\mathbf{D}_x}\hat{\mathbf{W}}_{\mathbf{D}_x})$ $\dot{\hat{\mathbf{W}}}_{\mathbf{C}_x} = -\mathbf{Q}_{\mathbf{C}_x}^{-1}(\mathbf{Z}_{\mathbf{C}_x}\mathbf{v}\mathbf{s}^T + \sigma_{\mathbf{C}_x}\hat{\mathbf{W}}_{\mathbf{C}_x})$ $\dot{\hat{\mathbf{W}}}_{\mathbf{g}_x} = -\mathbf{Q}_{\mathbf{g}_x}^{-1}(\mathbf{z}_{\mathbf{g}_x}\mathbf{s}^T + \sigma_{\mathbf{g}_x}\hat{\mathbf{W}}_{\mathbf{g}_x})$ $\dot{\hat{\mathbf{W}}}_{\mathbf{h}} = -\mathbf{Q}_{\mathbf{h}}^{-1}(\mathbf{z}_{\mathbf{h}}\mathbf{e}_\tau^T + \sigma_{\mathbf{h}}\hat{\mathbf{W}}_{\mathbf{h}})$ $\dot{\hat{\mathbf{W}}}_{\mathbf{f}} = -\mathbf{Q}_{\mathbf{f}}^{-1}(\mathbf{z}_{\mathbf{f}}\mathbf{e}_i^T + \sigma_{\mathbf{f}}\hat{\mathbf{W}}_{\mathbf{f}})$ (7.5-31)
Realization Issue	Need to know the regressor matrix, joint accelerations and their higher derivatives.	Does not need the information for the regressor matrix, joint accelerations, or their derivatives.

Example 7.2:

Consider flexible-joint robot in example 7.1 but with consideration of the motor dynamics. Actual values of link parameters are selected as $m_1 = m_2 = 0.5(kg)$, $l_1 = l_2 = 0.75(m)$, $l_{c1} = l_{c2} = 0.375(m)$, $I_1 = I_2 = 0.0234(kg\text{-}m^2)$, and $k_1 = k_2 = 100(\text{N-}m/\text{rad})$. Parameters for the actuator part are chosen as $j_1 = 0.02(kg\text{-}m^2)$, $j_2 = 0.01(kg\text{-}m^2)$, $b_1 = 5(\text{N-}m\text{-}sec/\text{rad})$, $b_2 = 4(\text{N-}m\text{-}sec/\text{rad})$, and $h_1 = h_2 = 10(\text{N-}m/\text{A})$. Electrical parameter for the actuator are $L_1 = L_2 = 0.025(\text{H})$, $r_1 = r_2 = 1(\Omega)$, $k_{b1} = k_{b2} = 1$ (Vol/rad/sec) (Chien and Huang 2007a). The stiffness of the constrained surface is assumed to be $k_w = 5000(N/m)$. In order to observe the effect of the actuator dynamics, the endpoint is required to track a $0.2m$ radius circle centered at $(0.8m, 1.0m)$ in 2 seconds which is much faster than the case in example 7.1. The initial condition for the generalized coordinate vector is at $\mathbf{q}(0) = \mathbf{\theta}(0) = [0.0022 \quad 1.5019 \quad 0 \quad 0]^T$, i.e., the endpoint is initially at $(0.8m, 0,75m)$. It is away from the desired initial endpoint position $(0.8m, 0,8m)$ for observation of the transient. The initial state for the reference model is $\mathbf{\tau}_r(0) = [8.1 \quad 1.4 \quad 0 \quad 0]^T$, which is the same as the initial state for the desired torque. The controller gain matrices are selected as

$$\mathbf{K}_d = \begin{bmatrix} 50 & 0 \\ 0 & 50 \end{bmatrix}, \mathbf{\Lambda} = \begin{bmatrix} 20 & 0 \\ 0 & 20 \end{bmatrix}, \text{ and } \mathbf{K}_c = \begin{bmatrix} 200 & 0 \\ 0 & 200 \end{bmatrix}.$$

The initial value for the desired current can be found by calculation as

$$\mathbf{i}_d(0) = \mathbf{i}(0) = [16.2 \quad 1.4]^T.$$

The matrices in the target impedance are picked as

$$\mathbf{M}_i = \begin{bmatrix} 0.5 & 0 \\ 0 & 0.5 \end{bmatrix}, \mathbf{B}_i = \begin{bmatrix} 100 & 0 \\ 0 & 100 \end{bmatrix}, \text{ and } \mathbf{K}_i = \begin{bmatrix} 1500 & 0 \\ 0 & 1500 \end{bmatrix}.$$

The 11-term Fourier series is selected as the basis function for the approximation. Therefore, $\hat{\mathbf{W}}_{\mathbf{D}_x}$ and $\hat{\mathbf{W}}_{\mathbf{C}_x}$ are in $\mathfrak{R}^{44 \times 2}$, while $\hat{\mathbf{W}}_{\mathbf{g}_x}$, $\hat{\mathbf{W}}_{\mathbf{h}}$, and $\hat{\mathbf{W}}_{\mathbf{f}}$ are in $\mathfrak{R}^{22 \times 2}$. The initial weighting vectors for the entries are assigned to be

$$\hat{\mathbf{w}}_{D_{x11}}(0) = [0.05 \quad 0 \quad \cdots \quad 0]^T \in \mathfrak{R}^{11 \times 1}$$

$$\hat{\mathbf{w}}_{D_{x12}}(0) = \hat{\mathbf{w}}_{D_{x21}}(0) = [-0.05 \quad 0 \quad \cdots \quad 0]^T \in \mathfrak{R}^{11 \times 1}$$

$$\hat{\mathbf{w}}_{D_{x22}}(0) = [0.1 \quad 0 \quad \cdots \quad 0]^T \in \mathfrak{R}^{11 \times 1}$$

$$\hat{\mathbf{w}}_{C_{x11}}(0) = [0.05 \quad 0 \quad \cdots \quad 0]^T \in \mathfrak{R}^{11 \times 1}$$

$$\hat{\mathbf{w}}_{C_{x12}}(0) = \hat{\mathbf{w}}_{C_{x21}}(0) = [-0.05 \quad 0 \quad \cdots \quad 0]^T \in \mathfrak{R}^{11 \times 1}$$

$$\hat{\mathbf{w}}_{C_{x22}}(0) = [0.1 \quad 0 \quad \cdots \quad 0]^T \in \mathfrak{R}^{11 \times 1}$$

$$\hat{\mathbf{w}}_{g_{x1}}(0) = \hat{\mathbf{w}}_{g_{x2}}(0) = [0 \quad 0 \quad \cdots \quad 0]^T \in \mathfrak{R}^{11 \times 1}$$

$$\hat{\mathbf{w}}_{h_1}(0) = \hat{\mathbf{w}}_{h_2}(0) = [0 \quad 0 \quad \cdots \quad 0]^T \in \mathfrak{R}^{11 \times 1}$$

$$\hat{\mathbf{w}}_{f_1}(0) = \hat{\mathbf{w}}_{f_2}(0) = [0 \quad 0 \quad \cdots \quad 0]^T \in \mathfrak{R}^{11 \times 1}$$

The gain matrices in the update law (31) are selected as $\mathbf{Q}_{\mathbf{D}_x}^{-1} = 0.001\mathbf{I}_{44}$, $\mathbf{Q}_{\mathbf{C}_x}^{-1} = 0.001\mathbf{I}_{44}$, $\mathbf{Q}_{\mathbf{g}_x}^{-1} = 100\mathbf{I}_{22}$, $\mathbf{Q}_{\mathbf{h}}^{-1} = 10\mathbf{I}_{22}$, and $\mathbf{Q}_{\mathbf{f}}^{-1} = 10000\mathbf{I}_{22}$. In this simulation, the approximation error is assumed to be neglected, and the σ-modification parameters are chosen as $\sigma_{(\cdot)} = 0$.

The simulation results are shown in Figure 7.10 to 7.20. Figure 7.10 shows the tracking performance of the robot endpoint and its desired trajectory in the Cartesian space. It is observed that the endpoint trajectory converges smoothly to the desired trajectory in the free space tracking and contacts compliantly in the constrained motion phase. Computation of the complex regressor is avoided in this strategy which greatly simplifies the design and implementation of the control law. Although the initial error is quite large, the transient state takes only about 0.2 seconds which can be justified from the joint space tracking history in Figure 7.11. The torque tracking performance is shown in Figure 7.12. It can be seen that the torque errors for both joints are small. Figure 7.13 presents the current tracking performance. It is observed that the strategy can give very small current error. The control voltages to the two motors are reasonable that can be verified in Figure 7.14. Figure 7.15 to 7.20 are the performance of function approximation. Although most parameters do not converge to their actual values, they still remain bounded as desired.

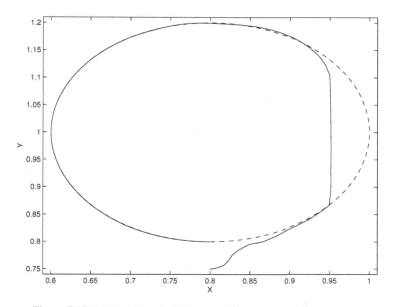

Figure 7.10 Robot endpoint tracking performance in the Cartesian space

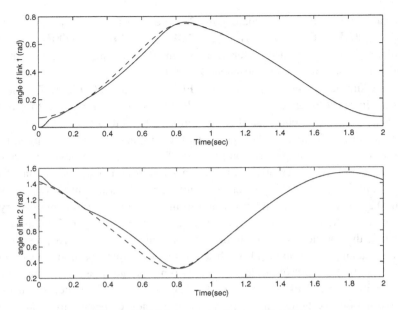

Figure 7.11 Joint space tracking performance

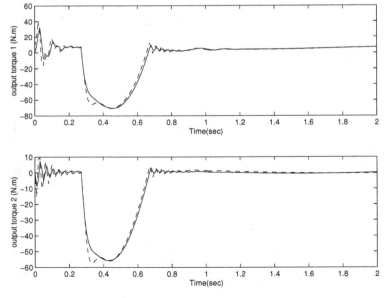

Figure 7.12 Tracking in the torque tracking loop

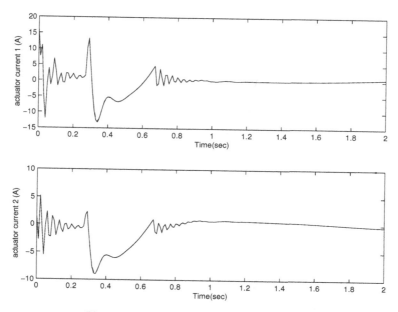

Figure 7.13 Tracking in the current tracking loop

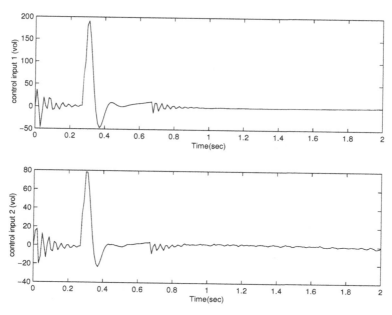

Figure 7.14 Control efforts

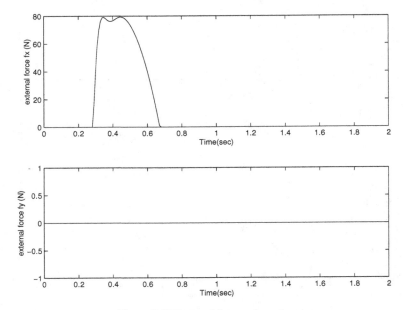

Figure 7.15 External force trajectories

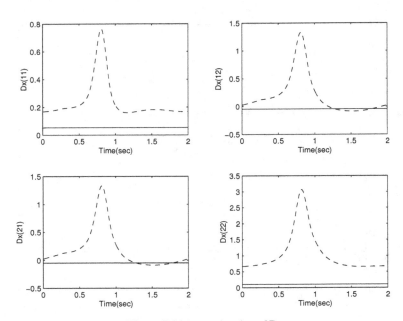

Figure 7.16 Approximation of $\mathbf{D}_x$

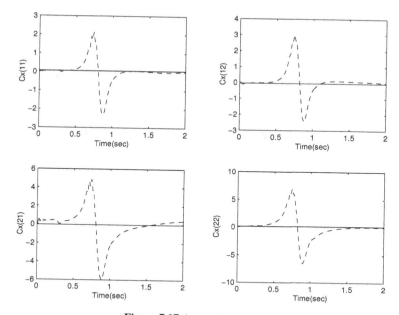

Figure 7.17 Approximation of $\mathbf{C}_x$

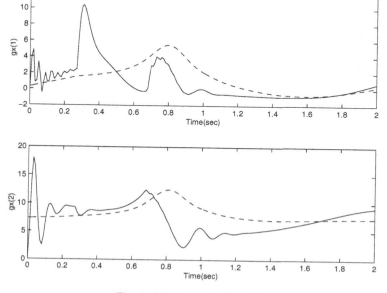

Figure 7.18 Approximation of $\mathbf{g}_x$

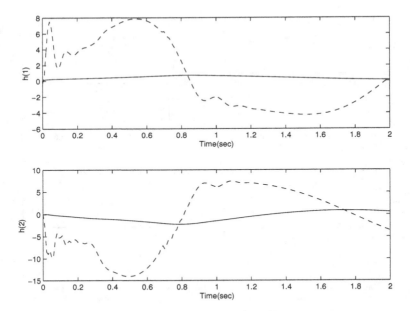

Figure 7.19 Approximation of **h**

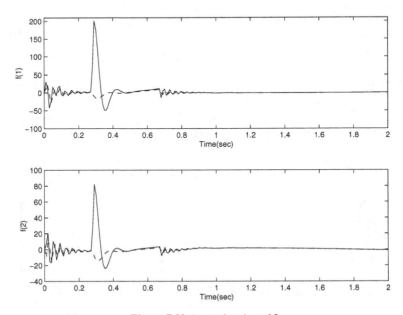

Figure 7.20 Approximation of **f**

7.6 Conclusions

In this chapter, we consider the case when the flexible-joint robot contacts with the environment. The adaptive impedance controllers are derived to give good performance in both free space tracking and constraint motion phase. Firstly, the MRC rule is utilized in Section 7.2 for the impedance control of a known FJR. To deal with the uncertainties, a regressor based adaptive controller is derived in Section 7.3. However, its implementation requires the knowledge of joint accelerations, the regressor matrix, and their higher order derivatives. Therefore, the control strategy is not practical. The regressor-free adaptive impedance controller is developed in Section 7.4 whose realization do not need the joint accelerations, the regressor matrix, or their higher order derivatives. The actuator dynamics is included in the system equation in Section 7.5. A regreesor-based adaptive impedance controller is then developed for EDFJR in Section 7.5.1. However, its realization still needs the joint accelerations, regressor matrix, and their derivatives. A regreesor-free adaptive impedance controller for EDFJR is then introduced in Section 7.5.2 and it is free from the information for joint accelerations or the regressor matrix. Several simulation results show that the regressor-free design can give good performance to an EDFJR operating in a compliant motion environment.

Appendix

Lemma A1:

Let $\mathbf{w}_i^T = [w_{i1} \quad w_{i2} \quad \cdots \quad w_{in}] \in \mathfrak{R}^{1 \times n}$, $i = 1, \ldots, m$ and $\mathbf{W}$ is a block diagonal matrix defined as $\mathbf{W} = diag\{\mathbf{w}_1, \mathbf{w}_2, \cdots, \mathbf{w}_m\} \in \mathfrak{R}^{mn \times m}$. Then, $Tr(\mathbf{W}^T \mathbf{W}) = \sum_{i=1}^{m} \|\mathbf{w}_i\|^2$.

Proof: The proof is straightforward as below:

$$\mathbf{W}^T \mathbf{W} = \begin{bmatrix} \mathbf{w}_1^T & \mathbf{0} & \cdots & \mathbf{0} \\ \mathbf{0} & \mathbf{w}_2^T & \cdots & \mathbf{0} \\ \vdots & \vdots & \ddots & \vdots \\ \mathbf{0} & \mathbf{0} & \cdots & \mathbf{w}_m^T \end{bmatrix} \begin{bmatrix} \mathbf{w}_1 & \mathbf{0} & \cdots & \mathbf{0} \\ \mathbf{0} & \mathbf{w}_2 & \cdots & \mathbf{0} \\ \vdots & \vdots & \ddots & \vdots \\ \mathbf{0} & \mathbf{0} & \cdots & \mathbf{w}_m \end{bmatrix}$$

$$= \begin{bmatrix} \mathbf{w}_1^T \mathbf{w}_1 & \mathbf{0} & \cdots & \mathbf{0} \\ \mathbf{0} & \mathbf{w}_2^T \mathbf{w}_2 & \cdots & \mathbf{0} \\ \vdots & \vdots & \ddots & \vdots \\ \mathbf{0} & \mathbf{0} & \cdots & \mathbf{w}_m^T \mathbf{w}_m \end{bmatrix}$$

$$= \begin{bmatrix} \|\mathbf{w}_1\|^2 & \mathbf{0} & \cdots & \mathbf{0} \\ \mathbf{0} & \|\mathbf{w}_2\|^2 & \cdots & \mathbf{0} \\ \vdots & \vdots & \ddots & \vdots \\ \mathbf{0} & \mathbf{0} & \cdots & \|\mathbf{w}_m\|^2 \end{bmatrix}$$

Therefore, we have $Tr(\mathbf{W}^T \mathbf{W}) = \sum_{i=1}^{m} \|\mathbf{w}_i\|^2$.

Lemma A2:

Suppose $\mathbf{w}_i^T = [w_{i1} \quad w_{i2} \quad \cdots \quad w_{in}] \in \mathfrak{R}^{1 \times n}$ and $\mathbf{v}_i^T = [v_{i1} \quad v_{i2} \quad \cdots \quad v_{in}] \in \mathfrak{R}^{1 \times n}$, $i = 1, \ldots, m$. Let $\mathbf{W}$ and $\mathbf{V}$ be block diagonal matrices that are defined as $\mathbf{W} = diag\{\mathbf{w}_1, \mathbf{w}_2, \cdots, \mathbf{w}_m\} \in \mathfrak{R}^{mn \times m}$ and $\mathbf{V} = diag\{\mathbf{v}_1, \mathbf{v}_2, \cdots, \mathbf{v}_m\} \in \mathfrak{R}^{mn \times m}$, respectively. Then, $Tr(\mathbf{V}^T \mathbf{W}) \le \sum_{i=1}^{m} \|\mathbf{v}_i\| \|\mathbf{w}_i\|$.

Proof: The proof is also straightforward:

$$
\mathbf{V}^T \mathbf{W} = \begin{bmatrix} \mathbf{v}_1^T & \mathbf{0} & \cdots & \mathbf{0} \\ \mathbf{0} & \mathbf{v}_2^T & \cdots & \mathbf{0} \\ \vdots & \vdots & \ddots & \vdots \\ \mathbf{0} & \mathbf{0} & \cdots & \mathbf{v}_m^T \end{bmatrix} \begin{bmatrix} \mathbf{w}_1 & \mathbf{0} & \cdots & \mathbf{0} \\ \mathbf{0} & \mathbf{w}_2 & \cdots & \mathbf{0} \\ \vdots & \vdots & \ddots & \vdots \\ \mathbf{0} & \mathbf{0} & \cdots & \mathbf{w}_m \end{bmatrix}
$$

$$
= \begin{bmatrix} \mathbf{v}_1^T \mathbf{w}_1 & \mathbf{0} & \cdots & \mathbf{0} \\ \mathbf{0} & \mathbf{v}_2^T \mathbf{w}_2 & \cdots & \mathbf{0} \\ \vdots & \vdots & \ddots & \vdots \\ \mathbf{0} & \mathbf{0} & \cdots & \mathbf{v}_m^T \mathbf{w}_m \end{bmatrix}
$$

Hence,

$$
Tr(\mathbf{V}^T \mathbf{W}) = \mathbf{v}_1^T \mathbf{w}_1 + \mathbf{v}_2^T \mathbf{w}_2 + \ldots + \mathbf{v}_m^T \mathbf{w}_m
$$
$$
\le \|\mathbf{v}_1\| \|\mathbf{w}_1\| + \|\mathbf{v}_2\| \|\mathbf{w}_2\| + \ldots + \|\mathbf{v}_m\| \|\mathbf{w}_m\|
$$
$$
= \sum_{i=1}^{m} \|\mathbf{v}_i\| \|\mathbf{w}_i\|.
$$

Lemma A3:

Let $\mathbf{W}$ be defined as in lemma A1, and $\tilde{\mathbf{W}}$ is a matrix defined as $\tilde{\mathbf{W}} = \mathbf{W} - \hat{\mathbf{W}}$, where $\hat{\mathbf{W}}$ is a matrix with proper dimension. Then

$$
Tr(\tilde{\mathbf{W}}^T \hat{\mathbf{W}}) \le \frac{1}{2} Tr(\mathbf{W}^T \mathbf{W}) - \frac{1}{2} Tr(\tilde{\mathbf{W}}^T \tilde{\mathbf{W}})
$$

Proof:

$$Tr(\tilde{\mathbf{W}}^T\hat{\mathbf{W}}) = Tr(\tilde{\mathbf{W}}^T\mathbf{W}) - Tr(\tilde{\mathbf{W}}^T\tilde{\mathbf{W}})$$

$$\leq \sum_{i=1}^{m}(\|\tilde{\mathbf{w}}_i\|\|\mathbf{w}_i\| - \|\tilde{\mathbf{w}}_i\|^2) \quad \text{(by lemma A1 and A2)}$$

$$= \frac{1}{2}\sum_{i=1}^{m}[\|\mathbf{w}_i\|^2 - \|\tilde{\mathbf{w}}_i\|^2 - (\|\tilde{\mathbf{w}}_i\| - \|\mathbf{w}_i\|)^2]$$

$$\leq \frac{1}{2}\sum_{i=1}^{m}(\|\mathbf{w}_i\|^2 - \|\tilde{\mathbf{w}}_i\|^2)$$

$$= \frac{1}{2}Tr(\mathbf{W}^T\mathbf{W}) - \frac{1}{2}Tr(\tilde{\mathbf{W}}^T\tilde{\mathbf{W}}) \quad \text{(by lemma A1)}$$

In the above lemmas, we consider properties of a block diagonal matrix. In the following, we would like to extend the analysis to a class of more general matrices.

Lemma A4:

Let $\mathbf{W}$ be a matrix in the form $\mathbf{W}^T = [\mathbf{W}_1^T \quad \mathbf{W}_2^T \quad \cdots \quad \mathbf{W}_p^T] \in \Re^{pmn \times m}$ where $\mathbf{W}_i = diag\{\mathbf{w}_{i1}, \mathbf{w}_{i2}, \cdots, \mathbf{w}_{im}\} \in \Re^{mn \times m}$, $i = 1, \ldots, p$, are block diagonal matrices with the entries of vectors $\mathbf{w}_{ij}^T = [w_{ij1} \quad w_{ij2} \quad \cdots \quad w_{ijn}] \in \Re^{1 \times n}$, $j = 1, \ldots, m$. Then, we may have $Tr(\mathbf{W}^T\mathbf{W}) = \sum_{i=1}^{p}\sum_{j=1}^{m}\|\mathbf{w}_{ij}\|^2$.

Proof:

$$\mathbf{W}^T\mathbf{W} = [\mathbf{W}_1^T \quad \cdots \quad \mathbf{W}_p^T]\begin{bmatrix} \mathbf{W}_1 \\ \vdots \\ \mathbf{W}_p \end{bmatrix}$$

$$= \mathbf{W}_1^T\mathbf{W}_1 + \cdots + \mathbf{W}_p^T\mathbf{W}_p$$

Hence, we may calculate the trace as

$$Tr(\mathbf{W}^T\mathbf{W}) = Tr(\mathbf{W}_1^T\mathbf{W}_1) + \cdots + Tr(\mathbf{W}_p^T\mathbf{W}_p)$$

$$= \sum_{j=1}^{m}\left\|\mathbf{w}_{1j}\right\|^2 + \cdots + \sum_{j=1}^{m}\left\|\mathbf{w}_{pj}\right\|^2 \qquad \text{(by lemma A1)}$$

$$= \sum_{i=1}^{p}\sum_{j=1}^{m}\left\|\mathbf{w}_{ij}\right\|^2$$

Lemma A5:

Let $\mathbf{V}$ and $\mathbf{W}$ be matrices defined in lemma A4, Then,
$$Tr(\mathbf{V}^T\mathbf{W}) \le \sum_{i=1}^{p}\sum_{j=1}^{m}\left\|\mathbf{v}_{ij}\right\|\left\|\mathbf{w}_{ij}\right\|.$$

Proof:

$$Tr(\mathbf{V}^T\mathbf{W}) = Tr(\mathbf{V}_1^T\mathbf{W}_1) + \cdots + Tr(\mathbf{V}_p^T\mathbf{W}_p)$$

$$\le \sum_{j=1}^{m}\left\|\mathbf{v}_{1j}\right\|\left\|\mathbf{w}_{1j}\right\| + \cdots + \sum_{j=1}^{m}\left\|\mathbf{v}_{pj}\right\|\left\|\mathbf{w}_{pj}\right\| \qquad \text{(by lemma A2)}$$

$$= \sum_{i=1}^{p}\sum_{j=1}^{m}\left\|\mathbf{v}_{ij}\right\|\left\|\mathbf{w}_{ij}\right\|$$

Lemma A6:

Let $\mathbf{W}$ be defined as in lemma A4, and $\tilde{\mathbf{W}}$ is a matrix defined as $\tilde{\mathbf{W}} = \mathbf{W} - \hat{\mathbf{W}}$, where $\hat{\mathbf{W}}$ is a matrix with proper dimension. Then

$$Tr(\tilde{\mathbf{W}}^T\hat{\mathbf{W}}) \le \frac{1}{2}Tr(\mathbf{W}^T\mathbf{W}) - \frac{1}{2}Tr(\tilde{\mathbf{W}}^T\tilde{\mathbf{W}})$$

Proof:

$$Tr(\tilde{\mathbf{W}}^T\hat{\mathbf{W}}) = Tr(\tilde{\mathbf{W}}^T\mathbf{W}) - Tr(\tilde{\mathbf{W}}^T\tilde{\mathbf{W}})$$

$$\le \sum_{i=1}^{p}\sum_{j=1}^{m}(\|\tilde{\mathbf{w}}_{ij}\|\|\mathbf{w}_{ij}\| - \|\tilde{\mathbf{w}}_{ij}\|^2) \quad \text{(by lemma A4 and A5)}$$

$$= \frac{1}{2}\sum_{i=1}^{p}\sum_{j=1}^{m}[\|\mathbf{w}_{ij}\|^2 - \|\tilde{\mathbf{w}}_{ij}\|^2 - (\|\tilde{\mathbf{w}}_{ij}\| - \|\mathbf{w}_{ij}\|)^2]$$

$$\le \frac{1}{2}\sum_{i=1}^{p}\sum_{j=1}^{m}(\|\mathbf{w}_{ij}\|^2 - \|\tilde{\mathbf{w}}_{ij}\|^2)$$

$$= \frac{1}{2}Tr(\mathbf{W}^T\mathbf{W}) - \frac{1}{2}Tr(\tilde{\mathbf{W}}^T\tilde{\mathbf{W}}) \quad \text{(by lemma A4)}$$

References

A. Abdallah, D. Dawson, P. Dorato and M. Jamishidi, "Survey of Robust Control for Rigid Robots," *IEEE Control System Magazine*, vol. 11, no. 2, pp. 24-30, 1991.

S. Ahmad, "Constrained Motion (Force/Position) Control of Flexible Joint Robots," *IEEE Transactions on System, Man, and Cybernetics*, vol. 23, 1993.

R. J. Anderson and M. W. Spong, "Hybrid Impedance Control of Robotic Manipulators," *IEEE Transactions on Robotics and Automation*, vol. 4, no. 5, pp. 549-556, 1988.

H. Asada and J-J. E. Slotine, Robot Analysis and Control, N.Y.: *Wiley*, 1986.

R. Carelli and R. Kelly, "An Adaptive Impedance Force Control for Robot Manipulators," *IEEE Transactions on Automatic Control*, vol. 36, no. 8, pp. 967-971, 1991.

Y. C. Chang and J. Shaw, "Low-frequency Vibration Control of a Pan/tilt Platform with Vision Feedback," *Journal of Sound and Vibration*, vol. 302, Issues 4-5, pp. 716-727, May 2007.

P. C. Chen and A. C. Huang, "Adaptive Tracking Control for a Class of Non-autonomous Systems Containing Time-varying Uncertainties with Unknown Bounds," *IEEE International Conference on Systems, Man and Cybernetics*, Hague, Netherlands, pp. 5870-5875, Oct. 2004.

P. C. Chen and A. C. Huang, "Adaptive Sliding Control of Active Suspension Systems based on Function Approximation Technique," *Journal of Sound and Vibration*, vol. 282, issue 3-5, pp. 1119-1135, April 2005a.

P. C. Chen and A. C. Huang, "Adaptive Multiple-surface Sliding Control of Hydraulic Active Suspension Systems Based on Function Approximation Technique," *Journal of Vibration and Control*, vol. 11, no. 5, pp. 685-706, 2005b.

P. C. Chen and A. C. Huang, "Adaptive Sliding Control of Active Suspension Systems with Uncertain Hydraulic Actuator Dynamics," *Vehicle System Dynamics*, vol. 44, no. 5, pp. 357-368, May 2006.

M. C. Chien and A. C. Huang, "Adaptive Impedance Control of Robot Manipulators based on Function Approximation Technique," *Robotica*, vol. 22, issue 04, pp. 395-403, August 2004.

M. C. Chien and A. C. Huang, "Regressor-Free Adaptive Impedance Control of Flexible-Joint Robots Using FAT," *American Control Conference*, 2006a.

M. C. Chien and A. C. Huang, "FAT-based adaptive control for flexible-joint robots without computation of the regressor matrix," *IEEE International Conference on Systems, Man and Cybernetics*, 2006b.

M. C. Chien and A. C. Huang, "Adaptive control of flexible-joint electrically-driven robot with time-varying uncertainties," *IEEE Transactions on Industrial Electronics*, vol. 54, no. 2, pp. 1032-1038, April 2007a.

M. C. Chien and A. C. Huang, "Adaptive control of electrically-driven robot without computation of Regressor matrix," *Journal of Chinese Institute of Engineers*, vol. 30, no. 5, pp. 855-862, July 2007b.

R. Colbaugh, H. Seraji and K. Glass, "Direct Adaptive Impedance Control of Manipulators," *Proceedings IEEE Conference on Decision and Control*, pp. 2410-2415, 1991.

R. Colbaugh, K. Glass and G. Gallegos, "Adaptive Compliant Motion Control of Flexible-joint Manipulators," *Proceedings American Control Conference*, Albuquerque, New Mexico, pp. 1873-1878, 1997.

W. E. Dixon, E. Zergeroglu, D. M. Dawson and M. W. Hannan, "Global Adaptive Partial State Feedback Tracking Control of Rigid-link Flexible-joint

Robots," *Proceedings IEEE/ASME International Conference on Advanced Intelligent Mechatronics*, pp. 281-286, Atlanta, USA, Sept. 1999.

S. S. Ge, "Adaptive Controller Design for Flexible Joint Manipulators," *Automatica*, vol. 32, no. 2, pp. 273-278, 1996.

S. S. Ge, T. H. Lee and C. J. Harris, Adaptive Neural Network Control of Robotic Manipulators, *World Scientific Publishing*, Singapore, 1998.

S. S. Ge, C. C. Hang, T. H. Lee and T. Zang, Stable Adaptive Neural Network Control, *Kluwer Academic*, Boston, 2001.

A. A. Goldenberg, "Implementation of Force and Impedance Control in Robot Manipulators," *Proceedings IEEE International Conference on Robotics and Automation*, pp. 1626-1632, 1988.

J. J. Gonzalez and G. R. Widmann, "A Force Commanded Impedance Control Scheme for Robots with Hard Nonlinearities," *IEEE Transactions on Control Systems Technology*, vol. 3, no. 4, pp. 398-408, 1995.

N. Hogan, "Impedance Control: an Approach to Manipulation: Part1-Theory, Part2-Implementation, Part3-an Approach to Manipulation," *ASME Journal of Dynamic Systems, Measurement, and Control*, vol. 107, pp. 1-24, 1985.

Y. R. Hu and G. Vukovich, "Position and Force Control of Flexible-joint Robots During Constrained Motion Tasks," *Mechanism and Machine Theory*, vol. 36, pp. 853-871, 2001.

A. C. Huang and Y. S. Kuo, "Sliding Control of Nonlinear Systems Containing Time-varying Uncertainties with Unknown Bounds," *International Journal of Control*, vol. 74, no. 3, pp. 252-264, 2001.

A. C. Huang and Y. C. Chen, "Adaptive Sliding Control for Single-Link Flexible-Joint Robot with Mismatched Uncertainties," *IEEE Transactions on Control Systems Technology*, vol. 12, no. 5, pp. 770-775, Sept. 2004a.

A. C. Huang and Y. C. Chen, "Adaptive Multiple-Surface Sliding Control for Non-Autonomous Systems with Mismatched Uncertainties," *Automatica*, vol. 40, issue 11, pp. 1939-1945, Nov. 2004b.

A. C. Huang, S. C. Wu and W. F. Ting, "An FAT-based Adaptive Controller for Robot Manipulators without Regressor Matrix: Theory and Experiments," *Robotica*, vol. 24, pp. 205-210, 2006.

A. C. Huang and K. K. Liao, "FAT-based Adaptive Sliding Control for Flexible Arms, Theory and Experiments," *Journal of Sound and Vibration*, vol. 298, issue 1-2, pp. 194-205, Nov. 2006.

S. K. Ider, "Force and Motion Trajectory Tracking Control of Flexible-joint Robots," *Mechanism and Machine Theory*, vol. 35, pp. 363-378, 2000.

K. P. Jankowski and H. A. ElMaraghy, "Nonlinear Decoupling for Position and Force Control of Constrained Robots with Flexible-joints," *Proceedings IEEE International Conference on Robotics and Automation*, Sacramento, California, pp. 1226-1231, April 1991.

H. Kawasaki, T. Bito and K. Kanzaki, "An Efficient Algorithm for the Model-based Adaptive Control of Robotic Manipulators," *IEEE Transactions on Robotics and Automation*, vol. 12, no. 3, pp. 496-501, 1996.

R. Kelly, R. Carelli, M. Amestegui and R. Ortega, "An Adaptive Impedance Control of Robot Manipulators," *Proceedings IEEE Conference on Robotics and Automation*, 572-557, 1987.

K. Kozlowski and P. Sauer, "The Stability of the Adaptive Tracking Controller of Rigid and Flexible Joint Robots," *Proceedings on the First Workshop on Robot Motion and Control*, pp. 85-93, Kiekrz, Poland, June 1999a.

K. Kozlowski and P. Sauer, "On Adaptive Control of Flexible Joint Manipulators: Theory and Experiments," *Proceedings IEEE International Symposium on Industrial Electronics*, vol. 3, pp. 1153-1158, Bled, Slovenia, July 1999b.

M. Krstic, I. Kanellakopoulos and P. Kokotovic, Nonlinear and Adaptive Control Design, *John Wiley & Sons, Inc.*, 1995.

K. Y. Lian, J. H. Jean and L. C. Fu, "Adaptive Force Control of Single-link Mechanism with Joint Flexibility," *IEEE Transactions on Robotics and Automation*, vol. 7, pp. 540-545, Aug. 1991.

Y. Liang, S. Cong and W. Shang, "Function Approximation-based Sliding Mode Adaptive Control," *Nonlinear Dynamics*, vol. 54, no. 3, pp. 223-230, Nov. 2008.

T. Lin and A. A. Goldenberg, "Robust Adaptive Control of Flexible Joint Robots with Joint Torque Feedback," *Proceedings IEEE International Conference on Robotics and Automation*, pp. 1229-1234, 1995.

T. Lin and A. A. Goldenberg, "A Unified Approach to Motion and Force Control of Flexible Joint Robots," *Proceedings IEEE International Conference on Robotics and Automation*, pp. 1115-1120, 1996.

T. Lin and A. A. Goldenberg, "On Coordinated Control of Multiple Flexible-joint Robots Holding a Constrained Object," *Proceedings IEEE International Conference on Robotics and Automation*, pp. 1490-1495, 1997.

W. S. Lu and Q. H. Meng, "Recursive Computation of Manipulator Regressor and its Application to Adaptive Motion Control of Robots," *IEEE Conference on Communication, Computation and Signal Processing*, pp. 170-173, 1991a.

W. S. Lu and Q. H. Meng, "Impedance Control with Adaptation for Robotic Manipulations," *IEEE Transactions on Robotics and Automation*, vol. 7, no. 3, pp. 408-415, 1991b.

W. S. Lu and Q. H. Meng, "Regressor Formulation of Robot Dynamics: Computation and Application," *IEEE Transactions on Robotics and Automation*, vol. 9, no. 3, pp. 323-333, Jun. 1993.

J. K Mills and G. J. Liu, "Robotic Manipulator Impedance Control of Generalized Contact Force and Position," *IEEE/RSJ International Workshop on Intelligent Robots and System*, pp. 1103-1108, 1991.

V. Mut, O. Nasisi, R. Carelli and B. Kuchen, "Tracking Adaptive Impedance Robot Control with Visual Feedback," *Robotica*, vol. 18, pp. 369-374, 2000.

R. Ortega and M. W. Spong, "Adaptive Motion Control of Rigid Robots: a Tutorial," *Proceeding of. 27th Conference on Decision and Control*, pp. 1575-1584, 1988.

C. Ott, A. A. Schaffer and G. Hirzinger, "Comparison of Adaptive and Non-adaptive Tracking Control Laws for a Flexible Joint Manipulator," *Proceedings IEEE/RSJ International Conference on Intelligent Robots and Systems*, Lausanne, Switzerland, pp. 2018-2024, Oct. 2000.

A. Ott, A. A. Schaffer, A. Kugi and G. Hirzinger, "Decoupling Based Cartesian Impedance Control of Flexible-joint Robots," *Proceedings IEEE International Conference on Robotics and Automation*, Taipei, Taiwan, pp. 3101-3107, Sept. 2003.

S. Ozgoli and H. D. Taghirad, "A Survey on the Control of Flexible Joint Robots," *Asian Journal of Control*, vol. 8, no. 4, pp. 1-15, Dec. 2006.

P. R. Pagilla and M. Tomizuka, "An Adaptive Output Feedback Controller for Robot Arms: Stability and Experiments," *Automatica*, vol. 37, no. 7, pp. 983-995, July 2001.

J. S. Park, Y. A. Jiang, T. Hesketh and D. J. Clements, "Trajectory Control of Manipulators Using Adaptive Sliding Mode Control," *Proceedings of IEEE SOUTHEASTCON*, pp. 142-146, 1994.

Z. Qu and J. Dorsey, "Robust Tracking Control of Robots by a Linear Feedback Law," *IEEE Transactions on Automatic Control*, vol. 36, no. 9, pp. 1081-1084, 1991.

M. H. Raibert and J. J. Craig, "Hybrid Position/Force Control of Manipulators," *ASME Journal of Dynamic Systems, Measurements and Control*, vol. 102, pp. 126-133, 1981.

C. E. Rohrs, L. S. Valavani, M. Athans and G. Stein, "Robustness of Continuous-time Adaptive Control Algorithms in the Presence of Unmodeled Dynamics," *IEEE Transactions on Automatic Control*, vol. 30, pp. 881-889, 1985.

N. Sadegh and R. Horowitz, "Stability and Robustness Analysis of a Class of Adaptive Controller for Robotic Manipulators," *International Journal of Robotics Research*, vol. 9, no. 3, pp. 74-92, 1990.

A. Schaffer, C. Ott., U. Frese and G. Hirzinger, "Cartesian Impedance Control of Redundant Robots: Recent Results with the DLR-light-weight-arms," *Proceedings IEEE International Conference on Robotics and Automation*, Taipei, Taiwan, pp. 3704-3709, Sept. 2003.

J.-J. E. Slotine and W. Li, "Adaptive Strategy in Constrained Manipulators," *Proceeding IEEE International Conference on Robotics and Automation*, pp. 595-601, 1987.

J.-J. E. Slotine and W. Li, "Adaptive Manipulator Control: a Case Study," *IEEE Transactions on Automatic Control*, vol. 33, no. 11, pp. 995-1003, 1988.

J.-J. E. Slotine and W. Li, Applied Nonlinear Control, NJ: *Prentice-Hall*, 1991.

Y. D. Song, "Adaptive Motion Tracking Control of Robot Manipulators: Non-regressor Based Approach," *IEEE International Conference on Robotics and Automation*, vol. 4, pp. 3008-3013, 1994.

M. W. Spong, "Adaptive Control of Flexible Joint Manipulators," *Systems and Control Letters*, vol. 13, pp. 15-21, 1989.

M. W. Spong, "Adaptive Control of Flexible Joint Manipulators: Comments on Two Papers," *Automatica*, vol. 31, no. 4, pp. 585-590, 1995.

M. W. Spong, "On the Force Control Problem for Flexible-joint Manipulators," *IEEE Transactions* on *Automatic Control*, vol. 34, pp. 107-111, 1989.

M. W. Spong and M. Vidyasagar, Robot Dynamics and Control, *Wiley*, 1989.

J. T. Spooner, M. Maggiore, R. Ordonez and K. M. Passino, Stable Adaptive Control and Estimation for Nonlinear Systems – Neural and Fuzzy Approximator Techniques, *Wiley*, New York, 2002.

I. Stakgold, Green's Functions and Boundary Value Problems, NY: *Wiley*, 1979.

C. Y. Su and Y. Stepanenko, "Adaptive Control for Constrained Robots Without Using Regressor," *IEEE International Conference on Robotics and Automation*, pp. 264-269, 1996.

Y. C. Tsai and A. C. Huang, "FAT based adaptive control for pneumatic servo system with mismatched uncertainties," *Mechanical Systems and Signal Processing*, vol. 22, no. 6, pp. 1263-1273, Aug. 2008a.

Y. C. Tsai and A. C. Huang, "Multiple-Surface Sliding Controller Design for Pneumatic Servo Systems," *Mechatronics*, vol. 18, Issue 9, pp. 506-512, Nov. 2008b.

F. Tyan and S. C. Lee, "An Adaptive Control of Rotating Stall and Surge of Jet Engines – A Function Approximation Approach," *44th IEEE Conference on Decision and Control and 2005 European Control Conference*, pp. 5498-5503, Dec. 2005.

D. Del Vecchio, R. Marino and P. Tomei, "Adaptive Learning Control for Robot Manipulators," *American Control Conference*, pp. 641-645, 2001.

J. H. Yang, "Adaptive Tracking Control for Manipulators with Only Position Feedback," *IEEE Canadian Conference on Electrical and Computer Engineering*, pp. 1740-1745, 1999.

W. Yim, "Adaptive Control of a Flexible Joint Manipulator," *Proceedings IEEE International Conference on Robotics and Automation*, pp. 3441-3446, Seoul, Korea, May 2001.

T. Yoshikawa, "Dynamic Hybrid Position/Force Control of Robot Manipulators, Description of Hand Constraints and Calculation of Joint Driving Force," *Proceedings IEEE International Conference on Robotics and Automation*, San Francisco, CA, pp. 1396-1398, April 1986.

T. Yoshikawa, "Force Control of Robot Manipulators," *Proceedings IEEE International Conference on Robotics and Automation*, pp. 220-226, 2000.

J. Yuan and Y. Stepanenko, "Adaptive PD Control of Flexible Joint Robots without using the High-order Regressor," *Proceedings of the 36th Midwest Symposium on Circuits and Systems*, pp. 389-393, 1993.

R. R. Y. Zhen and A. A. Goldenberg, "An Adaptive Approach to Constrained Robot Motion Control," *Proceedings IEEE International Conference on Robotics and Automation*, pp. 1833-1838, 1995.

Symbols, Definitions and Abbreviations

Abbreviations

RR	: Rigid Robot
RRE	: Rigid Robot interacting with Environment
EDRR	: Electrically-Driven Rigid Robot
EDRRE	: Electrically-Driven Rigid Robot interacting with Environment
FJR	: Flexible-Joint Robot
FJRE	: Flexible-Joint Robot interacting with Environment
EDFJR	: Electrically-Driven Flexible-Joint Robot
EDFJRE	: Electrically-Driven FJR interacting with Environment
PE	: Persistent Excitation
FAT	: Function Approximation Technique
MRC	: Model Reference Control
MRAC	: Model Reference Adaptive Control
SPR	: Strictly Positive Real
UUB	: Uniformly Ultimately Bounded

General Symbols and Definitions

a	: scalar (unbold lower case)
$\mathbf{a}$	: vector (bold lower case)
$\mathbf{A}$	: matrix (bold upper case)
$\mathbf{I}_n$	: $n \times n$ identity matrix
a_i	: i-th element of vector $\mathbf{a}$
a_{ij}	: (i,j)-th element of matrix $\mathbf{A}$
$\mathbf{a}^T$	: transpose of vector $\mathbf{a}$
$\mathbf{A}^T$	: transpose of matrix $\mathbf{A}$
$\mathbf{A}^{-1}$	: inverse of matrix $\mathbf{A}$

$Tr(\mathbf{A})$	: trace of matrix $\mathbf{A}$
$\hat{a}$	: estimation of scalar a
$\tilde{a}$	: error between a and $\hat{a}$
$\hat{\mathbf{a}}$	: estimation of vector $\mathbf{a}$
$\tilde{\mathbf{a}}$	: error between $\mathbf{a}$ and $\hat{\mathbf{a}}$
$\hat{\mathbf{A}}$	: estimation of matrix $\mathbf{A}$
$\tilde{\mathbf{A}}$	: error between $\mathbf{A}$ and $\hat{\mathbf{A}}$
$\lambda_i(\mathbf{A})$	: i-th eigenvalue of matrix $\mathbf{A}$
$\lambda_{\min}(\mathbf{A})$	: minimum eigenvalue of matrix $\mathbf{A}$
$\lambda_{\max}(\mathbf{A})$	: maximum eigenvalue of matrix $\mathbf{A}$
$|a|$	: absolute value of scalar a
$\|\mathbf{a}\|$	: norm of vector $\mathbf{a}$
$|\mathbf{A}|$	: determinant of matrix $\mathbf{A}$
$\|\mathbf{A}\|$	: norm of matrix $\mathbf{A}$
$\min\{.\}$	: minimum operation
$\sup(.)$	: least upper bound
$\operatorname{diag}\{\dots\}$	: diagonal matrix

Symbols and Definitions in Robot Model

l_i	: length of link i
m_i	: mass of link i
I_i	: moment of inertia of link i
$\mathbf{B}$	: actuator damping matrix
$\mathbf{B}_i$	: desired apparent damping
$\mathbf{C}$	: vector of centrifugal and Coriolis forces
$\mathbf{C}_x$	: $\mathbf{C}$ in the Cartesian space
$\mathbf{D}$	: inertia matrix
$\mathbf{D}_x$	: $\mathbf{D}$ in the Cartesian space
$\mathbf{F}_{ext}$	: external force
$\mathbf{g}$	: gravitational force vector
$\mathbf{g}_x$	: $\mathbf{g}$ in the Cartesian space
$\mathbf{H}$	: electro-mechanical conversion matrix
$\mathbf{i}$	: armature current vector
$\mathbf{i}_d$	: desired current trajectory
$\mathbf{J}$	: actuator inertia matrix
$\mathbf{J}_a$	: Jacobian matrix

$\mathbf{K}$	: joint stiffness matrix
$\mathbf{K}_b$	: back-emf matrix
$\mathbf{K}_i$	: desired apparent stiffness
$\mathbf{L}$	: electrical inductance matrix
$\mathbf{M}_i$	: desired apparent inertia
$\mathbf{p}$	: parameter vector
$\mathbf{p}_x$	: parameter vector in the Cartesian space
$\mathbf{q}$	: generalized coordinate vector
$\mathbf{q}_d$	: desired trajectory for $\mathbf{q}$
$\mathbf{R}$	: electrical resistance matrix
$\mathbf{x}$	: coordinate in the Cartesian space
$\mathbf{x}_d$	: desired trajectory for $\mathbf{x}$
$\mathbf{Y}$	: regressor matrix
θ	: actuator angle
τ	: control torque vector
τ_a	: actuator input torque vector
τ_t	: transmission torque

Symbols and Definitions in Controller Design

d	: disturbance
e	: error signal
$\mathbf{e}$	: error vector
$\mathbf{e}_i$	: current error vector
$\mathbf{e}_m$	: output tracking error vector
$\mathbf{e}_\tau$	: torque tracking error vector
k_w	: stiffness of the wall
$\mathbf{K}_d$	: gain matrix for velocity error
$\mathbf{K}_p$	: gain matrix for position error
$\mathbf{K}_\tau$	: conversion matrix
$\mathbf{P}, \mathbf{Q}, \mathbf{\Gamma}$	: positive matrices
s	: sliding surface variable
$\mathbf{s}$	: error vector
sat(.)	: saturation function
sgn(.)	: signum function
u	: control input signal
$\mathbf{u}$	: control input vector

$\mathbf{v}$	: a known signal vector
V	: Lyapunov function candidate
$\mathbf{w}$	: a vector of weightings
$\mathbf{W}$	: a matrix of weightings
$\mathbf{z}$	: a vector of basis functions
$\mathbf{Z}$	: a matrix of basis functions
α	: a constant
β	: number of basis function
ε	: approximation error
ε_i	: approximation error
ϕ	: thickness of the boundary layer
σ	: sigma modification constant
φ	: known signal vector
τ_{robust}	: robust term
Λ	: diagonal gain matrix

Index